THE
MATH
BOOK

BIG IDEAS

THE ART BOOK	THE LITERATURE BOOK
THE ASTRONOMY BOOK	THE MATHS BOOK
THE BIBLE BOOK	THE MEDICINE BOOK
THE BIOLOGY BOOK	THE MOVIE BOOK
THE BLACK HISTORY BOOK	THE MYTHOLOGY BOOK
THE BUSINESS BOOK	THE PHILOSOPHY BOOK
THE CHEMISTRY BOOK	THE PHYSICS BOOK
THE CLASSICAL MUSIC BOOK	THE POLITICS BOOK
THE CRIME BOOK	THE PSYCHOLOGY BOOK
THE ECOLOGY BOOK	THE RELIGIONS BOOK
THE ECONOMICS BOOK	THE SCIENCE BOOK
THE FEMINISM BOOK	THE SHAKESPEARE BOOK
THE HISTORY BOOK	THE SHERLOCK HOLMES BOOK
THE ISLAM BOOK	THE SOCIOLOGY BOOK
THE LAW BOOK	THE WORLD WAR II BOOK

SIMPLY EXPLAINED

THE MATH BOOK

DK LONDON

SENIOR ART EDITOR
Gillian Andrews

SENIOR EDITORS
Camilla Hallinan, Laura Sandford

US EDITOR
Jenny Wilson

ILLUSTRATIONS
James Graham

JACKET EDITOR
Emma Dawson

JACKET DESIGNER
Surabhi Wadhwa

JACKET DESIGN
DEVELOPMENT MANAGER
Sophia MTT

PRODUCER, PRE-PRODUCTION
Andy Hilliard

PRODUCER
Rachel Ng

MANAGING EDITOR
Gareth Jones

SENIOR MANAGING ART EDITOR
Lee Griffiths

ASSOCIATE PUBLISHING DIRECTOR
Liz Wheeler

ART DIRECTOR
Karen Self

DESIGN DIRECTOR
Philip Ormerod

PUBLISHING DIRECTOR
Jonathan Metcalf

DK DELHI

PROJECT ART EDITOR
Pooja Pipil

ART EDITOR
Mridushmita Bose

ASSISTANT ART EDITOR
Nobina Chakravorty

SENIOR EDITOR
Anita Kakar

EDITOR
Aadithyan Mohan

SENIOR JACKET DESIGNER
Suhita Dharamjit

JACKETS EDITORIAL COORDINATOR
Priyanka Sharma

SENIOR DTP DESIGNER
Harish Aggarwal

DTP DESIGNERS
Vijay Khandwal, Anita Yadav

PICTURE RESEARCHER
. Rituraj Singh

MANAGING JACKETS EDITOR
Saloni Singh

PICTURE RESEARCH MANAGER
Taiyaba Khatoon

PRE-PRODUCTION MANAGER
Balwant Singh

PRODUCTION MANAGER
Pankaj Sharma

SENIOR MANAGING EDITOR
Rohan Sinha

MANAGING ART EDITOR
Sudakshina Basu

original styling by
STUDIO 8

TOUCAN BOOKS

EDITORIAL DIRECTOR
Ellen Dupont

SENIOR DESIGNER
Thomas Keenes

SENIOR EDITOR
Dorothy Stannard

EDITORS
John Andrews, Tim Harris, Abigail Mitchell,
Rachel Warren Chadd

EDITORIAL ASSISTANTS
Christina Fleischer, Isobel Rodol, Gage Rull

ADDITIONAL TEXT
Marcus Weeks

EDITORIAL ADVISORS
Tom Le Bas, Robert Snedden

INDEXER
Marie Lorrimer

PROOFREADER
Richard Beatty

This American Edition, 2023
First American Edition, 2019
Published in the United States by DK Publishing
1745 Broadway, 20th Floor, New York, NY 10019
Copyright © 2019, 2023 Dorling Kindersley Limited
DK, a Division of Penguin Random House LLC
24 25 10 9 8 7 6 5 4 3
035–335337–Feb/2023

A catalog record for this book is available
from the Library of Congress.
ISBN 978-0-7440-7937-1

DK books are available at special discounts when purchased
in bulk for sales promotions, premiums, fund-raising, or
educational use. For details, contact: DK Publishing Special
Markets, 1745 Broadway, 20th Floor, New York, NY 10019
SpecialSales@dk.com

Printed and bound in China

For the curious
www.dk.com

MIX
Paper | Supporting
responsible forestry
FSC™ C018179

This book was made with Forest Stewardship
Council™ certified paper – one small step in
DK's commitment to a sustainable future.
For more information go to
www.dk.com/our-green-pledge

CONTRIBUTORS

KARL WARSI, CONSULTANT EDITOR

Karl Warsi taught mathematics in UK schools and colleges for many years. In 2000, he began publishing books on mathematics, creating bestselling textbook series for secondary-level students, both in the UK and worldwide. He is committed to inclusion in education, and the idea that people of all ages learn in different ways.

JAN DANGERFIELD

A lecturer and senior examiner in Further Mathematics, Jan Dangerfield is also a fellow of the UK's Chartered Institute of Educational Assessors and a Fellow of the Royal Statistical Society. She has been a member of the British Society for the History of Mathematics for more than 30 years.

HEATHER DAVIS

British author and educator Heather Davis has taught mathematics for 30 years. She has published textbooks for Hodder Education and managed publications for the UK's Association of Teachers of Mathematics. She presents courses for examination boards both in the UK and internationally and writes and presents enrichment activities for students.

JOHN FARNDON

A widely published author of popular books on science and nature, John Farndon has been shortlisted five times for the Royal Society's Young People's Science Book Prize, among other awards. He has written or collaborated on around 1,000 books on a range of subjects, including internationally acclaimed titles such as *The Oceans Atlas, Do You Think You're Clever?* and *Do Not Open*, and contributed to major books such as *Science* and *Science Year by Year*.

JONNY GRIFFITHS

After studying mathematics and education at Cambridge University, the Open University, and the University of East Anglia, Jonny Griffiths taught math at Paston Sixth Form College in Norfolk, UK, for over 20 years. In 2005–06, he was made a Gatsby Teacher Fellow for creating the popular mathematics website Risps. In 2016, he founded the competition Ritangle for students of mathematics.

TOM JACKSON

A writer for 25 years, Tom Jackson has written about 200 non-fiction books for adults and children and contributed to many more on a wide range of science and technology topics. They include *Numbers: How Counting Changed the World; Everything is Mathematical*, a book series with Marcus du Sautoy; and *Help Your Kids with Science* with Carol Vorderman.

MUKUL PATEL

Mukul Patel, who studied mathematics at Imperial College, London, writes and collaborates across many disciplines. He is the author of *We've Got Your Number*, a book on mathematics for children, and film scripts voiced by Tilda Swinton. He has also composed extensively for contemporary choreographers and designed sound installations for architects. He is currently investigating ethical issues in AI.

SUE POPE

A mathematics educator, Sue Pope is a long-standing member of the Association of Teachers of Mathematics and co-runs workshops on the history of mathematics in teaching at their conferences. Published widely, she recently co-edited *Enriching Mathematics in the Primary Curriculum*.

MATT PARKER, FOREWORD

Originally a math teacher from Australia, Matt Parker is a now a stand-up comedian, mathematics communicator, and a prominent math YouTuber on the Numberphile and Stand-up Maths channels, where his videos have had more than 100 million views. Matt performs live comedy with Festival of the Spoken Nerd and once calculated pi live in front of a sold-out Royal Albert Hall. He also presents television and radio programs for Discovery Channel and the BBC, and his 2019 book *Humble Pi: A Comedy of Maths Errors* topped the *Sunday Times* best-seller chart.

CONTENTS

THE MIDDLE AGES
500–1500

THE RENAISSANCE
1500–1680

THE ENLIGHTENMENT
1680–1800

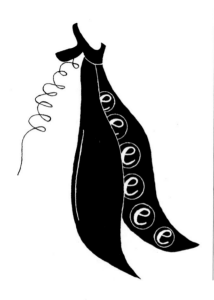

THE 19TH CENTURY
1800–1900

MODERN MATHEMATICS
1900–PRESENT

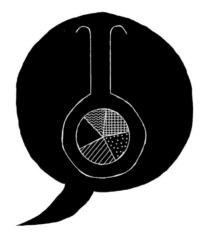

FOREWORD

Summarizing all of mathematics in one book is a daunting and indeed impossible task. Humankind has been exploring and discovering mathematics for millennia. Practically, we have relied on math to advance our species, with early arithmetic and geometry providing the foundations for the first cities and civilizations. And philosophically, we have used mathematics as an exercise in pure thought to explore patterns and logic.

As a subject, mathematics is surprisingly hard to pin down with one catch-all definition. "Mathematics" is not simply, as many people think, "stuff to do with numbers." That would exclude a huge range of mathematical topics, including much of the geometry and topology covered in this book. Of course, numbers are still very useful tools to understand even the most esoteric areas of mathematics, but the point is that they are not the most interesting aspect of it. Focusing just on numbers misses the forest for the threes.

For the record, my own definition of math as "the sort of things that mathematicians enjoy doing," while delightfully circular, is largely unhelpful. *Big Ideas Simply Explained* is actually not a bad definition. Mathematics could be seen as the attempt to find the simplest explanations for the biggest ideas. It is the endeavor of finding and summarizing patterns. Some of those patterns involve the practical triangles required to build pyramids and divide land; other patterns attempt to classify all of the 26 sporadic groups of abstract algebra. These are very different problems in terms of both usefulness and complexity, but both types of pattern have become the obsession of mathematicians throughout the ages.

There is no definitive way to organize all of mathematics, but looking at it chronologically is not a bad way to go. This book uses the historical journey of humans discovering math as a way to classify it and wrangle it into a linear progression, which is a valiant but difficult effort. Our current mathematical body of knowledge has been built up by a haphazard and diverse group of people across time and cultures.

So something like the short section on magic squares covers thousands of years and the span of the globe. Magic squares—arrangements of numbers where the sum in each row, column, and diagonal is always the same—are one of the oldest areas of recreational mathematics. Starting in the 9th century BCE in China, the story then bounces around via Indian texts from 100 CE, Arab scholars in the Middle Ages, Europe during the Renaissance, and finally modern Sudoku-style puzzles. Across a mere two pages this book has to cover 3,000 years of history ending with geomagic squares in 2001. And even in this small niche of mathematics, there are many magic square developments that there was simply not enough room to include. The whole book should be viewed as a curated tour of mathematical highlights.

Studying even just a sample of mathematics is a great reminder of how much humans have achieved. But it also highlights where mathematics could do better; things like the glaring omission of women from the history of mathematics cannot be ignored. A lot of talent has been squandered over the centuries, and a lot of credit has not been appropriately given. But I hope that we are now improving the diversity of mathematicians and encouraging all humans to discover and learn about mathematics.

Because going forward, the body of mathematics will continue to grow. Had this book been written a century earlier it would have been much the same up until about page 280. And then it would have ended. No ring theory from Emmy Noether, no computing from Alan Turing, and no six degrees of separation from Kevin Bacon. And no doubt that will be true again 100 years from now. The edition printed a century from now will carry on past page 325, covering patterns totally alien to us. And because anyone can do math, there is no telling who will discover this new math, and where or when. To make the biggest advancement in mathematics during the 21st century, we need to include all people. I hope this book helps inspire everyone to get involved.

Matt Parker

INTRODU

CTION

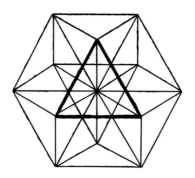

The history of mathematics reaches back to prehistory, when early humans found ways to count and quantify things. In doing so, they began to identify certain patterns and rules in the concepts of numbers, sizes, and shapes. They discovered the basic principles of addition and subtraction—for example, that two things (whether pebbles, berries, or mammoths) when added to another two invariably resulted in four things. While such ideas may seem obvious to us today, they were profound insights for their time. They also demonstrate that the history of mathematics is above all a story of discovery rather than invention. Although it was human curiosity and intuition that recognized the underlying principles of mathematics, and human ingenuity that later provided various means of recording and notating them, those principles themselves are not a human invention. The fact that $2 + 2 = 4$ is true, independent of human existence; the rules of mathematics, like the laws of physics, are universal, eternal, and unchanging. When mathematicians first showed that the angles of any triangle in a flat plane when added together come to 180°, a straight line, this was not their invention: they had simply discovered a fact that had always been (and will always be) true.

Early applications

The process of mathematical discovery began in prehistoric times, with the development of ways of counting things people needed to quantify. At its simplest, this was done by cutting tally marks in a bone or stick, a rudimentary but reliable means of recording numbers of things. In time, words and symbols were assigned to the numbers and the first systems of numerals began to evolve, a means of expressing operations such as acquisition of additional items, or depletion of a stock, the basic operations of arithmetic.

As hunter-gatherers turned to trade and farming, and societies became more sophisticated, arithmetical operations and a numeral system became essential tools in all kinds of transactions. To enable trade, stocktaking, and taxes in uncountable goods such as oil, flour, or plots of land, systems of measurement were developed, putting a numerical value on dimensions such as weight and length. Calculations also became more complex, developing the concepts of multiplication and division from addition and subtraction—allowing the area of land to be calculated, for example.

In the early civilizations, these new discoveries in mathematics, and specifically the measurement of objects in space, became the foundation of the field of geometry, knowledge that could be used in building and toolmaking. In using these measurements for practical purposes, people found that certain patterns were emerging, which could in turn prove useful. A simple but accurate carpenter's square can be made from a triangle with sides of three, four, and five units. Without that accurate tool and knowledge,

> It is impossible to be a mathematician without being a poet of the soul.
> **Sofya Kovalevskaya**
> **Russian mathematician**

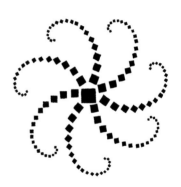

the roads, canals, ziggurats, and pyramids of ancient Mesopotamia and Egypt could not have been built. As new applications for these mathematical discoveries were found—in astronomy, navigation, engineering, bookkeeping, taxation, and so on—further patterns and ideas emerged. The ancient civilizations each established the foundations of mathematics through this interdependent process of application and discovery, but also developed a fascination with mathematics for its own sake, so-called pure mathematics. From the middle of the first millennium BCE, the first pure mathematicians began to appear in Greece, and slightly later in India and China, building on the legacy of the practical pioneers of the subject—the engineers, astronomers, and explorers of earlier civilizations.

Although these early mathematicians were not so concerned with the practical applications of their discoveries, they did not restrict their studies to mathematics alone. In their exploration of the properties of numbers, shapes, and processes, they discovered universal rules and patterns that raised metaphysical questions about the nature of the

> Geometry is knowledge of the eternally existent.
> **Pythagoras**
> **Ancient Greek mathematician**

cosmos, and even suggested that these patterns had mystical properties. Often mathematics was therefore seen as a complementary discipline to philosophy—many of the greatest mathematicians through the ages have also been philosophers, and vice versa—and the links between the two subjects have persisted to the present day.

Arithmetic and algebra

So began the history of mathematics as we understand it today—the discoveries, conjectures, and insights of mathematicians that form the bulk of this book. As well as the individual thinkers and their ideas, it is a story of societies and cultures, a continuously developing thread of thought from the ancient

civilizations of Mesopotamia and Egypt, through Greece, China, India, and the Islamic empire to Renaissance Europe and into the modern world. As it evolved, mathematics was also seen to comprise several distinct but interconnected fields of study.

The first field to emerge, and in many ways the most fundamental, is the study of numbers and quantities, which we now call arithmetic, from the Greek word *arithmos* ("number"). At its most basic, it is concerned with counting and assigning numerical values to things, but also the operations, such as addition, subtraction, multiplication, and division, that can be applied to numbers. From the simple concept of a system of numbers comes the study of the properties of numbers, and even the study of the very concept itself. Certain numbers—such as the constants π, e, or the prime and irrational numbers—hold a special fascination and have become the subject of considerable study.

Another major field in mathematics is algebra, which is the study of structure, the way that mathematics is organized, and therefore has some relevance in every other field. What marks algebra from arithmetic is the »

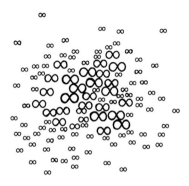

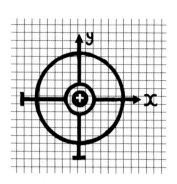

use of symbols, such as letters, to represent variables (unknown numbers). In its basic form, algebra is the study of the underlying rules of how those symbols are used in mathematics—in equations, for example. Methods of solving equations, even quite complex quadratic equations, had been discovered as early as the ancient Babylonians, but it was medieval mathematicians of the Islamic Golden Age who pioneered the use of symbols to simplify the process, giving us the word "algebra," which is derived from the Arabic *al-jabr*. More recent developments in algebra have extended the idea of abstraction into the study of algebraic structure, known as abstract algebra.

Geometry and calculus

A third major field of mathematics, geometry, is concerned with the concept of space, and the relationships of objects in space: the study of the shape, size, and position of figures. It evolved from the very practical business of describing the physical dimensions of things, in engineering and construction projects, measuring and apportioning plots of land, and astronomical observations for navigation and compiling

> In mathematics, the art of asking questions is more valuable than solving problems.
> **Georg Cantor**
> **German mathematician**

calendars. A particular branch of geometry, trigonometry (the study of the properties of triangles), proved to be especially useful in these pursuits. Perhaps because of its very concrete nature, for many ancient civilizations, geometry was the cornerstone of mathematics, and provided a means of problem-solving and proof in other fields.

This was particularly true of ancient Greece, where geometry and mathematics were virtually synonymous. The legacy of great mathematical philosophers such as Pythagoras, Plato, and Aristotle was consolidated by Euclid, whose principles of mathematics based on a combination of geometry

and logic were accepted as the subject's foundation for some 2,000 years. In the 1800s, however, alternatives to classical Euclidean geometry were proposed, opening up new areas of study, including topology, which examines the nature and properties not only of objects in space, but of space itself.

Since the Classical period, mathematics had been concerned with static situations, or how things are at any given moment. It failed to offer a means of measuring or calculating continuous change. Calculus, developed independently by Gottfried Leibniz and Isaac Newton in the 1600s, provided an answer to this problem. The two branches of calculus, integral and differential, offered a method of analyzing such things as the slope of curves on a graph and the area beneath them as a way of describing and calculating change.

The discovery of calculus opened up a field of analysis that later became particularly relevant to, for example, the theories of quantum mechanics and chaos theory in the 1900s.

Revisiting logic

The late 19th and early 20th centuries saw the emergence of another field of mathematics—the

foundations of mathematics. This revived the link between philosophy and mathematics. Just as Euclid had done in the 3rd century BCE, scholars including Gottlob Frege and Bertrand Russell sought to discover the logical foundations on which mathematical principles are based. Their work inspired a re-examination of the nature of mathematics itself, how it works, and what its limits are. This study of basic mathematical concepts is perhaps the most abstract field, a sort of meta-mathematics, yet an essential adjunct to every other field of modern mathematics.

New technology, new ideas

The various fields of mathematics—arithmetic, algebra, geometry, calculus, and foundations—are worthy of study for their own sake, and the popular image of academic mathematics is that of an almost incomprehensible abstraction. But applications for mathematical discoveries have usually been found, and advances in science and technology have driven innovations in mathematical thinking.

A prime example is the symbiotic relationship between mathematics and computers. Originally developed as a mechanical means of doing the "donkey work" of calculation to provide tables for mathematicians, astronomers and so on, the actual construction of computers required new mathematical thinking. It was mathematicians, as much as engineers, who provided the means of building mechanical, and then electronic computing devices, which in turn could be used as tools in the discovery of new mathematical ideas. No doubt, new applications for mathematical theorems will be found in the future too—and with numerous problems still unsolved, it seems that there is no end to the mathematical discoveries to be made.

The story of mathematics is one of exploration of these different fields, and the discovery of new ones. But it is also the story of the explorers, the mathematicians who set out with a definite aim in mind, to find answers to unsolved problems, or to travel into unknown territory in search of new ideas—and those who simply stumbled upon an idea in the course of their mathematical journey, and were inspired to see where it would lead. Sometimes the discovery would come as a game-changing revelation, providing a way into unexplored fields; at other times it was a case of "standing on the shoulders of giants," developing the ideas of previous thinkers, or finding practical applications for them.

This book presents many of the "big ideas" in mathematics, from the earliest discoveries to the present day, explaining them in layperson's language, where they came from, who discovered them, and what makes them significant. Some may be familiar, others less so. With an understanding of these ideas, and an insight into the people and societies in which they were discovered, we can gain an appreciation of not only the ubiquity and usefulness of mathematics, but also the elegance and beauty that mathematicians find in the subject. ∎

Mathematics, rightly viewed, possesses not only truth, but supreme beauty.
Bertrand Russell
British philosopher and mathematician

ANCIENT A
CLASSICAL
3500 BCE—500 CE

ND PERIODS

Different **quantities are denoted** on Sumerian clay tablets, prefiguring **a numerical system**.

c. 3500 BCE

c. 3000 BCE

The Sumerians introduce a **base-60 numerical** system, in which a **small cone denotes 1** and a **large cone denotes 60**.

The ancient Egyptians describe **methods of working** out **areas and volumes** and record them on the Rhind papyrus.

c. 1650 BCE

c. 530 BCE

Pythagoras **founds a school** where he teaches his **metaphysical beliefs and mathematical discoveries**, including Pythagoras's theorem.

Hippasus of Metapontum discovers **irrational numbers**, which cannot be expressed in fractions.

c. 430 BCE

c. 387 BCE

Plato founds his **Academy in Athens**—the sign over the entrance reads "Let no one ignorant of geometry enter here."

One of the **most influential** textbooks ever written, Euclid's *Elements*, contains **mathematical advances** including proof of the infinity of prime numbers.

c. 300 BCE

As early as 40,000 years ago, humans were making tally marks on wood and bone as a means of counting. They undoubtedly had a rudimentary sense of number and arithmetic, but the history of mathematics only properly began with the development of numerical systems in early civilizations. The first of these emerged in the fourth millennium BCE, in Mesopotamia, western Asia, home to the world's earliest agriculture and cities. Here, the Sumerians elaborated on the concept of tally marks, using different symbols to denote different quantities, which the Babylonians then developed into a sophisticated numerical system of cuneiform (wedge-shaped) characters. From about 1800 BCE, the Babylonians used elementary geometry and algebra to solve practical problems—such as building, engineering, and calculating land divisions—alongside the arithmetical skills they used to conduct commerce and levy taxes.

A similar story emerges in the slightly later civilization of the ancient Egyptians. Their trade and taxation required a sophisticated numerical system, and their building and engineering works relied on both a means of measurement and some knowledge of geometry and algebra. The Egyptians were also able to use their mathematical skills in conjunction with observations of the heavens to calculate and predict astronomical and seasonal cycles and construct calendars for the religious and agricultural year.

They established the study of the principles of arithmetic and geometry as early as 2000 BCE.

Greek rigor

The 6th century BCE onward saw a rapid rise in the influence of ancient Greece across the eastern Mediterranean. Greek scholars quickly assimilated the mathematical ideas of the Babylonians and Egyptians. The Greeks used a numerical system of base-10 (with ten symbols) derived from the Egyptians. Geometry in particular chimed with Greek culture, which idolized beauty of form and symmetry. Mathematics became a cornerstone of Classical Greek thinking, reflected in its art, architecture, and even philosophy. The almost mystical qualities of geometry and numbers inspired

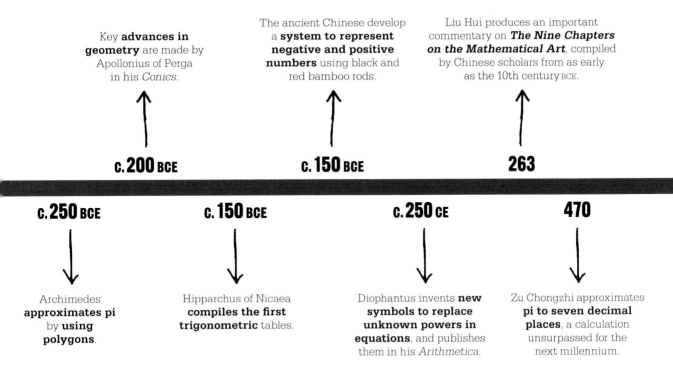

Key **advances in geometry** are made by Apollonius of Perga in his *Conics*.

The ancient Chinese develop a **system to represent negative and positive numbers** using black and red bamboo rods.

Liu Hui produces an important commentary on ***The Nine Chapters on the Mathematical Art***, compiled by Chinese scholars from as early as the 10th century BCE.

c. 200 BCE

c. 150 BCE

263

c. 250 BCE

c. 150 BCE

c. 250 CE

470

Archimedes **approximates pi** by **using polygons**.

Hipparchus of Nicaea **compiles the first trigonometric** tables.

Diophantus invents **new symbols to replace unknown powers in equations**, and publishes them in his *Arithmetica*.

Zu Chongzhi approximates **pi to seven decimal places**, a calculation unsurpassed for the next millennium.

Pythagoras and his followers to establish a cultlike community, dedicated to studying the mathematical principles they believed were the foundations of the Universe and everything in it.

Centuries before Pythagoras, the Egyptians had used a triangle with sides of 3, 4, and 5 units as a building tool to ensure corners were square. They had come across this idea by observation, and then applied it as a rule of thumb, whereas the Pythagoreans set about rigorously showing the principle, offering a proof that it is true for all right-angled triangles. It is this notion of proof and rigor that is the Greeks' greatest contribution to mathematics.

Plato's Academy in Athens was dedicated to the study of philosophy and mathematics, and Plato himself described the five Platonic solids (the tetrahedron, cube, octahedron, dodecahedron, and icosahedron). Other philosophers, notably Zeno of Elea, applied logic to the foundations of mathematics, exposing the problems of infinity and change. They even explored the strange phenomenon of irrational numbers. Plato's pupil Aristotle, with his methodical analysis of logical forms, identified the difference between inductive reasoning (such as inferring a rule of thumb from observations) and deductive reasoning (using logical steps to reach a certain conclusion from established premises, or axioms).

From this basis, Euclid laid out the principles of mathematical proof from axiomatic truths in his *Elements*, a treatise that was the foundation of mathematics for the next two millennia. With similar rigor, Diophantus pioneered the use of symbols to represent unknown numbers in his equations; this was the first step toward the symbolic notation of algebra.

A new dawn in the East

Greek dominance was eventually eclipsed by the rise of the Roman Empire. The Romans regarded mathematics as a practical tool rather than worthy of study. At the same time, the ancient civilizations of India and China independently developed their own numerical systems. Chinese mathematics in particular flourished between the 2nd and 5th centuries CE, thanks largely to the work of Liu Hui in revising and expanding the classic texts of Chinese mathematics. ∎

NUMERALS TAKE THEIR PLACES

POSITIONAL NUMBERS

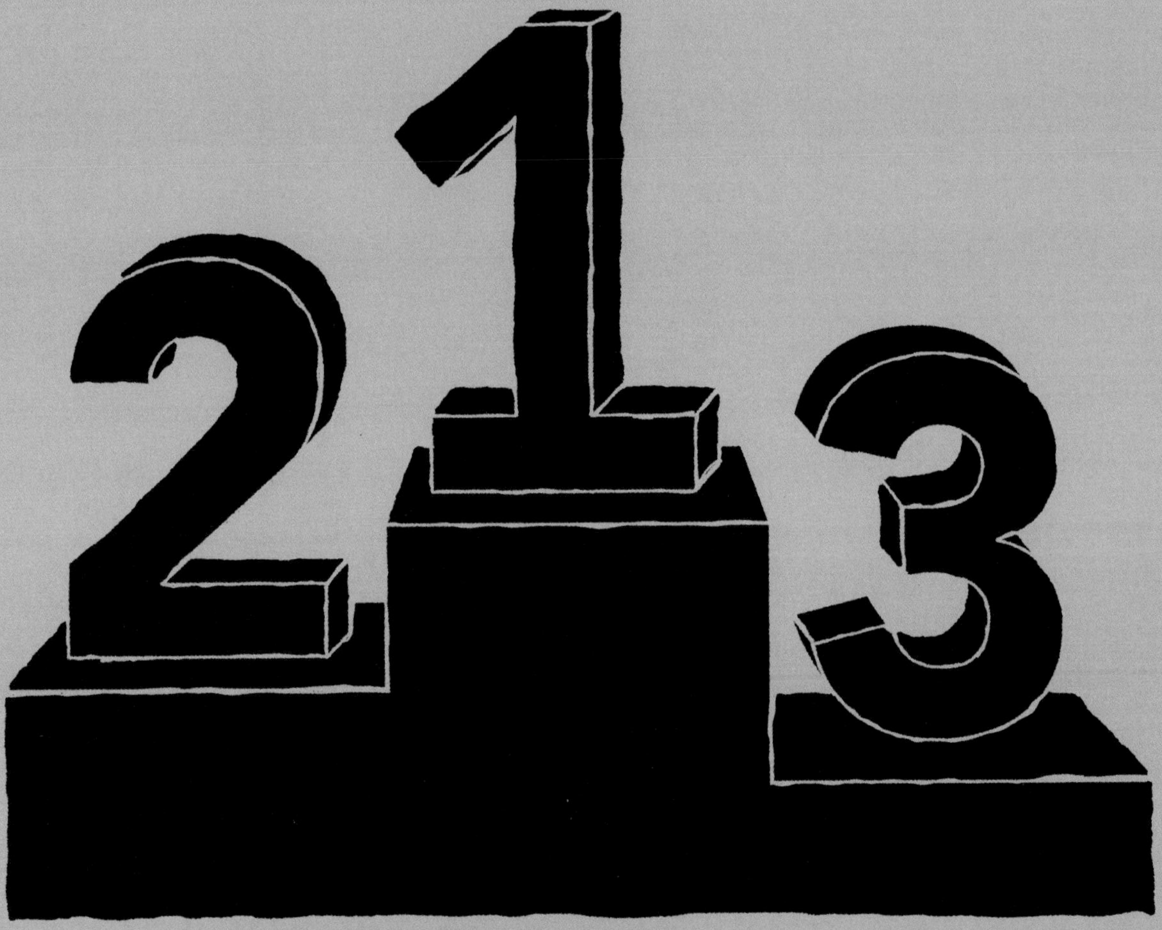

IN CONTEXT

KEY CIVILIZATION
Babylonians

FIELD
Arithmetic

BEFORE
40,000 years ago Stone Age people in Europe and Africa count using tally marks on wood or bone.

3500–3200 BCE Sumerians develop early calculation systems to measure land and to study the night sky.

3200–3000 BCE Babylonians use a small clay cone for 1 and a large cone for 60, along with a clay ball for 10, as their base-60 system evolves.

AFTER
2nd century CE The Chinese use an abacus in their base-10 positional number system.

7th century In India, Brahmagupta establishes zero as a number in its own right and not just as a placeholder.

It is given to us to calculate, to weigh, to measure, to observe; this is natural philosophy.
Voltaire
French philosopher

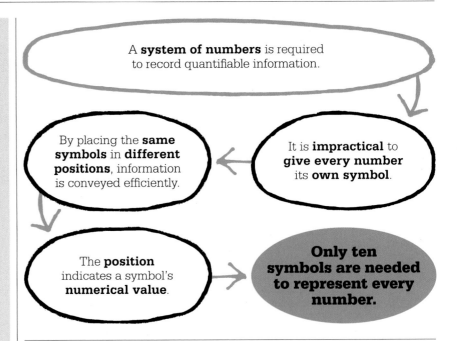

A **system of numbers** is required to record quantifiable information.

It is **impractical** to **give every number** its **own symbol**.

By placing the **same symbols** in **different positions**, information is conveyed efficiently.

The **position** indicates a symbol's **numerical value**.

Only ten symbols are needed to represent every number.

The first people known to have used an advanced numeration system were the Sumerians of Mesopotamia, an ancient civilization living between the Tigris and Euphrates rivers in what is present-day Iraq. Sumerian clay tablets from as early as the 4th millennium BCE include symbols denoting different quantities. The Sumerians, followed by the Babylonians, needed efficient mathematical tools in order to administer their empires.

What distinguished the Babylonians from neighbors such as Egypt was their use of a positional (place value) number system. In such systems, the value of a number is indicated both by its symbol and its position. Today, for instance, in the decimal system, the position of a digit in a number indicates whether its value is in ones (less than 10), tens, hundreds, or more. Such systems make calculation more efficient because a small set of symbols can represent a huge range of values. By contrast, the ancient Egyptians used separate symbols for ones, tens, hundreds, thousands, and above, and had no place value system. Representing larger numbers could require 50 or more hieroglyphs.

Using different bases

The Hindu–Arabic numeration that is employed today is a base-10 (decimal) system. It requires only 10 symbols—nine digits (1, 2, 3, 4, 5, 6, 7, 8, 9) and a zero as a placeholder. As in the Babylonian system, the position of a digit indicates its value, and the smallest value digit is always to the right. In a base-10 system, a two-digit number, such as 22, indicates $(2 \times 10^1) + 2$; the value of the 2 on the left is ten times that of the 2 on the right. Placing digits after the number 22 will create hundreds, thousands, and larger powers of 10. A symbol after a whole number (the standard notation now is a decimal point) can also separate it from its fractional parts, each representing a tenth of the place value of the preceding figure. The Babylonians worked with a more complex sexagesimal (base-60) number system that was probably inherited from the earlier Sumerians and is still used across the world today for measuring time, degrees in a circle (360° = 6 × 60), and geographic coordinates. Why they used 60 as a number base is still not known for sure. It may have been chosen because it can be divided by many other numbers—1, 2, 3, 4, 5, 6, 10, 12, 15, 20, and 30. The Babylonians also based their calendar year on the solar year (365.24 days); the number of days in a year was 360 (6 × 60) with additional days for festivals.

In the Babylonian sexagesimal system, a single symbol was used alone and repeated up to nine times to represent symbols for 1 to 9. For 10, a different symbol was used, placed to the left of the one symbol, and repeated two to five times in numbers up to 59. At 60 (60 × 1), the original symbol for one was reused but placed further to the left than the symbol for 1. Because it was a base-60 system, two such symbols signified 61, while three such symbols indicated 3,661, that is, 60 × 60 (60²) + 60 + 1.

The base-60 system had obvious drawbacks. It necessarily requires many more symbols than a base-10 system. For centuries, the sexagesimal system also had no »

The Babylonian sun-god Shamash awards a rod and a coiled rope, ancient measuring devices, to newly trained surveyors, on a clay tablet dating from around 1000 BCE.

Cuneiform

In the late 1800s, academics deciphered the "cuneiform" (wedge-shaped) markings on clay tablets recovered from Babylonian sites in and around Iraq. Such marks, denoting letters and words as well as an advanced number system, were etched in wet clay with either end of a stylus. Like the Egyptians, the Babylonians needed scribes to administer their complex society, and many of the tablets bearing mathematical records are thought to be from training schools for scribes.

A great deal has now been discovered about Babylonian mathematics, which extended to multiplication, division, geometry, fractions, square roots, cube roots, equations, and other forms, because— unlike Egyptian papyrus scrolls—the clay tablets have survived well. Several thousand, mostly dating from between 1800 and 1600 BCE, are housed in museums around the world.

Cuneiform, a word derived from the Latin *cuneus* ("wedge") to describe the shape of the symbols, was inscribed into wet clay, stone, or metal.

1		11		21		31		41		51	
2		12		22		32		42		52	
3		13		23		33		43		53	
4		14		24		34		44		54	
5		15		25		35		45		55	
6		16		26		36		46		56	
7		17		27		37		47		57	
8		18		28		38		48		58	
9		19		29		39		49		59	
10		20		30		40		50		60	

The Babylonian base-60 number system was built from two symbols—the single unit symbol, used alone and combined for numbers 1 to 9, and the 10 symbol, repeated for 20, 30, 40, and 50.

place value holders, and nothing to separate whole numbers from fractional parts. By around 300 BCE, however, the Babylonians used two wedges to indicate no value, much as we use a placeholder zero today; this was possibly the earliest use of zero.

Other counting systems

In Mesoamerica, on the other side of the world, the Mayan civilization developed its own advanced numeration system in the 1st millennium BCE—apparently in complete isolation. Theirs was a base-20 (vigesimal) number system, which probably evolved from a simple counting method using fingers and toes. In fact, base-20 number systems were used across the world, in Europe, Africa, and Asia. Language often contains remnants of this system. For example, in French, 80 is expressed as *quatre-vingt* (4 × 20); Welsh and Irish also express some numbers as multiples of 20, while in English a score is 20. In the Bible, for instance, Psalm 90 talks of a human lifespan being "threescore years and ten" or as great as "fourscore years."

From around 500 BCE until the 16th century when Hindu–Arabic numbers were officially adopted in China, the Chinese used rod numerals to represent numbers. This was the first decimal place value system. By alternating quantities of vertical rods with horizontal rods, this system could indicate ones, tens, hundreds, thousands, and more powers of 10, much as the decimal system does today. For example, 45 was written with four horizontal bars representing 4×10^1 (40) and five vertical bars for 5×1 (5). However, four vertical rods followed by five vertical rods indicated 405 (4×100, or 10^2) + 5×1—the absence of horizontal rods meant there were no tens in the number. Calculations were carried out by manipulating the rods on a counting board. Positive and negative numbers were represented by red and black rods respectively or different cross sections (triangular and rectangular). Rod numerals are still used occasionally in China, just as Roman numerals are sometimes used in Western society.

The Chinese place value system is reflected in the Chinese abacus (suanpan). Dating back to at least 200 BCE, it is one of the oldest bead-counting devices, although the Romans used something similar. The Chinese version, which is still used today, has a central bar and a varying number of vertical wires to separate ones from tens, hundreds, or more. In each column, there are two beads above the bar worth five each and five beads below the bar worth one each.

The Japanese adopted the Chinese abacus in the 14th century and developed their own abacus, the soroban, which has one bead worth five above the central bar

The Babylonian and Assyrian civilizations have perished… yet Babylonian mathematics is still interesting, and the Babylonian scale of 60 is still used in astronomy.
G. H. Hardy
British mathematician

> The fact that we work in 10s as opposed to any other number is purely a consequence of our anatomy. We use our ten fingers to count.
> **Marcus du Sautoy**
> **British mathematician**

and four beads each worth one below the bar in each column. Japan still uses the soroban today: there are even contests in which young people demonstrate their ability to perform soroban calculations mentally, a skill known as *anzan*.

Modern numeration

The Hindu–Arabic decimal system used throughout the world today has its origins in India. In the 1st to 4th centuries CE, the use of nine symbols along with zero was developed to allow any number

Ebisu, the Japanese god of fishermen and one of the seven gods of fortune, uses a soroban to calculate his profits in *The Red Snapper's Dream* by Utagawa Toyohiro.

to be written efficiently, through the use of place value. The system was adopted and refined by Arab mathematicians in the 9th century. They introduced the decimal point, so that the system could also express fractions of whole numbers.

Three centuries later, Leonardo of Pisa (Fibonacci) popularized the use of Hindu–Arabic numerals in Europe through his book *Liber Abaci* (1202). Yet the debate about whether to use the new system rather than Roman numerals and traditional counting methods lasted for several hundred years, before its adoption paved the way for modern mathematical advances.

With the advent of electronic computers, other number bases became important—particularly binary, a number system with base 2. Unlike the base-10 system with its 10 symbols, binary has just two: 1 and 0. It is a positional system, but instead of multiplying by 10, each column is multiplied by

2, also expressed as 2^1, 2^2, 2^3 and upward. In binary, the number 111 means $1 \times 2^2 + 1 \times 2^1 + 1 \times 2^0$, that is $4 + 2 + 1$, or 7 in our decimal number system.

In binary, as in all modern number systems whatever their base, the principles of place value are always the same. Place value—the Babylonian legacy—remains a powerful, easily understood, and efficient way to represent large numbers. ∎

The Dresden Codex, the oldest surviving Mayan book, dating from the 13th or 14th century, illustrates Mayan number symbols and glyphs.

Mayan numeral system

The Mayans, who lived in Central America from around 2000 BCE, used a base-20 (vigesimal) number system from around 1000 BCE to perform astronomical and calendar calculations. Like the Babylonians, they used a calendar of 360 days plus festivals, to make 365.24 days based on the solar year; their calendars helped them work out the growing cycles of crops.

The Mayan system employed symbols: a dot representing one and a bar representing five. By using combinations of dots over

bars they could generate numerals up to 19. Numbers larger than 19 were written vertically, with the lowest numbers at the bottom, and there is evidence of Mayan calculations up to hundreds of millions. An inscription from 36 BCE shows that they used a shell-shaped symbol to denote zero, which was widely used by the 4th century.

The Mayans' number system was in use in Central America until the Spanish conquests in the 16th century. Its influence, however, never spread further.

THE SQUARE AS THE HIGHEST POWER

QUADRATIC EQUATIONS

IN CONTEXT

KEY CIVILIZATIONS
Egyptians (c. 2000 BCE),
Babylonians (c. 1600 BCE)

FIELD
Algebra

BEFORE
c. 1800 BCE The Berlin papyrus
records a quadratic equation
solved in ancient Egypt.

AFTER
7th century CE The Indian
mathematician Brahmagupta
solves quadratic equations
using only positive integers.

10th century CE Egyptian
scholar Abu Kamil Shuja ibn
Aslam uses negative and
irrational numbers to solve
quadratic equations.

1545 Italian mathematician
Gerolamo Cardano publishes
his *Ars Magna*, setting out the
rules of algebra.

Quadratic equations are those involving unknown numbers to the power of 2 but not to a higher power; they contain x^2 but not x^3, x^4, and so on. One of the main rudiments of mathematics is the ability to use equations to work out solutions to real-world problems. Where those problems involve areas or paths of curves such as parabolas, quadratic equations become very useful, and describe physical phenomena, such as the flight of a ball or a rocket.

Ancient roots
The history of quadratic equations extends across the world. It is likely that these equations first arose

See also: Irrational numbers 44–45 ▪ Negative numbers 76–79 ▪ Diophantine equations 80–81 ▪ Zero 88–91 ▪ Algebra 92–99 ▪ The binomial theorem 100–01 ▪ Cubic equations 102–05 ▪ Imaginary and complex numbers 128–31

Quadratic **equations** contain the **power of 2**, so are used when calculating with **two dimensions**.

The **number of dimensions** is equal to the **maximum number of real solutions** an equation has.

There is a maximum of **two real solutions** for a **quadratic equation**, **three** for a **cubic equation**, and so on.

If a quadratic equation, or any equation, is **set equal to zero** (e.g. $x^2 + 3x + 2 = 0$), the **solutions are called roots**.

In a quadratic equation, these **two roots** are the points where a quadratic curve **crosses** the **x-axis** on a graph.

Der Berliner Papyrus 6619.

The Berlin papyrus was copied and published by German Egyptologist Hans Schack-Schackenburg in 1900. It contains two mathematical problems, one of which is a quadratic equation.

from the need to subdivide land for inheritance purposes, or to solve problems involving addition and multiplication.

One of the oldest surviving examples of a quadratic equation comes from the ancient Egyptian text known as the Berlin papyrus (c. 1800 BCE). The problem contains the following information: the area of a square of 100 cubits is equal to that of two smaller squares. The side of one of the smaller squares is equal to one half plus a quarter of the side of the other. In modern notation, this translates into two simultaneous equations: $x^2 + y^2 = 100$ and $x = (1/2 + 1/4)y$ $= 3/4\,y$. These can be simplified to

the quadratic equation $(3/4\,y)^2 + y^2 = 100$ to find the length of a side on each square.

The Egyptians used a method called "false position" to determine the solution. In this method, the mathematician selects a convenient number that is usually easy to calculate, then works out what the solution to the equation would be using that number. The result shows how to adjust the number to give the correct solution the equation. For example, in the Berlin papyrus problem, the simplest length to use for the larger of the two small squares is 4, because the problem deals with quarters. For the side of the smallest square,

3 is used because this length is $3/4$ of the side of the other small square. Two squares created using these false position numbers would have areas of 16 and 9 respectively, which when added together give a total area of 25. This is only $1/4$ of 100, so the areas must be quadrupled to match the Berlin papyrus equation. The lengths therefore must be doubled from the false positions of 4 and 3 to reach the solutions: 8 and 6.

Other early records of quadratic equations are found in Babylonian clay tablets, where the diagonal of a square is given to five decimal places. The Babylonian tablet YBC 7289 (c. 1800–1600 BCE) shows a method of working out the »

The quadratic formula is a way to solve quadratic equations. By modern convention, quadratic equations include a number, a, multiplied by x^2; a number, b, multiplied by x; and a number, c, on its own. The illustration below shows how the formula uses a, b, and c to find the value of x. Quadratic equations often equal 0, because this makes them easy to work out on a graph; the x solutions are the points where the curve crosses the x axis.

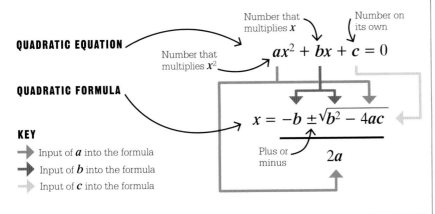

QUADRATIC EQUATION
Number that multiplies x^2

Number that multiplies x
Number on its own

$$ax^2 + bx + c = 0$$

QUADRATIC FORMULA

$$x = \frac{-b \pm \sqrt{b^2 - 4ac}}{2a}$$

Plus or minus

KEY
→ Input of a into the formula
→ Input of b into the formula
⇢ Input of c into the formula

multiplied together make a positive number. While $\sqrt{2} \times \sqrt{2} = 2$, it is also true that $-\sqrt{2} \times -\sqrt{2} = 2$.

In 1545, Italian scholar Gerolamo Cardano published his *Ars Magna* (*The Great Art, or the Rules of Algebra*) in which he explored the problem: "What pair of numbers have a sum of ten and product of 40?" He found that the problem led to a quadratic equation which, when he completed the square, gave $\sqrt{(-15)}$. No numbers available to mathematicians at the time gave a negative number when multiplied by themselves, but Cardano suggested suspending belief and working with the square root of negative 15 to find the equation's two solutions. Numbers such as $\sqrt{(-15)}$ would later be known as "imaginary" numbers.

quadratic equation $x^2 = 2$ by drawing rectangles and trimming them down into squares. In the 7th century CE, Indian mathematician Brahmagupta wrote a formula for solving quadratic equations that could be applied to equations in the form $ax^2 + bx = c$. Mathematicians at the time did not use letters or symbols, so he wrote his solution in words, but it was similar to the modern formula shown above.

In the 8th century, Persian mathematician al-Khwarizmi employed a geometric solution for quadratic equations known as completing the square. Until the 10th century, geometric methods were were often used, as quadratic equations were used to solve real-world problems involving land rather than abstract algebraic challenges.

Negative solutions
Indian, Persian, and Arab scholars thus far had used only positive numbers. When solving the equation $x^2 + 10x = 39$, they gave the solution as 3. However, this is one of two correct solutions to the problem; −13 is the other. If x is −13, $x^2 = 169$ and $10x = -130$. Adding a negative number gives the same result as subtracting its equivalent positive number, so $169 + -130 = 169 - 130 = 39$.

In the 10th century, Egyptian scholar Abu Kamil Shuja ibn Aslam made use of negative numbers and algebraic irrational numbers (such as the square root of 2) as both solutions and coefficients (numbers multiplying an unknown quantity). By the 1500s, most mathematicians accepted negative solutions and were comfortable with surds (irrational roots – those that cannot be expressed exactly as a decimal). They had also started using numbers and symbols, rather than writing equations in words. Mathematicians now utilized the plus or minus symbol, $\pm$, in solving quadratic equations. With the equation $x^2 = 2$, the solution is not just $x = \sqrt{2}$ but $x = \pm\sqrt{2}$. The plus or minus symbol is included because two negative numbers

Structure of equations
Modern quadratic equations usually look like $ax^2 + bx + c = 0$. The letters a, b, and c represent known numbers, while x represents the unknown number. Equations contain variables (symbols for numbers that are unknown), coefficients, constants (those that do not multiply variables), and operators (symbols such as the plus and equals sign). Terms are the parts separated by operators;

Politics is for the present, but an equation is for eternity.
Albert Einstein

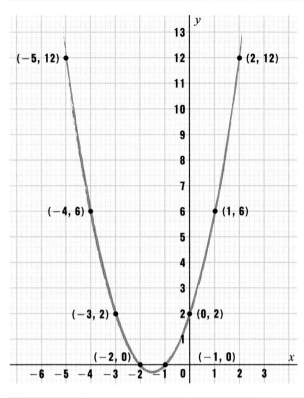

A graph of the quadratic function $y = ax^2 + bx + c$ creates a U-shaped curve called a parabola. This graph plots the points (in black) of the quadratic function where $a = 1$, $b = 3$, and $c = 2$. This expresses the quadratic equation $x^2 + 3x + 2 = 0$. The solutions for x are where $y = 0$ and the curve crosses the x axis. These are -2 and -1.

Practical applications

Although they were initially used for working out geometric problems, today quadratic equations are important in many aspects of mathematics, science, and technology. Projectile flight, for example, can be modeled with quadratic equations. An object thrown up into the air will fall back down again as a result of gravity. The quadratic function can predict projectile motion—the height of the object over time. Quadratic equations are used to model the relationship between time, speed, and distance, and in calculations with parabolic objects such as lenses. They can also be used to forecast profits and loss in the world of business. Profit is based on total revenue minus production cost; companies create a quadratic equation known as the profit function with these variables to work out the optimal sale prices to maximize profits.

they can be a number or variable, or a combination of both. The modern quadratic equation has four terms: ax^2, bx, c, and 0.

Parabolas

A function is a group of terms that defines a relationship between variables (often x and y). The quadratic function is generally written as $y = ax^2 + bx + c$, which, on a graph, produces a curve called a parabola (see above). When real (not imaginary) solutions to $ax^2 + bx + c = 0$ exist, they will be the roots—the points where the parabola crosses the x axis. Not all parabolas cut the x axis in two places. If the parabola touches the x axis only once, this means that there are coincident roots (the solutions are equal to each other). The simplest equation of this form is $y = x^2$. If the parabola does not touch or cross the x axis, there are no real roots.

Parabolas prove useful in the real world because of their reflective. properties. Satellite dishes are parabolic for this reason. Signals received by the dish will reflect off the parabola and be directed to one single point—the receiver. ∎

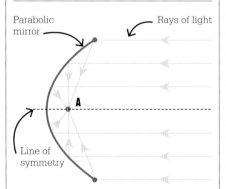

Parabolic objects have special reflective properties. With a parabolic mirror, any ray of light parallel to its line of symmetry will reflect off the surface to the same fixed point (A).

Quadratic equations are used by military specialists to model the trajectory of projectiles fired by artillery—such as this MIM-104 Patriot surface-to-air missile, commonly used by the US Army.

THE ACCURATE RECKONING FOR INQUIRING INTO ALL THINGS
THE RHIND PAPYRUS

IN CONTEXT

KEY CIVILIZATION
Ancient Egyptians
(c. 1650 BCE)

FIELD
Arithmetic

BEFORE
c. 2480 BCE Stone carvings record flood levels on the River Nile, measured in cubits— about 20½ in (52 cm)—and palms—about 3 in (7.5 cm).

c. 1850 BCE The Moscow papyrus provides solutions to 25 mathematical problems, including the calculation of the surface area of a hemisphere and the volume of a pyramid.

AFTER
c. 1800 BCE The Berlin papyrus is produced. It shows that the ancient Egyptians used quadratic equations.

6th century BCE The Greek scientist Thales travels to Egypt and studies its mathematical theories.

The Rhind papyrus in the British Museum in London provides an intriguing account of mathematics in ancient Egypt. Named after Scottish antiquarian Alexander Henry Rhind, who purchased the papyrus in Egypt in 1858, it was copied from earlier documents by a scribe, Ahmose, more than 3,500 years ago. It measures 12½ in (32 cm) by 78½ in (200 cm) and includes 84 problems concerned with arithmetic, algebra, geometry, and measurement. The problems, recorded in this and other ancient Egyptian artifacts such as the earlier Moscow papyrus, illustrated techniques for working out areas, proportions, and volumes.

The Eye of Horus, an Egyptian god, was a symbol of power and protection. Parts of it were also used to denote fractions whose denominators were powers of 2. The eyeball, for example, represents ¼, while the eyebrow is ⅛.

Representing concepts

The Egyptian number system was the first decimal system. It used strokes for single digits and a different symbol for each power of 10. The symbols were then repeated to create other numbers. A fraction was shown as a number with a dot above it. The Egyptian concept of a fraction was closest to a unit fraction—that is, $^1/_n$, where **n** is a whole number. When a fraction was doubled, it had to be rewritten as one unit fraction added to another unit fraction; for example, $^2/_3$ in modern notation would be ½ + ⅙ in Egyptian notation (not ⅓ + ⅓ because the Egyptians did not allow repeats of the same fraction).

The 84 problems in the Rhind papyrus illustrate the mathematical methods in common use in ancient Egypt. Problem 24, for example, asks what quantity, if added to its seventh part, becomes 19. This translates as $x + ^x/_7 = 19$. The approach applied to problem 24 is known as "false position." This technique—used well into the Middle Ages—is based on trial and improvement, choosing

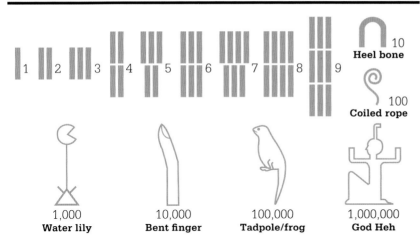

Ancient Egyptians used vertical lines to denote the numbers 1 to 9. Powers of 10, particularly those inscribed on stone, were depicted as hieroglyphs—picture symbols.

the simplest, or "false," value for a variable and adjusting the value using a scaling factor (the required quantity divided by the result).

In the workings for problem 24, one-seventh is easiest to find for the number 7, so 7 is used first as a "false" value for the variable. The result of the calculation—7 plus $^7/_7$ (or 1)—is 8, not 19, so a scaling factor is needed. To find how far the guess of 7 is from the required quantity, 19 is divided by 8 (the "false" answer). This produces a result of $2 + ^1/_4 + ^1/_8$ (not $2^3/_8$, as Egyptian multiplication was based on doubling and halving fractions), which is the scaling factor that should be applied. So 7 (the original "false" value) is multiplied by $2 + ^1/_4 + ^1/_8$ (the scaling factor) to give the quantity $16 + ^1/_2 + ^1/_8$ (or $16^5/_8$).

Many problems in the papyrus deal with working out shares of commodities or land. Problem 41 asks for the volume of a cylindrical store with a diameter of 9 cubits and a height of 10 cubits. The method finds the area of a square whose side length is $^8/_9$ of the diameter, then multiplies this by the height. The figure of $^8/_9$ is used as an approximation for the proportion of the area of a square that would be taken up by a circle if it were drawn within the square. This method is used in problem 50 to find the area of a circle: subtract $^1/_9$ from the diameter of the circle, and find the area of the square with the resulting side length.

Level of accuracy

Since the Ancient Greeks, the area of a circle has been found by multiplying the square of its radius (r^2) with the number pi (π), written as πr^2. The ancient Egyptians had no concept of pi, but the calculations in the Rhind papyrus show that they were close to its value. Their circle area calculation—with the circle diameter as twice the radius ($2r$)—can be expressed as $(^8/_9 \times 2r)^2$, which, simplified, is $^{256}/_{81} r^2$, giving an equivalent for pi of $^{256}/_{81}$. As a decimal, this is about 0.6 percent greater than the true value of pi. ∎

Instruction books

The Rhind and Moscow papyri are the most complete mathematical documents to survive from the height of the ancient Egyptian civilization. They were painstakingly copied by scribes well versed in arithmetic, geometry, and mensuration (the study of measurements) and are likely to have been used for training of other scribes. Although they captured probably the most advanced mathematical knowledge of the time, they were not seen as works of scholarship. Instead, they were instruction manuals for use in trade, accounting, construction, and other activities that involved measurement and calculation.

Egyptian engineers, for example, used mathematics in the building of pyramids. The Rhind papyrus includes a calculation for the slope of a pyramid using the *seked*—a measure for the horizontal distance traveled by a slope for each drop of 1 cubit. The steeper the side of a pyramid, the fewer the *sekeds*.

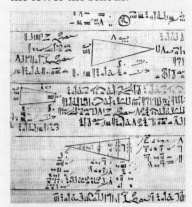

The Rhind papyrus scribe used the hieratic system of writing numerals. This cursive style was more compact and practical than drawing complex hieroglyphs.

THE SUM IS THE SAME IN EVERY DIRECTION

MAGIC SQUARES

IN CONTEXT

KEY CIVILIZATION
Ancient Chinese

FIELD
Number theory

BEFORE
9th century BCE The Chinese
I Ching (*Book of Changes*) lays
out trigrams and hexagrams of
numbers for use in divination.

AFTER
1782 Leonhard Euler writes
about Latin squares in his
*Recherches sur une nouvelle
espèce de carrés magiques*
(*Investigations on a new type
of magic square*).

1979 The first Sudoku-style
puzzle is published by Dell
Magazines in New York.

2001 British electronics
engineer Lee Sallows
invents magic squares called
"geomagic squares," which
contain geometric shapes
rather than numbers.

A **magic square** is a **square grid**—three-by-three or higher—in which distinct integers have been placed in each cell.

In each **row, column, and diagonal**, the sum of the numbers **will be the same**.

This **sum** is the **magic total**.

There are thousands of ways in which to arrange the numbers 1 to 9 in a three-by-three grid. Only eight of these produce a magic square, where the sum of the numbers in each row, column, and diagonal—the magic total—is the same. The sum of the numbers 1 to 9 is 45, as is the sum of all three rows or columns. The magic total, therefore, is $\frac{1}{3}$ of 45, or 15. In fact, there is really just one combination of numbers in a magic square. The other seven are rotations of this combination.

Ancient origins

Magic squares are probably the earliest example of "recreational mathematics." Their exact origin is unknown, but the first known reference, in the Chinese legend of *Lo Shu* (*Scroll of the river Lo*), dates from 650 BCE. In the legend, a turtle appears to the great King Yu as he faces a devastating flood. The markings on the turtle's back form a magic square, with numbers from 1 to 9 represented by circular dots. Because of this legend, the arrangement of odd and even numbers (even numbers are always in the corners of the square) were believed to have magical properties and was used as a good luck talisman through the ages.

As ideas from China spread along trade routes such as the Silk Road, other cultures became interested in magic squares. Magic squares are discussed in Indian texts dating from 100 CE,

and *Brihat-Samhita* (c. 550 CE), a book of divination, includes the first recorded magic square in India, used to measure out quantities of perfume. Arab scholars, who created a vital link between the learning of ancient civilizations and the European Renaissance, introduced magic squares to Europe in the 14th century.

Different-sized squares

The number of rows and columns in a magic square is called its order. For example, a three-by-three magic square is said to have an order of three. An order-two magic square does not exist because it would only work if all the numbers were identical. As the orders increase, so do the quantities of magic squares. Order four produces 880 magic squares—with a magic total of 34. There are hundreds of millions of order-five magic squares, while the quantity of order-six magic squares has not yet been calculated.

An order-four magic square appears beneath the bell in *Melencolia I* by the German artist Albrecht Dürer and wittily includes the engraving's date of 1514.

Magic squares have been an enduring source of fascination for mathematicians. The 15th-century Italian mathematician Luca Pacioli, author of *De viribus quantitatis* (*On the Power of Numbers*), collected magic squares. In 18th-century Switzerland, Leonhard Euler also became interested in them, and devised a form that he named Latin squares. The rows and columns in a Latin square contain figures or symbols that appear only once in each row and column.

One derivation of the Latin square—Sudoku—has become a popular puzzle. Devised in the US in the 1970s (where it was called Number Place), Sudoku took off in Japan in the 1980s, acquiring its now-familiar name, which means "single digits." A Sudoku puzzle is a nine-by-nine Latin square with the added restriction that subdivisions of the square must also contain all nine numbers. ▪

The most magically magical of any magic square ever made by a magician.
Benjamin Franklin
Talking about a magic square that he discovered

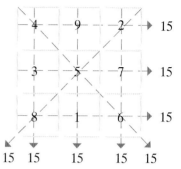

The Lo Shu magic square has a magic total of 15.

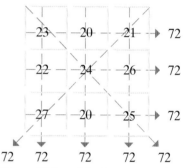

Here, 19 is added to each of the numbers in the Lo Shu square; the magic total is 72.

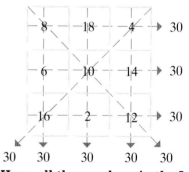

Here, all the numbers in the Lo Shu square have been doubled; the magic total is 30.

Once you have one magic square, you can add the same quantity to every number in the square and still end up with a magic square. Similarly, if you multiply all the numbers by the same quantity, you still have a magic square.

NUMBER
IS THE CAUSE OF
GODS AND DAEMONS

PYTHAGORAS

The 6th-century BCE Greek philosopher Pythagoras is also antiquity's most famous mathematician. Whether or not he was responsible for all the many achievements attributed to him in math, science, astronomy, music, and medicine, there is no doubt that he founded an exclusive community that lived for the pursuit of mathematics and philosophy, and regarded numbers as the sacred building blocks of the Universe.

Angles and symmetry

The Pythagoreans were masters of geometry and knew that the sum of the three angles of a triangle (180°) is equal to the sum of two right angles (90° + 90°), a fact which two centuries later was described by Euclid as the triangle postulate. Pythagoras's followers were also aware of some of the regular polyhedra; these are the perfectly symmetrical three-dimensional shapes (such as the cube) that were later known as the Platonic solids.

Pythagoras himself is primarily associated with the formula that describes the relationship between the sides of a right-angled triangle.

Thales of Miletus, one of the Seven Sages of ancient Greece, possibly inspired the younger Pythagoras with his geometrical and scientific ideas. They may have met in Egypt.

Widely known as Pythagoras's theorem, it states that $a^2 + b^2 = c^2$, where c is the longest side of the triangle (the hypotenuse), and a and b represent the other two, shorter sides that are adjacent

The smallest, or most primitive, of the Pythagorean triples is a triangle with side lengths 3, 4, and 5. As this graphic shows, 9 plus 16 equals 25.

Pythagorean triples

The sets of three integers that solve the equation $a^2 + b^2 = c^2$ are known as Pythagorean triples, although their existence was known long before Pythagoras. Around 1800 BCE, the Babylonians recorded sets of Pythagorean numbers on the Plimpton 322 clay tablet; these show that triples become more spread out as the number line progresses. The Pythagoreans developed methods for finding sets of triples, and also proved that there are an infinite number of such sets. After many

of Pythagoras's schools were destroyed in a 6th-century BCE political purge, Pythagoreans emigrated to other parts of southern Italy, spreading their knowledge of triples across the ancient world. Two centuries later, Euclid developed a formula to generate triples: $a = m^2 - n^2$, $b = 2mn$, $c = m^2 + n^2$. With certain exceptions, m and n can be any two integers, such as 7 and 4, which produce the triple 33, 56, 65 ($33^2 + 56^2 = 65^2$). The formula dramatically sped up the process of finding new Pythagorean triples.

See also: Irrational numbers 44–45 ▪ The Platonic solids 48–49 ▪ Syllogistic logic 50–51 ▪ Calculating pi 60–65 ▪ Trigonometry 70–75 ▪ The golden ratio 118–23 ▪ Projective geometry 154–55

The graphic below demonstrates why the Pythagorean equation ($a^2 + b^2 = c^2$) works. Within a large square there are four right-angled triangles of equal size (sides labeled a, b, and c). They are arranged so that a tilted square is formed in the middle, by the hypotenuses (c sides) of the four triangles.

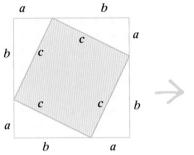

The large square, with area A, has a side length of ($a + b$). Its area is therefore equal to ($a + b$)^2. $A = (a+b)(a+b)$.

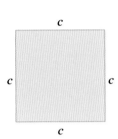

The smaller, tilted square inside the large square has an area of c^2.

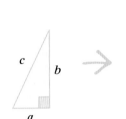

Each triangle has an area of $^{ab}/_2$ (the base a multiplied by the height b and divided by 2). The total area of all four triangles is $^{4ab}/_2 = 2ab$.

The total area of the tilted square plus the triangles is equal to the area of the large square (A).
$A = c^2 + 2ab$

A is equal to ($a+b$)($a+b$): $(a+b)(a+b) = c^2 + 2ab$

Expand the parentheses (multiply each term in the first parentheses by each term in the second). Add it all together: $a^2 + b^2 + 2ab = c^2 + 2ab$

Subtract $2ab$ from each side: $a^2 + b^2 = c^2$

---

to the right angle. For example, a right-angled triangle with two shorter sides of lengths 3in and 4in will have a hypotenuse of length 5in. The length of this hypotenuse is found because $3^2 + 4^2 = 5^2$ ($9 + 16 = 25$). Such sets of whole-number solutions to the equation $a^2 + b^2 = c^2$ are known as Pythagorean triples. Multiplying the triple 3, 4, and 5 by 2 produces another Pythagorean triple: 6, 8, and 10 ($36 + 64 = 100$). The set 3, 4, 5 is called a "primitive" Pythagorean triple because its components share no common divisor larger than 1. The set 6, 8, 10 is not primitive as its components share the common divisor 2.

There is good evidence that the Babylonians and the Chinese were well aware of the mathematical relationship between sides of a right-angled triangle centuries before Pythagoras's birth. However, Pythagoras is believed to have been the first to prove the truth of the formula that states this relationship, and its validity for all right-angled triangles, which is why the theorem takes his name.

Journeys of discovery

Pythagoras was well-traveled, and the ideas he absorbed from other countries undoubtedly fueled his mathematical inspiration. Hailing from Samos, which was not far from Miletus in western Anatolia (present-day Turkey), he may have studied at the school of Thales of Miletus under the philosopher Anaximander. He embarked on his travels at the age of 20, and spent many years away. He is thought to have visited Phoenicia, Persia, Babylon, and Egypt, and may also have reached

India. The Egyptians knew that a triangle with sides of 3, 4, and 5 (the first Pythagorean triple) would produce a right angle, so their surveyors used ropes of these lengths to construct perfect right angles for their building projects. Observing this method firsthand may have encouraged »

Reason is immortal, all else is mortal.
Pythagoras

Proving **every** instance of a conjecture (an unproven theorem) would take **forever**.

Instead, mathematicians try to prove the **underlying theorem**.

Once the theorem has been proved, **the truth of every instance follows**.

Pythagoras's theorem is a clear example of this process, as it proves that the sides of every right-angled triangle follow the rule $a^2 + b^2 = c^2$.

Pythagoras to study and prove the underlying mathematical theorem. In Egypt, Pythagoras may also have met Thales of Miletus, a keen geometrician, who calculated the heights of pyramids and applied deductive reasoning to geometry.

A Pythagorean community

After 20 years of traveling, Pythagoras eventually settled in Croton (now Crotone), southern Italy, a city with a large Greek population. There, he established the Pythagorean brotherhood— a community to whom he could teach both his mathematical and philosophical beliefs. Women were welcome in the brotherhood, and

Strength of mind rests in sobriety; for this keeps your reason unclouded by passion.
Pythagoras

formed a significant part of its 600 members. When they joined, members were obliged to give all their possessions and wealth to the brotherhood, and also swore to keep its mathematical discoveries secret. Under Pythagoras's leadership, the community gained considerable political influence.

As well as his theorem, Pythagoras and his close-knit community made numerous other advances in mathematics, but carefully guarded that knowledge. Among their discoveries were polygonal numbers: these, when represented by dots, can form the shapes of regular polygons. For example, 4 is a polygonal number as 4 dots can form a square, and 10 is a polygonal number as 10 dots can form a triangle with 4 dots at the base, 3 dots on the next row, 2 on the next, and 1 dot at the top of the triangle ($4 + 3 + 2 + 1 = 10$).

Two millennia after Pythagoras, in 1638, Pierre de Fermat enlarged on this idea when he asserted that any number could be written as the sum of up to k k-gonal numbers; in other words, every single number is the sum of up to 3 triangular numbers, up to 4 square numbers, or up to 5 pentagonal numbers, and so on. For example, 19 can

be written as the sum of three triangular numbers: $1 + 3 + 15 = 19$. Fermat could not prove this conjecture; it was only in 1813 that French mathematician Augustin-Louis Cauchy completed the proof.

Fascinated by numbers

Another type of number that excited Pythagoras was the perfect number. It was so called because it is the exact sum of all the divisors less than itself. The first perfect number is 6, as its divisors 1, 2, and 3 add up to 6. The second is 28 ($1 + 2 + 4 + 7 + 14 = 28$), the third 496, and the fourth 8,128.

The finest type of man gives himself up to discovering the meaning and purpose of life itself… this is the man I call a philosopher.
Pythagoras

I have often admired the
mystical way of Pythagoras,
and the secret
magick of numbers.
Sir Thomas Browne
English polymath

There was no practical value in identifying such numbers, but their quirkiness and the beauty of their patterns fascinated Pythagoras and his brotherhood.

By contrast, Pythagoras was said to have an overwhelming fear and disbelief of irrational numbers, numbers that cannot be expressed as fractions of two integers, the most famous example being π. Such numbers had no place among the well-ordered integers and fractions by which Pythagoras claimed the Universe was governed. One story suggests that his fear of irrational numbers drove his followers to drown a fellow Pythagorean—Hippasus—for revealing their existence when attempting to find $\sqrt{2}$.

Pythagoras's reputation for ruthlessness is also highlighted in a story about a member of the brotherhood who was executed for publicly disclosing that the Pythagoreans had discovered a new regular polyhedron. The new

In *The School of Athens*, painted by Raphael in 1509–11 for the Vatican in Rome, Pythagoras is shown with a book, surrounded by scholars eager to learn from him.

shape was formed from 12 regular pentagons, and known as the dodecahedron—one of the five Platonic solids. Pythagoreans revered the pentagon, and their symbol was the pentagram, a five-pointed star with a pentagon at its center. Breaking the brotherhood's rule of secrecy by revealing their knowledge of the dodecahedron would therefore have been an especially heinous crime, punishable by death.

An integrated philosophy

In ancient Greece, mathematics and philosophy were considered complementary subjects and were studied together. Pythagoras is credited with coining the term "philosopher," from the Greek *philos* ("love") and *sophos* ("wisdom"). For Pythagoras and his successors, the duty of a philosopher was the pursuit of wisdom.

Pythagoras's own brand of philosophy integrated spiritual ideas with mathematics, science, and reasoning. Among his beliefs was the idea of metempsychosis, which he may have encountered on his travels in Egypt or elsewhere in the Middle East. This held that souls are immortal and at death transmigrate to occupy a new body. In Athens two centuries later, Plato was entranced by the idea and included it in many of his »

Pythagoras noticed that **numerical patterns** are found in **music** and **shapes**.

Some families of numbers are **polygonal**; when represented by dots, they create **regular polygons**.

The **ratios** of the lengths of **lyre strings** are related to the sequence of **notes in a musical scale**.

A hammer double the weight of a second hammer will produce a note **an octave lower**.

Numbers and the ratios between them **govern shapes and the sounds made by musical instruments and tools**.

dialogues. Later, Christianity, too, embraced the idea of a division between body and soul; and Pythagoras's ideas would become a core part of Western thought.

Importantly for mathematics, Pythagoras also believed that everything in the Universe related to numbers and obeyed mathematical rules. Certain numbers were endowed with characteristics and spiritual significance in what amounted to a kind of number worship, and Pythagoras and his followers sought mathematical patterns in everything around them.

Numbers in harmony

Music was of great importance to Pythagoras. He is said to have considered it a holy science, rather than something simply to be used for entertainment. It was a unifying element in his concept of Harmonia, the joining together of the cosmos and the psyche. This may be why he is credited with discovering the link between mathematical ratios and harmony. It is said that, while walking past a blacksmith's forge, he noticed that different notes were produced when hammers of unequal weight were struck against equal lengths of metal. If the weights of the hammers were in exact and particular proportions, their resulting notes were harmonic.

The hammers in the forge had individual weights of 6, 8, 9, and 12 units. Those weighing 6 and 12 units sounded the same notes at different pitches; in today's music terminology they would be said to be an octave apart. The frequency of the note produced by the hammer of weight 6 was double that of the hammer weighing 12, which corresponds with the ratio of their weights. The hammers of weights 12 and 9 produced a harmonious sound—a perfect fourth—as their weights were in the ratio 4:3. The notes made by the hammers of weights 12 and 8 were also harmonious—a perfect fifth—as their weights were in the ratio

Pythagoras was reputedly an excellent lyre player. This drawing of ancient Greek musicians illustrates two members of the lyre family— the trigonon (left) and the cithara.

3:2. In contrast, the hammers of weights 9 and 8 were dissonant, as 9:8 is not a simple mathematical ratio. By noticing that harmonious musical notes were connected to numerical ratios, Pythagoras was the first to uncover the relationship between mathematics and music.

Creating a musical scale

Although scholars have questioned the story of the forge, Pythagoras is also widely credited with another musical discovery. He is said to have experimented with notes produced by lyre strings of different lengths. He found that while a vibrating string produces a note with frequency f, halving the length of the string produces a note an octave higher, with frequency $2f$. When Pythagoras used the same ratios that produced harmoniously sounding hammers, and applied them to vibrating strings, he similarly produced notes in harmony with one another. Pythagoras then constructed a musical scale, starting with one note and the note an octave above it, filling in the notes between using perfect fifths.

This scale was used until the 1700s, when it was replaced by the even-tempered scale, in which the notes between the two octaves are more evenly spaced. Although the

Pythagorean scale worked well for music lying within one octave, it was not suited for more modern music, which was written in different keys and extended across several octaves.

While there have been many different types of musical scales in use by different cultures, the long tradition of Western music dates back to the Pythagoreans and their quest to understand the relationship between music and mathematical proportions.

The legacy of Pythagoras

Pythagoras's status as the most famous mathematician from antiquity is justified by his contributions to geometry, number theory, and music. His ideas were not always original, but the rigor with which he and his followers developed them, using axioms and logic to build a system of mathematics, was a fine legacy for those who succeeded him. ∎

There is geometry in the humming of the strings, there is music in the spacing of the spheres.
Pythagoras

Pythagoras

Pythagoras was born around 570 BCE on the Greek island of Samos in the eastern Aegean Sea. His ideas have influenced many of the greatest scholars in history, from Plato to Nicolaus Copernicus, Johannes Kepler, and Isaac Newton. Pythagoras is thought to have traveled widely, assimilating ideas from scholars in Egypt and elsewhere in the Middle East before establishing his community of around 600 people in Croton, southern Italy, around 518 BCE. This ascetic brotherhood required its members to live for intellectual pursuits, while following strict rules of diet and clothing. It is from this time onward that his theorem and other discoveries were probably set down, although no records remain. At the age of 60, Pythagoras is said to have married a young member of the community, Theano, and perhaps had two or three children. Political upheaval in Croton led to a revolt against the Pythagoreans. Pythagoras may have been killed when his school was set on fire, or shortly afterward. He is said to have died around 495 BCE.

A REAL NUMBER THAT IS NOT RATIONAL
IRRATIONAL NUMBERS

IN CONTEXT

KEY FIGURE
Hippasus (5th century BCE)

FIELD
Number systems

BEFORE
19th century BCE Cuneiform inscriptions show that the Babylonians constructed right-angled triangles and understood their properties.

6th century BCE In Greece, the relationship between the side lengths of a right-angled triangle is discovered, and is later attributed to Pythagoras.

AFTER
400 BCE Theodorus of Cyrene proves the irrationality of the square roots of the nonsquare numbers between 3 and 17.

4th century BCE The Greek mathematician Eudoxus of Cnidus establishes a strong mathematical foundation for irrational numbers.

Any number that can be expressed as a ratio of two integers—a fraction, a decimal that either ends or repeats in a recurring pattern, or a percentage—is said to be a rational number. All whole numbers are rational as they can be shown as fractions divided by 1. Irrational numbers, however, cannot be expressed as a ratio of two numbers

Hippasus, a Greek scholar, is believed to have first identified irrational numbers in the 5th century BCE, as he worked on geometrical problems. He was familiar with Pythagoras's theorem, which states that the square of the hypotenuse in a right-angled triangle is equal to the sum of the squares of the other two sides. He applied the theorem to a right-angled triangle that has both shorter sides equal to 1. As $1^2 + 1^2 = 2$, the length of the hypotenuse is the square root of 2.

Hippasus realized, however, that the square root of 2 could not be expressed as the ratio of two whole numbers—that is, it could not be written as a fraction, as there is no rational number that can be multiplied by itself to produce precisely 2. This makes the square

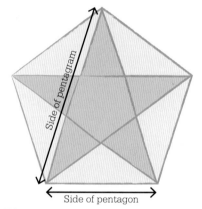

Hippasus may have encountered irrational numbers while exploring the relationship between the length of the side of a pentagon and one side of a pentagram formed inside it. He found that it was impossible to express it as a ratio between two whole numbers.

root of 2 an irrational number, and 2 itself is termed nonsquare or square-free. The numbers 3, 5, 7, and many others are similarly nonsquare and in each case, their square root is irrational. By contrast, numbers such as 4 (2^2), 9 (3^2), and 16 (4^2) are square numbers, with square roots that are also whole numbers and therefore rational.

The concept of irrational numbers was not readily accepted, although later Greek and Indian

See also: Positional numbers 22–27 ▪ Quadratic equations 28–31 ▪ Pythagoras 36–43 ▪ Imaginary and complex numbers 128–31 ▪ Euler's number 186–91

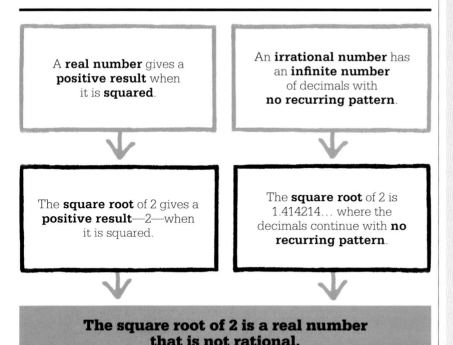

A **real number** gives a **positive result** when it is **squared**.

An **irrational number** has an **infinite number** of decimals with **no recurring pattern**.

The **square root** of 2 gives a **positive result**—2—when it is squared.

The **square root** of 2 is 1.414214… where the decimals continue with **no recurring pattern**.

The square root of 2 is a real number that is not rational.

Hippasus

Details of Hippasus's early life are sketchy, but it is thought that he was born in Metapontum, in Magna Graecia (now southern Italy), around 500 BCE. According to the philosopher Iamblichus, who wrote a biography of Pythagoras, Hippasus was a founder of a Pythagorean sect called the Mathematici, which fervently believed that all numbers were rational.

Hippasus is usually credited with discovering irrational numbers, an idea that would have been considered heresy by the sect. According to one story, Hippasus drowned when his fellow Pythagoreans threw him over the side of a boat in disgust. Another story suggests that a fellow Pythagorean discovered irrational numbers, but Hippasus was punished for telling the outside world about them. The year of Hippasus's death is not known but is likely to have been in the 5th century BCE.

Key works

5th century BCE *Mystic Discourse*

mathematicians explored their properties. In the 9th century, Arab scholars used them in algebra.

In decimal terms

The positional decimal system of Hindu–Arabic numeration allowed further study of irrational numbers, which can be shown as an infinite series of digits after the decimal point with no recurring pattern. For example, 0.1010010001… with an extra zero between each successive pair of 1s, continuing indefinitely, is an irrational number. Pi (π), which is the ratio of the circumference of a circle to its diameter, is irrational. This was proved in 1761 by Johann Heinrich Lambert—earlier estimations of π had been 3 or $^{22}/_7$.

Between any two rational numbers, another rational number can always be found. The average of the two numbers will also be rational, as will the average of that number and either of the original numbers. Irrational numbers can also be found between any two rational numbers. One method is to change a digit in a recurring sequence. For example, an irrational number can be found between the recurring numbers 0.124124… and 0.125125… by changing 1 to 3 in the second cycle of 124, to give 0.124324…, and doing so again at the fifth, then ninth cycle, increasing the gap between the replacement 3s by one cycle each time.

One of the great challenges of modern number theory has been establishing whether there are more rational or irrational numbers. Set theory strongly indicates that there are many more irrational numbers than rational numbers, even though there are infinite numbers of each. ▪

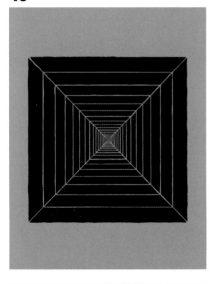

THE QUICKEST RUNNER CAN NEVER OVERTAKE THE SLOWEST
ZENO'S PARADOXES OF MOTION

IN CONTEXT

KEY FIGURE
Zeno of Elea (c. 495–430 BCE)

FIELD
Logic

BEFORE
Early 5th century BCE The Greek philosopher Parmenides founds the Eleatic school of philosophy in Elea, a Greek colony in southern Italy.

AFTER
350 BCE Aristotle produces his treatise *Physics*, in which he draws on the concept of relative motion to refute Zeno's paradoxes.

1914 British philosopher Bertrand Russell, who described Zeno's paradoxes as immeasurably subtle, states that motion is a function of position with respect to time.

Zeno of Elea belonged to the Eleatic school of philosophy that flourished in ancient Greece in the 5th century BCE. In contrast to the pluralists, who believed that the Universe could be divided into its constituent atoms, Eleatics believed in the indivisibility of all things.

Zeno wrote 40 paradoxes to show the absurdity of the pluralist view. Four of these—the dichotomy paradox, Achilles and the tortoise, the arrow paradox, and the stadium paradox—address motion. The dichotomy paradox shows the absurdity of the pluralist view that motion can be divided. A body moving a certain distance, it says, would have to reach the halfway point before it arrived at the end, and in order to reach that halfway mark, it would first have to reach the quarter-way mark, and so on ad infinitum. Because the body has to pass through an infinite number of points, it would never reach its goal.

In the paradox of Achilles and the tortoise, Achilles, who is 100 times faster than the tortoise, gives the creature a head start of 100 meters in a race. At the sound

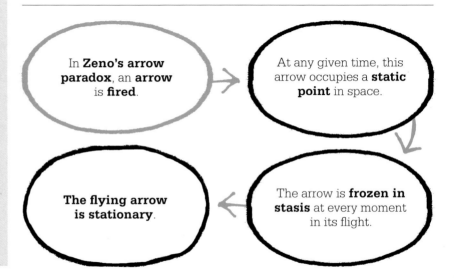

In **Zeno's arrow paradox**, an **arrow** is **fired**.

At any given time, this arrow occupies a **static point** in space.

The arrow is **frozen in stasis** at every moment in its flight.

The flying arrow is stationary.

See also: Pythagoras 36–43 ▪ Syllogistic logic 50–51 ▪ Calculus 168–75 ▪ Transfinite numbers 252–53 ▪ The logic of mathematics 272–73 ▪ The infinite monkey theorem 278–79

of the starting signal, Achilles runs 100 meters to reach the tortoise's starting point, while the tortoise runs 1 meter, giving it a 1 meter lead. Undeterred, Achilles runs another meter; however, in the same time, the tortoise runs one-hundredth of a meter, so it is still in the lead. This continues, and Achilles never catches up.

The stadium paradox concerns three columns of people, each containing an equal number of people; one group is at rest, while the other two run past each other at the same speed in opposite directions. According to the paradox, a person in one moving group can pass two people in the other moving group in a fixed time, but only one person in the stationary group. The paradoxical conclusion is that half a given time is equivalent to double that time.

Over the centuries, many mathematicians have refuted the paradoxes. The development of calculus allowed mathematicians to deal with infinitesimal quantities without resulting in contradiction. ∎

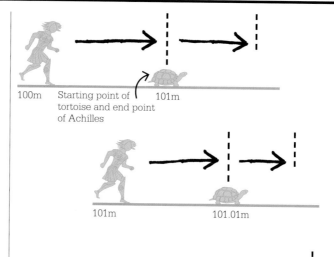

100m   Starting point of tortoise and end point of Achilles   101m

101m   101.01m

101.01m   101.0101m

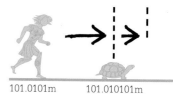

101.0101m   101.010101m

The paradox of Achilles and the tortoise maintains that a fast object, such as Achilles, will never catch up with a slow one, such as a tortoise. Achilles will get closer to the tortoise, but never actually overtake it.

Zeno of Elea

Zeno of Elea was born around 495 BCE in the Greek city of Elea (now Velia, in southern Italy). At a young age, he was adopted by the philosopher Parmenides, and was said to have been "beloved" by him. Zeno was inducted into the school of Eleatic thought, founded by Parmenides. At the age of around 40, Zeno traveled to Athens, where he met Socrates. Zeno introduced the Socratic philosophers to Eleatic ideas.

Zeno was renowned for his paradoxes, which contributed to the development of mathematical rigor. Aristotle later described him as the inventor of the dialectical method (a method starting from two opposing viewpoints) of logical argument. Zeno collected his arguments in a book, but this did not survive. The paradoxes are known from Aristotle's treatise *Physics*, which lists nine of them.

Although little is known of Zeno's life, the ancient Greek biographer Diogenes claimed he was beaten to death for trying to overthrow the tyrant Nearchus. In a clash with Nearchus, Zeno is reported to have bitten off the man's ear.

THEIR COMBINATIONS GIVE RISE TO ENDLESS COMPLEXITIES

THE PLATONIC SOLIDS

IN CONTEXT

KEY FIGURE
Plato (c. 428–348 BCE)

FIELD
Geometry

BEFORE
6th century BCE Pythagoras
identifies the tetrahedron,
cube, and dodecahedron.

4th century BCE Theaetetus,
an Athenian contemporary
of Plato, discusses the
octahedron and icosahedron.

AFTER
c. 300 BCE Euclid's *Elements*
fully describes the five regular
convex polyhedra.

1596 German astronomer
Johannes Kepler proposes
a model of the Solar System,
explaining it geometrically
in terms of Platonic solids.

1735 Leonhard Euler devises
a formula that links the
faces, vertices, and edges
of polyhedra.

A regular **polygon** has
equal angles
and equal **sides**.

Only **five solids** (3-D shapes)
have identical vertices and
faces that are all identical
regular polygons.

These **five solids**
are the **tetrahedron,
cube, octahedron,
dodecahedron**, and
icosahedron.

They are known as the
Platonic solids.

The perfect symmetry of
the five Platonic solids was
probably known to scholars
long before the Greek philosopher
Plato popularized the forms in his
dialogue *Timaeus*, written in
c. 360 BCE. Each of the five regular
convex polyhedra—3-D shapes with
flat faces and straight edges—has
its own set of identical polygonal
faces, the same number of faces
meeting at each vertex, as well as
equilateral sides, and same-sized
angles. Theorizing on the nature of
the world, Plato assigned four of the
shapes to the classical elements:
the cube (also known as a regular
hexahedron) was associated with
earth; the icosahedron with water;
the octahedron with air; and the
tetrahedron with fire. The 12-faced
dodecahedron was associated with
the heavens and its constellations.

Composed of polygons

Only five regular polyhedra are
possible—each one created either
from identical equilateral triangles,
squares, or regular pentagons, as
Euclid explained in Book XIII of
his *Elements*. To create a Platonic
solid, a minimum of three identical
polygons must meet at a vertex,
so the simplest is a tetrahedron—

The Platonic solids

A tetrahedron has four triangular faces.

A cube has six square faces.

An octahedron has eight triangular faces.

A dodecahedron has 12 pentagonal faces.

An icosahedron has 20 triangular faces.

a pyramid made up of four equilateral triangles. Octahedra and icosahedra are also formed with equilateral triangles, while cubes are created from squares, and dodecahedra are constructed with regular pentagons.

Platonic solids also display duality: the vertices of one polyhedron correspond to the faces of another. For example, a cube, which has six faces and eight vertices, and an octahedron (eight faces and six vertices) form a dual pair. A dodecahedron (12 faces and 20 vertices), and an icosahedron (20 faces and 12 vertices) form another dual pair. Tetrahedra, which have four faces and four vertices, are said to be self-dual.

Shapes in the Universe?

Like Plato, later scholars sought Platonic solids in nature and the Universe. In 1596, Johannes Kepler reasoned that the positions of the six planets then known (Mercury, Venus, Earth, Mars, Jupiter, and Saturn) could be explained in terms of the Platonic solids. Kepler later acknowledged he was wrong, but his calculations led him to discover that planets have elliptical orbits.

In 1735, Swiss mathematician Leonhard Euler noted a further property of Platonic solids, later shown to be true for all polyhedra. The sum of the vertices (V) minus the number of edges (E) plus the number of faces (F) always equals 2, that is, $V - E + F = 2$.

It is also now known that Platonic solids are indeed found in nature—in certain crystals, viruses, gases, and the clustering of galaxies. ▪

Plato

Born around 428 BCE to wealthy Athenian parents, Plato was a student of Socrates, who was also a family friend. Socrates' execution in 399 BCE deeply affected Plato and he left Greece to travel. During this period his discovery of the work of Pythagoras inspired a love of mathematics. Returning to Athens, in 387 BCE he founded the Academy, inscribing over its entrance the words "Let no one ignorant of geometry enter here." Teaching mathematics as a branch of philosophy, Plato emphasized the importance of geometry, believing that its forms—especially the five regular convex polyhedra— could explain the properties of the Universe. Plato found perfection in mathematical objects, believing they were the key to understanding the differences between the real and the abstract. He died in Athens around 348 BCE.

Key works

c. 375 BCE *The Republic*
c. 360 BCE *Philebus*
c. 360 BCE *Timaeus*

DEMONSTRATIVE KNOWLEDGE MUST REST ON NECESSARY BASIC TRUTHS
SYLLOGISTIC LOGIC

IN CONTEXT

KEY FIGURE
Aristotle (384–322 BCE)

FIELD
Logic

BEFORE
6th century BCE Pythagoras
and his followers develop a
systematic method of proof
for geometric theorems.

AFTER
c. 300 BCE Euclid's *Elements*
describes geometry in terms of
logical deduction from axioms.

1677 Gottfried Leibniz
suggests a form of symbolic
notation for logic, anticipating
the development of
mathematical logic.

1854 George Boole publishes
The Laws of Thought, his
second book on algebraic logic.

1884 *The Foundations of
Arithmetic* by German
mathematician Gottlob Frege
examines the logical principles
underpinning mathematics.

I n Classical Greece, there was
no clear distinction between
mathematics and philosophy;
the two were considered
interdependent. For philosophers,
one important principle was the
formulation of cogent arguments
that followed a logical progression
of ideas. The principle was based
on Socrates' dialectal method of
questioning assumptions to expose
inconsistencies and contradictions.
Aristotle, however, did not find this
model entirely satisfactory, so he
set about determining a systematic
structure for logical argument.
First, he identified the different
kinds of proposition that can be
used in logical arguments, and how
they can be combined to reach a
logical conclusion. In *Prior Analytics*,
he describes the propositions as
being of broadly four types, in the
form of "all S are P," "no S are P,"
"some S are P," and "some S are not
P," where S is a subject, such as

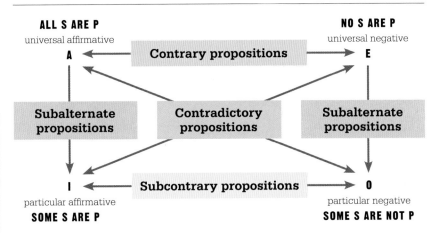

In the Square of Opposition, S is a subject, such as "sugar," and P a predicate,
such as "sweet." A and O are contradictory, as are E and I (if one is true, the other
is false, and vice versa). A and E are contrary (both cannot be true but both can
be false); I and O are subcontrary: both can be true but both cannot be false. I is
a subaltern of A and O is a subaltern of E. In syllogistic logic, this means that if
A is true, I must be true, but that if I is false, A must be false as well.

See also: Pythagoras 36–43 ▪ Zeno's paradoxes of motion 46–47 ▪ Euclid's *Elements* 52–57 ▪ Boolean algebra 242–47 ▪ The logic of mathematics 272–73

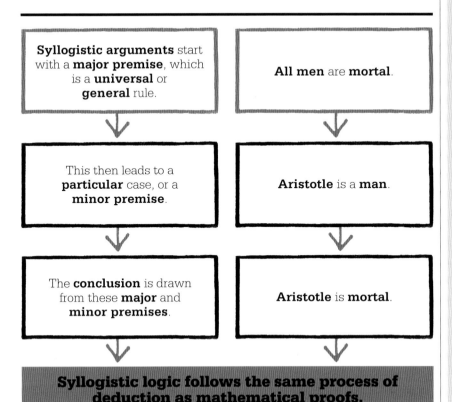

Syllogistic arguments start with a **major premise**, which is a **universal** or **general** rule.

All men are **mortal**.

This then leads to a **particular** case, or a **minor premise**.

Aristotle is a **man**.

The **conclusion** is drawn from these **major** and **minor premises**.

Aristotle is **mortal**.

Syllogistic logic follows the same process of deduction as mathematical proofs.

Aristotle

The son of a physician at the Macedonian court, Aristotle was born in 384 BCE, in Stagira, Chalkidiki. At the age of about 17, he left to study at Plato's Academy in Athens, where he excelled. Soon after Plato's death, anti-Macedonian prejudice forced him to leave Athens. He continued his academic work in Assos (now in Turkey). In 343 BCE, Philip II recalled him to Macedonia to head the school at the court; one of his students was Philip's son, later known as Alexander the Great.

In 335 BCE, Aristotle returned to Athens and founded the Lyceum, a rival institution to the Academy. In 323 BCE, after Alexander's death, Athens again became fiercely anti-Macedonian, and Aristotle retired to his family estate in Chalcis, on Euboea. He died there in 322 BCE.

Key works

c. 350 BCE *Prior Analytics*
c. 350 BCE *Posterior Analytics*
c. 350 BCE *On Interpretation*
335–323 BCE *Nichomachean Ethics*
335–323 BCE *Politics*

sugar, and P the predicate—a quality, such as sweet. From just two such propositions an argument can be constructed and a conclusion deduced. This is, in essence, the logical form known as the syllogism: two premises leading to a conclusion. Aristotle identified the structure of syllogisms that are logically valid, those where the conclusion follows from the premises, and those that are not, where the conclusion does not follow from the premises, providing a method for both constructing and analyzing logical arguments.

Seeking a rigorous proof

Implicit in his discussion of valid syllogistic logic is the process of deduction, working from a general rule in the major premise, such as "All men are mortal," and a particular case in the minor premise, such as "Aristotle is a man," to reach a conclusion that necessarily follows—in this case, "Aristotle is mortal." This form of deductive reasoning is the foundation of mathematical proofs.

Aristotle notes in *Posterior Analytics* that, even in a valid syllogistic argument, a conclusion cannot be true unless it is based on premises accepted as true, such as self-evident truths or axioms. With this idea, he established the principle of axiomatic truths as the basis for a logical progression of ideas—the model for mathematical theorems from Euclid onward. ∎

THE WHOLE IS GREATER THAN THE PART

EUCLID'S *ELEMENTS*

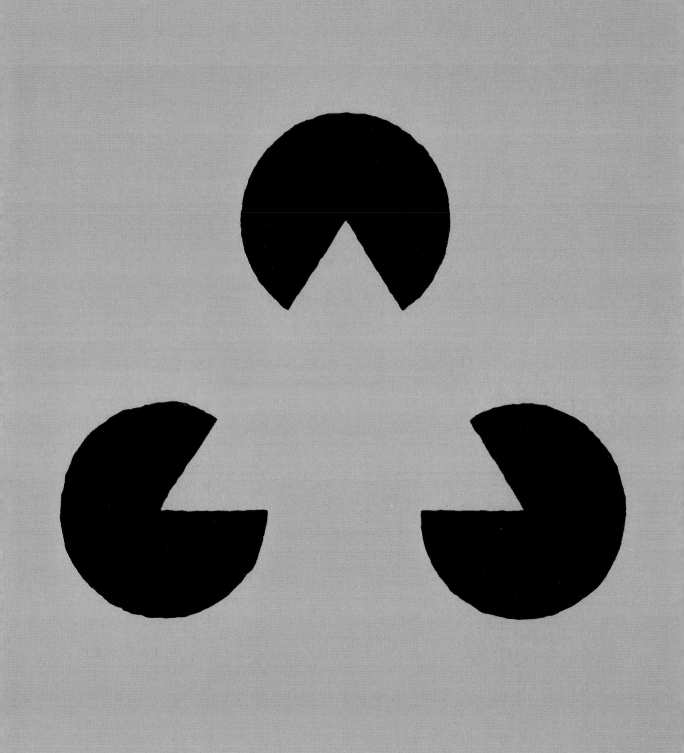

IN CONTEXT

KEY FIGURE
Euclid (c. 300 BCE)

FIELD
Geometry

BEFORE
c. 600 BCE The Greek philosopher, mathematician, and astronomer Thales of Miletus deduces that the angle inscribed inside a semicircle is a right angle. This becomes Proposition 31 of Euclid's *Elements*.

c. 440 BCE The Greek mathematician Hippocrates of Chios writes the first systematically organized geometry textbook, *Elements*.

AFTER
c. 1820 Mathematicians such as Carl Friedrich Gauss, János Bolyai, and Nicolai Ivanovich Lobachevsky begin to move toward hyperbolic non-Euclidean geometry.

uclid's *Elements* has a strong claim for being the most influential mathematical work of all time. It dominated human conceptions of space and number for more than 2,000 years and was the standard geometrical textbook until the start of the 1900s.

Euclid lived in Alexandria, Egypt, in around 300 BCE, when the city was part of the culturally rich Greek-speaking Hellenistic world that flourished around the Mediterranean Sea. He would have written on papyrus, which is not very durable; all that remains of his work are the copies, translations, and commentaries made by later scholars.

Collection of works

The *Elements* is a collection of 13 books that range widely in subject matter. Books I to IV tackle plane geometry—the study of flat surfaces. Book V addresses the idea of ratio and proportion, inspired by the thinking of the Greek mathematician and astronomer Eudoxus of Cnidus. Book VI contains more advanced plane geometry. Books VII to IX

There is
no royal road
to geometry.
Euclid

are devoted to number theory and discuss the properties and relationships of numbers. The long and difficult Book X deals with incommensurables. Now known as irrational numbers, these numbers cannot be expressed as a ratio of integers. Books XI to XIII examine three-dimensional solid geometry.

Book XIII of the *Elements* is actually attributed to another author—Athenian mathematician and disciple of Plato, Theaetetus, who died in 369 BCE. It covers the five regular convex solids—the tetrahedron, cube, octahedron, dodecahedron, and icosahedron, which are often called the Platonic

---

Euclid

Details of Euclid's date and place of birth are unknown and knowledge of his life is scant. It is thought that he studied at the Academy in Athens, which had been founded by Plato. In the 5th century CE, the Greek philosopher Proclus wrote in his history of mathematicians that Euclid taught at Alexandria during the reign of Ptolemy I Soter (323–285 BCE).

Euclid's work covers two areas: elementary geometry and general mathematics. In addition to the *Elements*, he wrote about perspective, conic sections, spherical geometry, mathematical astronomy, number theory, and the importance of mathematical rigor. Several of the works attributed to Euclid have been lost, but at least five have survived to the 21st century. It is thought that Euclid died between the mid-4th century and the mid-3rd century BCE.

Key works

Elements
Conics
Catoptrics
Phaenomena
Optics

See also: Pythagoras 36–43 ▪ The Platonic solids 48–49 ▪ Syllogistic logic 50–51 ▪ Conic sections 68–69 ▪ The problem of maxima 142–43 ▪ Non-Euclidean geometries 228–29

solids—and is the first recorded example of a classification theorem (one that itemizes all possible figures given certain limitations).

Euclid is known to have written an account of conic sections, but this work has not survived. Conic sections are figures formed from the intersection of a plane and a cone and they may be circular, elliptical, or parabolic in shape.

World of proof

The title of Euclid's work has a particular meaning that reflects his mathematical approach. In the 1900s, British mathematician John Fauvel maintained that the meaning of the Greek word for "element," *stoicheia*, changed over time, from "a constituent of a line," such as an olive tree in a line of trees, to "a proposition used to prove another," and eventually evolved to mean "a starting point for many other theorems." This is the sense in which Euclid used it. In the 5th century CE, the philosopher Proclus talked of an element as "a letter of an alphabet," with combinations of letters creating words in the same way that combinations of axioms— statements that are self-evidently true—create propositions.

Logical deductions

Euclid was not writing in a vacuum; he built upon foundations laid by a number of influential Greek mathematicians who came before him. Thales of Miletus, Hippocrates, and Plato (among others) had all begun to move toward the mathematical mindset that Euclid so brilliantly formalized: the world of proof. It is this that makes Euclid unique; his writings are the earliest surviving example of fully axiomatized mathematics. He identified certain basic facts and progressed from there to statements that were sound logical deductions (propositions). Euclid also managed to assemble all the mathematical knowledge of his day, and organize it into a mathematical structure where the

This opening page of Euclid's *Elements* shows illuminated Latin text with diagrams and comes from the first printed edition, produced in Venice in 1482.

logical relationships between the various propositions were carefully explained.

Euclid faced a Herculean task when he attempted to systematize the mathematics that lay before him. In devising his axiomatic system, he began with 23 definitions for terms such as point, line, surface, circle, and diameter. He then put forward five postulates: any two points can be joined with a straight line segment; any straight line segment can be extended to infinity; given any straight line segment, a circle can be drawn having the segment as its radius and one endpoint as its center; all right angles are equal to one another; and a postulate about parallel lines (see p.56).

He then went on to add five axioms, or common notions; if $A = B$ and $B = C$, then $A = C$; if $A = B$ and $C = D$, $A + C = B + D$; »

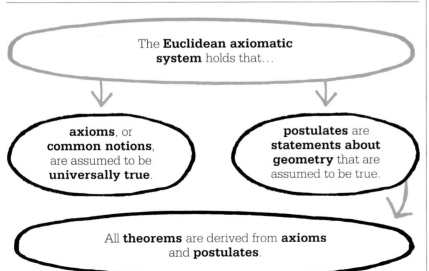

The **Euclidean axiomatic system** holds that…

axioms, or **common notions**, are assumed to be **universally true**.

postulates are **statements about geometry** that are assumed to be true.

All **theorems** are derived from **axioms** and **postulates**.

Euclid's five postulates

1. Any two points can be joined with a straight line segment.

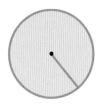

2. Any straight line segment can be extended to infinity.

3. Given a center and a radius, a circle can always be drawn with this center and this radius.

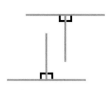

4. All right angles are equal to one another.

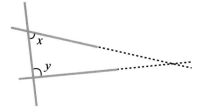

5. If $x + y$ is less than two right angles, then the lines must eventually meet on one side.

---

if $A = B$ and $C = D$, then $A − C = B − D$; if A coincides with B, then A and B are equal; and the whole of A is greater than part of A.

To prove Proposition 1 (see opposite), Euclid drew a line with endpoints labeled A and B (see below). Taking each endpoint as a center, he then drew two intersecting circles, so that each had the radius AB. This used his third postulate. Where the circles met, he called that point C, and he could draw two more lines AC and BC, calling on his first postulate. The radius of the two circles is the same, so $AC = AB$ and $BC = AB$; this means that $AC = BC$, which is Euclid's first axiom (things that are equal to the same thing are also equal to one another). It follows that $AB = BC = CA$, meaning that he had drawn an equilateral triangle on AB.

In Latin translations of *Elements,* deductions end with the letters QEF (*quod erat faciendum,* meaning "which was to be [and has been] done." Logical proofs end with QED (*quod erat demonstrandum,* meaning "which was to be [and has been] demonstrated").

The equilateral triangle construction is a good example of Euclid's method. Each step has to be justified by reference to the definitions, the postulates, and the axioms. Nothing else can be taken as obvious, and intuition is regarded as potentially suspect.

Euclid's very first proposition was criticized by later writers. They noted, for instance, that Euclid did not justify or explain the existence of C, the point of intersection of the two circles. Although apparent, it is not mentioned in his preliminary assumptions. Postulate 5 talks about a point of intersection, but that is between two lines, and not two circles. Similarly, one of the definitions describes a triangle as a plane figure bounded by three lines, which all lie in that plane. However, it seems that Euclid did not explicitly show that the lines AB, BC, and CA lie in the same plane.

Postulate 5 is also known as the "parallel postulate" because it can be used to prove properties of parallel lines. It says that if a straight line crossing two straight lines (A, B) creates interior angles on one side that total less than two right angles (180°), lines A and B will eventually cross on that side, if extended indefinitely. Euclid did not use it until Proposition 29, in which he stated that one condition for a straight line crossing two parallel lines was that the interior

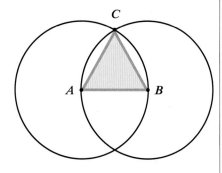

To construct an equilateral triangle, for Proposition 1, Euclid drew a line and centered a circle on its endpoints, here A and B. By drawing a line from each endpoint to C, where the circles intersect, he created a triangle with sides AB, AC, and BC of equal length.

Geometry is knowledge of what always exists.
Plato

angles on the same side were equal to two right angles. The fifth postulate is more elaborate than the other four, and Euclid himself seems to have been wary of it.

A vital part of any axiomatic system is to have enough axioms, and postulates in the case of Euclid, to derive every true proposition, but to avoid superfluous axioms that can be derived from others. Some asked whether the parallel postulate could be proved as a proposition using Euclid's common notions, definitions, and the other four postulates; if it could, the fifth was unnecessary. Euclid's contemporaries and later scholars made unsuccessful attempts to construct such a proof. Finally, in the 1800s, the fifth postulate was ruled both necessary for Euclid's geometry and independent of his other four postulates.

Beyond Euclidean geometry

The *Elements* also examines spherical geometry, an area explored by two of Euclid's successors, Theodosius of Bithynia and Menelaus of Alexandria. While Euclid's definition of "a point" addresses a point on the plane, a point can also be understood as a point on a sphere.

This raises the question of how Euclid's five postulates can be applied to the sphere. In spherical geometry, almost all the axioms look different from the postulates set out in Euclid's *Elements*. The *Elements* gave rise to what is called Euclidean geometry; spherical geometry is the first example of a non-Euclidean geometry. The parallel postulate is not true for spherical geometry, where all pairs of lines have points in common, nor for hyperbolic geometry, where they can meet infinite numbers of times. ∎

The first 16 propositions in Book 1	
Proposition 1	On a given finite straight line, to construct an equilateral triangle.
Proposition 2	To place at a given point (as an extremity) a straight line equal to a given straight line.
Proposition 3	Given two unequal straight lines, to cut off from the greater a straight line equal to the less.
Proposition 4	If two sides of one triangle are equal in length to two sides of another triangle, and if the angles contained by each pair of equal sides are equal, then the base of one triangle will equal the base of the other, the two triangles will be of equal area, and the remaining angles in one triangle will be equal to those in the other triangle.
Proposition 5	In an isosceles triangle, the angles at the base are equal to one another, and, if the equal straight lines are extended below the base, the angles under the base will also be equal to one another.
Proposition 6	If in a triangle two angles are equal to one another, the sides separated from the third side by these angles will also be equal.
Proposition 7	Given two straight lines constructed on a straight line (from its extremities) and meeting in a point, there cannot be constructed on the same straight line (from its extremities), and on the same side of it, two other straight lines meeting in another point and equal to the former two respectively, namely each to that which starts at the same extremity.
Proposition 8	If two sides of one triangle are equal in length to two sides of another triangle, and the base of one triangle is equal to the base of the other, the angles of the two triangles will also be equal.
Proposition 9	To bisect a given rectilineal angle.
Proposition 10	To bisect a given finite straight line.
Proposition 11	To draw a straight line at right angles to a given straight line from a given point on it.
Proposition 12	To a given infinite straight line, from a given point which is not on it, to draw a perpendicular straight line.
Proposition 13	If a straight line set up on a straight line makes angles, it will make either two right angles or angles equal to two right angles.
Proposition 14	If with any straight line, and at a point on it, two straight lines not lying on the same side and meeting at the point make adjacent angles equal to two right angles, the two straight lines will be in a straight line with one another.
Proposition 15	If two straight lines cut one another, they make the vertical angles equal to one another.
Proposition 16	In any triangle, if one of the sides is extended, the angle between the triangle and the extended side is greater than any of the angles inside the triangle.

COUNTING WITHOUT NUMBERS

THE ABACUS

IN CONTEXT

KEY CIVILIZATION
Ancient Greeks (c. 300 BCE)

FIELD
Number systems

BEFORE
c. 18,000 BCE In Central Africa, numbers are recorded on bone as carved marks.

c. 3000 BCE South American Indïans record numbers by tying knots in string.

c. 2000 BCE The Babylonians continue developing positional numbers.

AFTER
1202 Leonardo of Pisa (Fibonacci) commends the Hindu–Arabic number system in *Liber Abaci*.

1621 In England, William Oughtred invents the slide rule, which simplifies the use of logarithms.

1972 Hewlett Packard invents an electronic scientific calculator for personal use.

T he abacus is a counting device and calculator that has been in use since ancient times. It comes in many forms, but all of them work on the same principles: values of different sizes are represented by "counters" arranged in columns or rows.

Early abaci

The word "abacus" may hint at its origins. It is a Latin word derived from the ancient Greek, *abax*, which means "slab" or "board"— a surface that would have been covered in sand and used as a drawing board. The oldest surviving abacus is the Salamis Tablet, a marble slab made c. 300 BCE that is etched with horizontal lines. Pebbles were placed on these lines to count out values. The bottom line represented 0 to 4; the line above counted 5s, and the lines above that 10s, 50s, and so on. The tablet was discovered on the Greek island of Salamis in 1846.

Some scholars believe that the Salamis Tablet was actually Babylonian. The Greek *abax* may have come from the Phoenician or Hebrew word for "dust" (*abaq*) and

The Soroban Championship

Japanese schoolchildren still use the soroban (Japanese abacus) in mathematics lessons as a way of developing mental arithmetic skills. The soroban is also used for far more complex calculations. Expert soroban users can usually do such calculations more quickly than someone punching the values into an electronic calculator.

Every year, the best abacists from across Japan take part in the Soroban Championship. They are tested on their speed and accuracy in a knockout system similar to a spelling bee. One of the highlights of the event is Flash Anzan™, a feat of mental arithmetic in which the players imagine operating an abacus to add 15 three-digit numbers—no physical abacus is allowed. The contestants watch the numbers appear on a big screen, flashing by faster with each round. The 2017 world record for Flash Anzan was 15 numbers added together in 1.68 seconds.

See also: Positional numbers 22–27 ▪ Pythagoras 36–43 ▪ Zero 88–91 ▪ Decimals 132–37 ▪ Calculus 168–75

The suanpan shown here is set to the number 917,470,346. The suanpan is traditionally a 2:5 abacus—each column has two "heaven" beads, each with a value of 5, and 5 "earth" beads, each with a value of 1, giving a potential value of 15 units. This allows for calculations involving the Chinese base-16 system, which uses 15 units rather than the 9 used in the decimal system. Numbers can be added together by entering the units of one number, starting from the right, then adjusting the beads as further numbers are entered. For subtraction, the units of the first number are entered, then bead values are adjusted downward in each column as further subtracted numbers are entered.

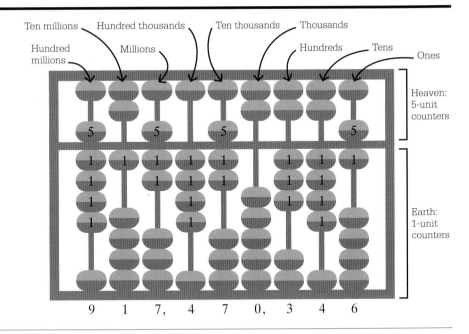

may refer to far older counting tables developed in Mesopotamian civilizations, where counters were set out on grids drawn in sand. The Babylonian positional number system, developed c. 2000 BCE, may have been inspired by the abacus.

The Romans upgraded the Greek counting table into a device that greatly simplified calculations. The horizontal rows of the Greek abacus became vertical columns in the Roman abacus, in which were set small pebbles—or *calculi* in Latin, from which we get the word "calculation."

A type of abacus was also in use in the pre-Columbian civilizations of Central America. Based on a five-digit vigesimal, or base-20, counting system, it used corn kernels threaded on strings to represent numbers. No device has survived, but scholars think that the ancient Olmec people invented it 3,000 years ago. By about 1000 CE, the Aztec people knew it as the *nepohualtzintzin*—the "personal accounts counter"—and wore it on the wrist as a bracelet.

Double base

Around the 2nd century CE, abaci had become a common tool in China. The Chinese abacus, or suanpan, matched the design of the Roman version, but rather than use pebbles set in a metal frame, it employed wooden counters on rods—the template for modern abaci. Whether the Roman or Chinese abaci came first is unclear, but their similarities may be a coincidence, inspired by the way people count using the five fingers of one hand. Both abaci have two decks—the lower deck counting to five, and the upper deck counting the fives.

A female personification of Arithmetic judges a contest between the Roman mathematician Boëthius, who uses numbers, and the Greek Pythagoras, who uses a counting board.

By the second millennium CE, the suanpan and its counting methods were becoming widespread across Asia. In the 1300s, it was exported to Japan, where it was called the soroban. This was slowly refined and by the 1900s, the soroban was a 1:4 abacus (with 1 upper bead on each rod, and 4 lower beads). ▪

EXPLORING PI IS LIKE EXPLORING THE UNIVERSE

CALCULATING PI

IN CONTEXT

KEY FIGURE
Archimedes
(c. 287–c. 212 BCE)

FIELD
Number theory

BEFORE
c. 1650 BCE The Rhind papyrus, written by Middle Kingdom Egyptian scribes as a mathematics guide, includes estimates of the value of π.

AFTER
5th century CE In China, Zu Chongzhi calculates π to seven decimal places.

1671 Scottish mathematician James Gregory develops the arctangent method for computing π. Gottfried Leibniz makes the same discovery in Germany three years later.

2019 In Japan, Emma Haruka Iwao uses a cloud computing service to calculate π to more than 31 trillion decimal places.

The fact that pi (π)—the ratio of the circumference of a circle to its diameter, roughly given as 3.141—cannot be expressed exactly as a decimal no matter how many decimal places are calculated has fascinated mathematicians for centuries. Welsh mathematician William Jones was the first to use the Greek letter π to represent the number in 1706, but its importance for calculating the circumference and area of a circle and the volume of a sphere has been understood for millennia.

Ancient texts

Determining pi's exact value is not straightforward and the quest continues to find pi's decimal representation to as many places as possible. Two of the earliest estimates for π are given in the ancient Egyptian documents known as the Rhind and Moscow papyri. The Rhind papyrus, thought to have been intended for trainee scribes, describes how to calculate the volumes of cylinders and pyramids and also the area of a circle. The method used to find the area of a circle was to find the area of a square with sides that are $8/9$ of the circle's diameter. Using this method implies that π is

> Pi is not merely the ubiquitous factor in high school geometry problems; it is stitched across the whole tapestry of mathematics.
> **Robert Kanigel**
> **American science writer**

approximately 3.1605 calculated to four decimal places, which is just 0.6 per cent greater than the most accurate known value of π.

In ancient Babylon, the area of a circle was found by multiplying the square of the circumference by $1/12$, implying that the value of π was 3. This value appears in the Bible (1 Kings 7:23): "And he made the Sea of cast bronze, ten cubits from one brim to the other; it was completely round. Its height was five cubits, and a line of thirty cubits measured its circumference."

In c. 250 BCE, the Greek scholar Archimedes developed an algorithm for determining the value of π based

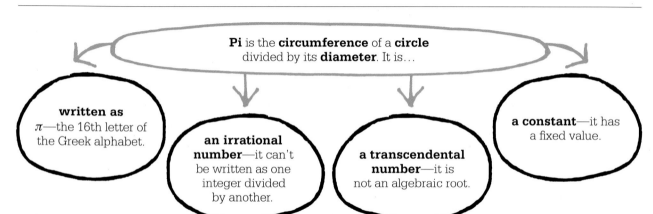

Pi is the **circumference** of a **circle** divided by its **diameter**. It is…

written as
π—the 16th letter of the Greek alphabet.

an irrational number—it can't be written as one integer divided by another.

a transcendental number—it is not an algebraic root.

a constant—it has a fixed value.

See also: The Rhind papyrus 32–33 ▪ Irrational numbers 44–45 ▪ Euclid's *Elements* 52–57 ▪ Eratosthenes' sieve 66–67 ▪ Zu Chongzhi 83 ▪ Calculus 168–75 ▪ Euler's number 186–91 ▪ Buffon's needle experiment 202–03

on constructing regular polygons that exactly fit within (inscribed), or enclosed (circumscribed), a circle. He calculated upper and lower limits for π by using Pythagoras's theorem—that the area of the square of the hypoteneuse (the side opposite the right angle) in a right-angled triangle is equal to the sum of the areas of the squares of the other two sides—to establish the relationship between the lengths of the sides of regular polygons when the number of sides was doubled. This enabled him to extend his algorithm to 96-sided polygons. Determining the area of a circle using a polygon with many sides had been proposed at least 200 years before Archimedes, but he was the first person to consider polygons that were both inscribed and circumscribed.

Squaring the circle

Another method for estimating π, "squaring the circle," was a popular challenge for mathematicians in ancient Greece. It involved constructing a square with the

Although polygons had long been used to estimate the circumference of circles, Archimedes was the first to use inscribed (inside the circle) and circumscribed (outside the circle) regular polygons to find upper and lower limits for π.

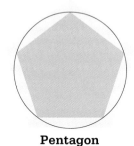
Pentagon

Hexagon

Octagon

same area as a given circle. Using only a pair of compasses and a straight edge, the Greeks would superimpose a square on a circle and then use their knowledge of the area of a square to approximate to the area of a circle. The Greeks were not successful with this method, and in the 1800s, squaring the circle was proved to be impossible, due to π's irrational nature. This is why attempts to achieve an impossible task are sometimes known as "squaring the circle." »

The works of Archimedes are, without exception, works of mathematical exposition.
Thomas L. Heath
Historian and mathematician

Archimedes

Born in c. 287 BCE in Syracuse, Sicily, the Greek polymath Archimedes excelled as a mathematician and engineer, and is also remembered for his "eureka" moment, when he realized that the volume of water displaced by an object is equal to the volume of that object. Among his claimed inventions is the Archimedes' screw, a revolving screw-shaped blade in a cylinder, which pushes water up a gradient.

In mathematics, he used practical approaches to establish the ratio of the volumes of a cylinder, sphere, and cone with

the same maximum radius and height to be 3:2:1. Many consider Archimedes to be a pioneer of calculus, which was not developed until the 1600s. He was killed by a Roman soldier during the Siege of Syracuse in 212 BCE, despite orders that his life be spared.

Key works

c. 250 BCE *On the Measurement of a Circle*
c. 225 BCE *On the Sphere and the Cylinder*
c. 225 BCE *On Spirals*

Another way mathematicians have attempted to square the circle is to slice it into sections and rearrange them into a rectangular shape (see below). The area of the rectangle is $r \times \frac{1}{2}(2\pi r) = r \times \pi r = \pi r^2$ (where r is the radius of the circle and $2\pi r$ is its circumference). The area of a circle is also πr^2. The smaller the segments used, the closer the shape is to a rectangle.

The quest spreads

More than 300 years after the death of Archimedes, Ptolemy (c. 100–170 CE) determined π to be 3:8:30 (base-60), that is, $3 + \frac{8}{60} + \frac{30}{3,600} = 3.1416$, which is just 0.007 percent greater than the closest known value of π. In China, 3 was often used as the value of π, until $\sqrt{10}$ became common from the 2nd century CE. The latter is 2.1 percent greater than π. In the 3rd century, Wang Fau stated that a circle with a circumference of 142 had a diameter of 45—that is $\frac{142}{45} = 3.15$, just 1.4 percent more than π—while Liu Hui used a 3,072-sided polygon to estimate π as 3.1416. In the 5th century, Zu Chongzhi and his son used a 24,576-sided polygon to calculate π as $\frac{355}{113} = 3.14159292$, a level of accuracy (to seven decimal places) not achieved in Europe until the 1500s.

> There is no end with pi. I would love to try with more digits.
> **Emma Haruka Iwao**
> **Japanese computer scientist**

In India, the mathematician–astronomer Aryabhata included a method for obtaining π in his *Aryabhatiyam* astronomical treatise of 499 CE: "Add 4 to 100, multiply by 8, and then add 62,000. By this rule the calculation of the circumference of a circle with a diameter of 20,000 can be approached." This works out as $[8(100 + 4) + 62,000] \div 20,000 = 62,832 \div 20,000 = 3.1416$.

Brahmagupta (c. 598–668 CE) derived square root approximations of π using regular polygons with 12, 24, 48, and 96 sides: $\sqrt{9.65}$, $\sqrt{9.81}$, $\sqrt{9.86}$, and $\sqrt{9.87}$ respectively. Having established that $\pi^2 = 9.8696$ to four decimal places, he simplified these calculations to $\pi = \sqrt{10}$.

During the 9th century, Arab mathematician al-Khwarizmi used $3\frac{1}{7}$, $\sqrt{10}$, and $\frac{62,832}{20,000}$ as values for π, attributing the first value to Greece and the other two to India. English cleric Adelard of Bath translated al-Khwarizmi's work in the 12th century, renewing an interest in the search for π in Europe. In 1220, Leonardo of Pisa (Fibonacci), who popularized Hindu-Arabic numerals in his book *Liber Abaci* (*The Book of Calculation*), 1202, computed π to be $\frac{864}{275} = 3.141$, a small improvement on Archimedes's approximation, but not as accurate as the calculations of Ptolemy, Zu Chongzhi, or Aryabhata. Two centuries later, Italian polymath Leonardo da Vinci (1452–1519) proposed making a rectangle whose length was the same as a circle's circumference and whose height was half its radius to determine the area of the circle.

Archimedes' method used in ancient Greece for calculating π was still being used in the late 16th century. In 1579, French mathematician François Viète used 393 regular polygons each with 216 sides to calculate π to 10 decimal places. In 1593, Flemish mathematician Adriaan van Roomen (Romanus) used a polygon

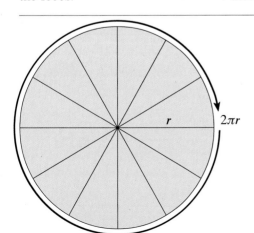

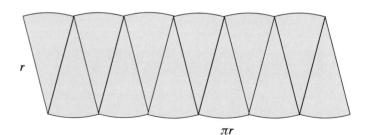

By arranging the segments of a circle in a near-rectangular shape, it can be shown that the area of a circle is πr^2. The height of the "rectangle" is approximately equal to the radius r of the circle, and the width is half of the circumference (half of $2\pi r$, which is πr).

The perimeter to height ratio of the Great Pyramid of Giza, in Egypt, is almost exactly π, which might suggest that ancient Egyptian architects were aware of the number.

with 230 sides to compute π to 17 decimal places; three years later, German–Dutch professor of mathematics Ludolph van Ceulen calculated π to 35 decimal places.

The development of arctangent series by Scottish astronomer–mathematician James Gregory in 1671, and independently by Gottfried Leibniz in 1674, provided a new approach for finding π. An arctangent (arctan) series is a way of determining the angles in a triangle from knowledge of the length of its sides, and involves radian measure, where a full turn is 2π radians (equivalent to 360°).

Unfortunately, hundreds of terms are needed to compute π to even a few decimal places using this series. Many mathematicians attempted to find more efficient methods to calculate π using arctan, including Leonhard Euler in the 1700s. Then, in 1841, British mathematician William Rutherford computed 208 digits of π using arctan series.

The advent of calculators and electronic computers in the 1900s made finding the digits of π much easier. In 1949, 2,037 digits of π were calculated in 70 hours. Four years later, it took around 13 minutes to compute 3,089 digits. In 1961, American mathematicians Daniel Shanks and John Wrench used arctan series to compute 100,625 digits in under eight hours. In 1973, French mathematicians Jean Guillaud and Martin Bouyer achieved 1 million decimal places, and in 1989, a billion decimal places were computed by Ukrainian–American brothers David and Gregory Chudnovsky.

In 2016, Peter Trueb, a Swiss particle physicist, used the y-cruncher software to calculate π to 22.4 trillion digits. A new world record was set when computer scientist Emma Haruka Iwao calculated π to more than 31 trillion decimal places in March 2019. ∎

Applying pi

Space scientists constantly use π in their calculations. For example, the length of orbits at different altitudes above a planet's surface can be worked out by using the basic principle that if the diameter of a circle is known, its circumference can be calculated by multiplying by π. In 2015, NASA scientists applied this method to compute the time it took the spacecraft Dawn to orbit Ceres, a dwarf planet in the asteroid belt between Mars and Jupiter.

When scientists at NASA's Jet Propulsion Laboratory in California wanted to know how much hydrogen might be available beneath the surface of Europa, one of Jupiter's moons, they estimated the hydrogen produced in a given unit area by first calculating Europa's surface area, which is $4\pi r^2$, as it is for any sphere. Since they knew Europa's radius, calculating its surface area was easy.

It is also possible to work out the distance traveled during one rotation of Earth by a person standing at a point on its surface using π, providing the latitude of the person's position is known.

Astrophysicists use π in their calculations to determine the orbital paths and characteristics of planetary bodies such as Saturn.

WE SEPARATE THE NUMBERS AS IF BY SOME SIEVE

ERATOSTHENES' SIEVE

In addition to calculating Earth's circumference and the distances from Earth to the Moon and Sun, the Greek polymath Eratosthenes devised a method for finding prime numbers. Such numbers, divisible only by 1 and themselves, had intrigued mathematicians for centuries. By inventing his "sieve" to eliminate nonprimes—using a number grid and crossing off multiples of 2, 3, 5, and above—Eratosthenes made prime numbers considerably more accessible.

Prime numbers have exactly two factors: 1 and the number itself. The Greeks understood the importance of primes as the building blocks of all positive integers. In his *Elements*, Euclid stated many properties of both composite numbers (integers above one that can be made by multiplying other integers) and primes. These included the fact that every integer can be written as a product of prime numbers or is itself a prime. A few decades later, Eratosthenes developed his

---

Eratosthenes developed his **"sieve"** as a **method** to **speed up** the process of **finding prime numbers**.

Numbers are written out in a table.

Multiples of prime numbers are **systematically** crossed out.

The **method** leaves you with a **grid** where **prime numbers** are **clearly** identified.

See also: Mersenne primes 124 ▪ The Riemann hypothesis 250–51 ▪ The prime number theorem 260–61 ▪ Finite simple groups 318–19

Eratosthenes' method starts with a table of consecutive numbers. First, 1 is crossed out. Then all multiples of 2 are crossed out except 2 itself. The same is then done for multiples of 3, 5, and 7. Multiples of any number higher than 7 are already crossed out, since 8, 9, and 10 are composites of 2, 3, and 5.

☐ Prime numbers

◩ 1 and composite numbers

1	2	3	4	5	6	7	8	9	10
11	12	13	14	15	16	17	18	19	20
21	22	23	24	25	26	27	28	29	30
31	32	33	34	35	36	37	38	39	40
41	42	43	44	45	46	47	48	49	50
51	52	53	54	55	56	57	58	59	60
61	62	63	64	65	66	67	68	69	70
71	72	73	74	75	76	77	78	79	80
81	82	83	84	85	86	87	88	89	90
91	92	93	94	95	96	97	98	99	100

method, which can be extended to uncover all primes. Using a number grid for 1 to 100 (see above), it is clear that 1 is not a prime number as its only factor is 1. The first prime number—and also the only even prime—is 2. As all other even numbers are divisible by 2, they cannot be primes, so all other primes must be odd. The next prime, 3, has only two factors, so all the other multiples of 3 cannot be primes. The number 4 (2 × 2) has already had its multiples removed, since they are all even. The next prime is 5, so all other multiples of 5 cannot be prime. The number 6 and all its multiples have been removed from the list of potential primes, as they are even multiples of 3. The next prime is 7, and removing its multiples eliminates 49, 77, and 91. All the multiples of 9 have gone, as they are multiples of 3, and all the multiples of 10 have been removed, because they are the even multiples of 5. The multiples of 11 up to 100 have already been removed, and so on for all successive numbers. There are only 25 prime numbers up to 100—starting with 2, 3, 5, 7, and 11, and ending with 97—all identified by simply removing every multiple of 2, 3, 5, and 7.

The search continues

Prime numbers attracted the attention of mathematicians from the 1600s onward, when figures such as Pierre de Fermat, Marin Mersenne, Leonhard Euler, and Carl Friedrich Gauss probed further into their properties.

Even in the age of computers, determining whether a large number is prime remains highly challenging. Public key cryptography—the use of two large primes to encrypt a message—is the basis of all internet security. If hackers ever do figure out a simple way of determining the prime factorization of very large numbers, a new system will need to be devised. ∎

Eratosthenes

Born around 276 BCE in Cyrene, a Greek city in Libya, Eratosthenes studied in Athens and became a mathematician, astronomer, geographer, music theorist, literary critic, and poet. He was the chief librarian at the Library of Alexandria, the greatest academic institution of the ancient world. He is known as the father of geography for founding and naming the subject as an academic discipline and developing much of the geographical language used today.

Eratosthenes also recognized that Earth is a sphere and calculated its circumference by comparing the angles of elevation of the Sun at noon at Aswan in southern Egypt and at Alexandria in the north of the country. In addition, he produced the first world map that featured meridian lines, the Equator, and even polar zones. He died around 194 BCE.

Key works

Mensuram orae ad terram (*On the Measurement of the Earth*) *Geographika* (*Geography*)

A GEOMETRICAL TOUR DE FORCE

CONIC SECTIONS

IN CONTEXT

KEY FIGURE
Apollonius of Perga
(c. 262–190 BCE)

FIELD
Geometry

BEFORE
c. 300 BCE Euclid's 13-volume *Elements* sets out the propositions that form the basis of plane geometry.

c. 250 BCE In *On Conoids and Spheroids*, Archimedes deals with the solids created by the revolution of conic sections about their axes.

AFTER
c. 1079 CE Persian polymath Omar Khayyam uses intersecting conics to solve algebraic equations.

1639 In France, 16-year-old Blaise Pascal asserts that where a hexagon is inscribed in a circle, the opposite sides of the hexagon meet at three points on a straight line.

O f the many pioneering mathematicians produced by ancient Greece, Apollonius of Perga was one of the most brilliant. He began studying mathematics after Euclid's great work *Elements* had emerged and he employed the Euclidian method of taking "axioms"—statements taken to be true—as starting points for further reasoning and proofs.

Apollonius wrote on many subjects, including optics (how light rays travel) and astronomy, as well as geometry. Much of his work survives only in fragments, but his most influential, *Conics*, is relatively

I have sent my son… to bring you… the second book of my *Conics*. Read it carefully and communicate it to such others as are worthy of it.
Apollonius of Perga

intact. It was written in eight volumes, of which seven survive: books 1–4 in Greek, and books 5–7 in Arabic. The work was designed to be read by mathematicians already well versed in geometry.

A new geometry
Early Greek mathematicians such as Euclid focused on the line and circle as the purest geometric forms. Apollonius viewed these in three-dimensional terms: if a circle is combined with all lines that emanate from it, above or below its plane, and those lines pass through the same fixed point—the vertex—a cone is created. By slicing that cone in different ways, a series of curves, known as conic sections, can be produced.

In *Conics*, Apollonius expounded in minute detail this new world of geometric construction, studying and defining the properties of conic sections. He based his workings on the assumption of two cones joined at the same vertex, with the area of their circular bases potentially stretching to infinity. To three of the conic sections he gave the names ellipse, parabola, and hyperbola. An ellipse occurs when a plane intersects a cone on a slant.

See also: Euclid's *Elements* 52–57 ▪ Coordinates 144–51 ▪ The area under a cycloid 152–53 ▪ Projective geometry 154–55
▪ The complex plane 214–15 ▪ Non-Euclidean geometries 228–29 ▪ Proving Fermat's last theorem 320–23

A parabola emerges if the cut is parallel to the edge of the cone, and a hyperbola results when the plane is vertical. Although he saw the circle as one of the four conic sections, it is really an ellipse with the plane perpendicular to the axis of the cone.

Paving the way for others

In his description of these four geometric objects, Apollonius used no algebraic formulae and no numbers. However, his view of a conic curve as a set of ordered parallel lines emanating from an axis looked toward the later creation of coordinate system geometry. He did not achieve the kind of precision that would come 1,800 years later with the work of French mathematicians René Descartes and Pierre de Fermat, but he did get close to coordinate representations of his conic curves. Some things held Apollonius back: he did not use negative numbers, nor did he explicitly work with zero. So while the two-dimensional Cartesian geometry developed by Descartes worked across four

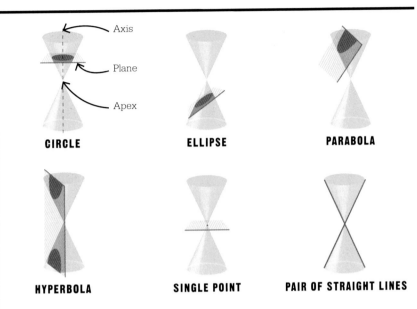

CIRCLE **ELLIPSE** **PARABOLA**

HYPERBOLA **SINGLE POINT** **PAIR OF STRAIGHT LINES**

When a plane intersects a cone, it creates a conic section. As well as the sections described by Apollonius, this can be a single point, where the plane cuts across the apex (top vertex), or straight lines cutting through the apex at an angle.

quadrants—with both positive and negative coordinates—Apollonius effectively worked in just one.

Apollonius's studies inspired many of the advances in geometry seen in the Islamic world during the Middle Ages. His work was then rediscovered in Europe during the Renaissance, leading mathematicians to develop the analytic geometry that helped to fuel the scientific revolution. ▪

[Conic sections are] the necessary key with which to attain the knowledge of the most important laws of nature.
Alfred North Whitehead
British mathematician

Apollonius of Perga

Little is known about the life of Apollonius. He was born in c.262 BCE in Perga, a center for the worship of the goddess Artemis, in southern Anatolia (now part of Turkey). After crossing the Mediterranean to Egypt, he was taught by Euclidean scholars in the great cultural city of Alexandria.

It is thought that all eight volumes of *Conics* were compiled while Apollonius was in Egypt. The first volumes produced little that was not known to Euclid, but the later works were a significant advance in geometry.

Beyond his work with conic sections, Apollonius is credited with estimating the value of pi more accurately than his contemporary Archimedes, and with being the first to state that a spherical mirror does not focus the sun's rays, while a parabolic mirror does.

Key work

c.200 BCE *Conics*

THE ART
OF MEASURING
TRIANGLES

TRIGONOMETRY

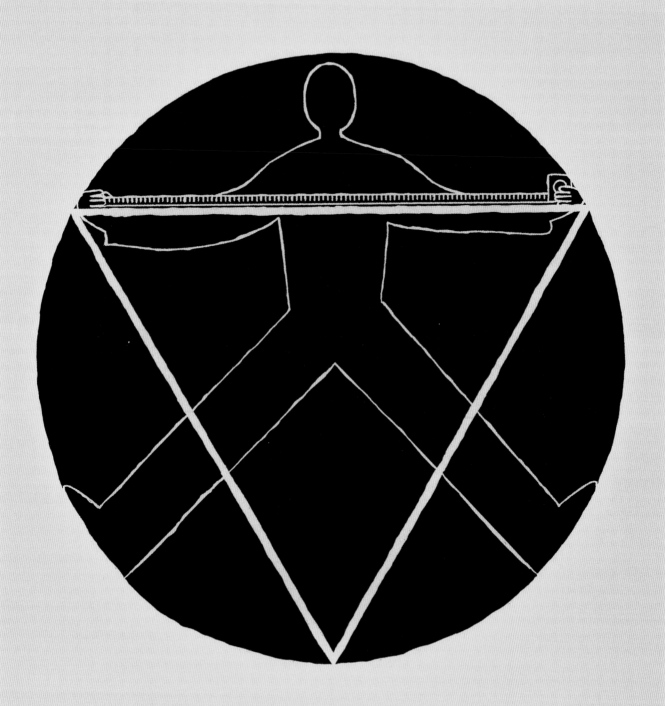

IN CONTEXT

KEY FIGURE
Hipparchus (c. 190–120 BCE)

FIELD
Geometry

BEFORE
c. 1800 BCE The Babylonian Plimpton 322 tablet contains a list of Pythagorean triples, long before Pythagoras devised his formula $a^2 + b^2 = c^2$.

c. 1650 BCE The Egyptian Rhind papyrus includes a method for calculating the slope of a pyramid.

6th century BCE In ancient Greece, Pythagoras discovers his theorem relating to the geometry of triangles.

AFTER
500 CE In India, the first trigonometric tables are used.

1000 CE In the Islamic world, mathematicians are using all the various ratios between the sides and angles of triangles.

Trigonometry is the study of the **relationship** between the **sides and angles of triangles**.

The **three angles** in any triangle **add** up to **180°**.

If two angles are known, the **third angle** can be determined.

The ratios of the sides of a right-angled triangle are called **trigonometric ratios**.

If the length of **one side** of a triangle is **known** and its **angles** are **known**, the **length of the other sides** can be **determined**.

Trigonometry, a term based on the Greek words for "triangle" and "measure," is of immense importance in both the historical development of mathematics and in the modern world. Trigonometry is one of the most useful of all the mathematical disciplines, enabling people to navigate the world, to understand electricity, and to measure the height of mountains.

Since antiquity, civilizations have appreciated the need for right angles in architecture. This led mathematicians to analyze the properties of right-angled triangles: all right-angled triangles contain two shorter sides (which may or may not be of equal length) and a diagonal, or hypotenuse, which is longer than either of the others; all triangles contain three angles; and right-angled triangles have one angle of 90°.

The Plimpton tablet
In the early 1900s, an examination of triangles, dating back to around 1800 BCE, was discovered on an ancient Babylonian clay tablet. The tablet, bought by American publisher George Plimpton in 1923 and known as Plimpton 322, is etched with numerical information relating to right-angled triangles. Its exact significance is debated, but the information appears to include Pythagorean triples (three positive numbers representing the lengths of sides of a right-angled triangle), alongside another set of numbers that resemble the ratios of the squares of sides. The tablet's original purpose is unknown, but it may have been used as a practical manual for measuring dimensions.

Even if he did not invent it, Hipparchus is the first person of whose systematic use of trigonometry we have documentary evidence.
Sir Thomas Heath
British historian of mathematics

See also: The Rhind papyrus 32–33 ▪ Pythagoras 36–43 ▪ Euclid's *Elements* 52–57 ▪ Imaginary and complex numbers 128–31 ▪ Logarithms 138–41 ▪ Pascal's triangle 156–61 ▪ Viviani's triangle theorem 166 ▪ Fourier analysis 216–17

At around the same time as the ancient Babylonians, Egypt's mathematicians were developing an interest in geometry. This was driven not just by their monumental building program, but also by the annual flooding of the Nile River, which required them to mark out the areas of fields each time the floods subsided. Egyptian interest is evident in the Rhind papyrus, a scroll that contains a set of tables relating to fractions. One of these tables poses the question: "If a pyramid is 250 cubits high and the side of its base is 360 cubits long, what is its *seked*?" The word *seked* means slope, so the problem is purely trigonometrical.

Hipparchus sets out rules

Influenced by Babylonian theories on angles, the ancient Greeks developed trigonometry as a branch of mathematics that was governed by definite rules rather than the tables of numbers relied on by the earliest mathematicians. In the 2nd century BCE, the astronomer and mathematician Hipparchus,

generally regarded as the founder of trigonometry, was particularly interested in triangles inscribed within circles and spheres, and the relationship between angles and lengths of chords (straight lines drawn between two points on a circle—or on any curve). Hipparchus compiled what was effectively the first true table of trigonometric values.

Ptolemy's contribution

Around 300 years later, in the Egyptian city of Alexandria, the gifted Greco-Roman polymath Claudius Ptolemaeus, better known as Ptolemy, wrote a mathematical treatise called the *Syntaxis Mathematikos* (later renamed the *Almagest* by Islamic scholars). In this work, Ptolemy further developed the ideas of Hipparchus on triangles and chords of circles, building formulae that would allow the prediction of the position of the Sun and other "heavenly bodies" based on the assumption of circular orbits around Earth. Ptolemy, like the mathematicians before him,

In the medieval period, astrolabes applied trigonometric principles to measure the position of celestial bodies. Hipparchus is credited with inventing the device.

used the Babylonian system of numbers known as the sexagesimal system, based on the number 60.

Ptolemy's work was developed further in India, where the growing discipline of trigonometry was regarded as part of astronomy. The Indian mathematician Aryabhata »

Hipparchus

Hipparchus was born in Nicaea (now Iznik in Turkey) in 190 BCE. Although little is known of his life, he achieved fame as an astronomer from the studies he carried out while living on the island of Rhodes. His findings were immortalized in Ptolemy's *Almagest*, where he is described as "a lover of truth."

The only work of Hipparchus to survive was his commentary on the *Phaenomena* of the poet Aratus and the mathematician and astronomer Eudoxus, criticizing the inaccuracy of their descriptions of constellations.

Hipparchus's most notable contribution to astronomy was his work *Sizes and Distances* (now lost, but used by Ptolemy), on the orbits of the Sun and Moon, which enabled him to calculate the dates of the equinoxes and solstices. He also compiled a star catalogue, which may be the one used by Ptolemy in *Almagest*. Hipparchus died in 120 BCE.

Key work

2nd century BCE *Sizes and Distances*

Types of trigonometry

a = **opposite**
b = **adjacent**
c = **hypotenuse**

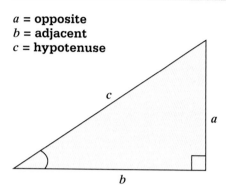

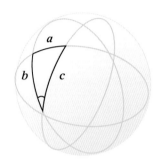

Planar trigonometry is the study of triangles on a plane (a flat, 2-D surface). Planar trigonometry is used by architects, for example, to ensure that buildings are stable, and by physicists to model motion.

Spherical trigonometry is the study of triangles on a sphere (a curved, 3-D surface). It is used by astronomers to calculate the positions of celestial bodies, and in navigation to calculate latitude and longitude.

(474–550 CE) pursued the study of chords to produce the first table of what is now known as the sine function (all the possible values of sine/cosine ratios for determining the unknown length of the side of a triangle when the lengths of the hypotenuse—the triangle's longest side—and the side opposite the angle are known).

In the 7th century CE, another great Indian mathematician and astronomer, Brahmagupta, made

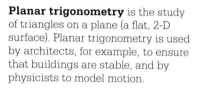

Trigonometry, like other branches of mathematics, was not the work of any one man, or nation.
Carl Benjamin Boyer
American historian of mathematics

his own contributions to geometry and trigonometry, including what is now known as Brahmagupta's formula. This is used to find the area of cyclic quadrilaterals, which are four-sided shapes inscribed within a circle. This area can also be found with a trigonometric method if the quadrilateral is split into two triangles.

Islamic trigonometry
Brahmagupta had already created a table of sine values, but in the 9th century CE, Persian astronomer and mathematician Habash al-Hasib ("Habash the Calculator") produced some of the first sine, cosine, and tangent tables to calculate the angles and sides of triangles. Around the same time, al-Battani (Albatenius) developed Ptolemy's work on the sine function and applied it to astronomical calculations. He recorded highly accurate observations of celestial objects from Raqqah, Syria. The motivation among Arab scholars for developing trigonometry was

not just for astronomy, but also for religious purposes, since it was important that Muslims knew the position of the holy city of Mecca from anywhere in the world.

In the 12th century CE, Indian mathematician and astronomer Bhaskara II invented the study of spherical trigonometry. This explores triangles and other shapes on the surface of a sphere rather than on a plane.

In later centuries, trigonometry became invaluable in navigation as well as astronomy. Bhaskara II's work, along with the ideas in Ptolemy's *Almagest*, were valued by the Islamic scholars of the medieval world, who had begun studying trigonometry well before Bhaskara II.

Aid to astronomy
Along with the developments in trigonometry, there was a gradual and corresponding shift in the way people viewed the heavens. From passively observing and recording the patterns in the movement of celestial bodies, scholars began to model that movement mathematically so that they could predict future astronomical events with ever greater accuracy. The

A trigonometric table is a small table by the use of which we can obtain knowledge of all geometrical dimensions and motions in space.
John Napier

study of trigonometry purely as an aid to astronomy persisted well into the 1500s, when new developments in Europe began to gain momentum. *De Triangulis Omnimodis* (*On Triangles of all Kinds*) was published in 1533. Written by German mathematician Johannes Müller von Königsberg, known as Regiomontanus, it was a compendium of all known theorems for finding sides and angles of both planar (2-D) and spherical triangles (those formed on the surface of a 3-D sphere). The publication of this work marked a turning point for trigonometry. It was no longer merely a branch of astronomy, but a key component of geometry.

Trigonometry was to develop even further; although geometry was its natural home, it was also increasingly applied to solve algebraic equations. French mathematician François Viète showed how algebraic equations could be solved using trigonometric functions, in conjunction with the new system of imaginary numbers that had been invented by Italian mathematician Rafael Bombelli in 1572.

At the end of the 1500s, Italian physicist and astronomer Galileo Galilei used trigonometry to model the trajectories of projectiles on which gravity was acting. The same equations are still used to project the motion of rockets and missiles into the atmosphere today. Also in the 1500s, Dutch cartographer and mathematician Gemma Frisius used trigonometry to determine distances, thus enabling accurate maps to be created for the first time.

New developments

Developments in trigonometry gathered pace in the 1600s. Scottish mathematician John Napier's discovery of logarithms in 1614 enabled the compilation of accurate sine, cosine, and tangent tables. In 1722, Abraham de Moivre, a French mathematician,

A network of triangulation stations such as this stone "trig point" in Wales was launched by the Ordnance Survey in 1936 to accurately map the island of Great Britain.

went a step further than Viète and showed how trigonometric functions could be used in the analysis of complex numbers. The latter comprised a real part and an imaginary part, and were to be of great significance in the development of mechanical and electrical engineering. Leonhard Euler used de Moivre's findings to derive the "most elegant equation in mathematics": $e^{i\pi} + 1 = 0$, also known as Euler's identity.

In the 1700s, Joseph Fourier applied trigonometry to his research into different forms of waves and vibrations. The "Fourier trigonometry series" has been used widely in scientific fields such as optics, electromagnetism, and, more recently, quantum mechanics. From its early beginnings, when the Babylonians and ancient Egyptians pondered the lengths of shadows cast by a stick in the ground, through architecture and astronomy to modern applications, trigonometry has formed a part of the language of mathematics in modeling the Universe. ∎

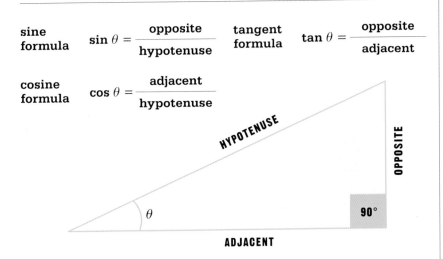

| sine formula | $\sin \theta = \dfrac{\text{opposite}}{\text{hypotenuse}}$ | tangent formula | $\tan \theta = \dfrac{\text{opposite}}{\text{adjacent}}$ |

| cosine formula | $\cos \theta = \dfrac{\text{adjacent}}{\text{hypotenuse}}$ |

HYPOTENUSE

OPPOSITE

θ

90°

ADJACENT

To find the unknown angle (θ) in a right-angled triangle, the sine formula is used when the lengths of the opposite (opposite θ) and the hypotenuse are known; the cosine formula is used when the lengths of the adjacent and hypotenuse are known; and the tangent formula is used when the lengths of the opposite and adjacent are known.

NUMBERS CAN BE LESS THAN NOTHING

NEGATIVE NUMBERS

IN CONTEXT

KEY CIVILIZATION
Ancient Chinese
(c. 1700 BCE–c. 600 CE)

FIELD
Number systems

BEFORE
c. 1000 BCE In China, bamboo rods are first used to denote numbers, including negatives.

AFTER
628 CE The Indian mathematician Brahmagupta provides rules for arithmetic with negative numbers.

1631 In *Practice of the Art of Analysis*, published 10 years after his death, British mathematician Thomas Harriot accepts negative numbers in algebraic notation.

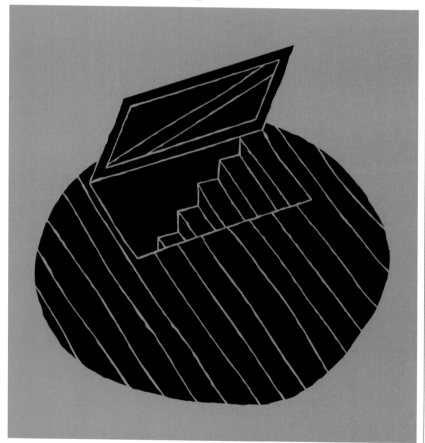

While practical notions of negative quantities were used from ancient times, particularly in China, negative numbers took far longer to be accepted within mathematics. Ancient Greek thinkers and many later European mathematicians regarded negative numbers—and the concept of something being less than nothing—as absurd. Only in the 1600s did European mathematicians begin to fully accept negative numbers.

Chinese rod system

The earliest ideas of negative quantities seem to have arisen in commercial accounting: the seller received money for what had been sold (a positive quantity), and the buyer spent the same amount,

See also: Positional numbers 22–27 ▪ Diophantine equations 80–81
▪ Zero 88–91 ▪ Algebra 92–99 ▪ Imaginary and complex numbers 128–31

In the Chinese rod numeral system, red indicates positive numbers, while black indicates negative numbers. To make the number being represented as clear as possible, horizontal and vertical symbols are used alternately—for example, the number 752 would use a vertical 7, then a horizontal 5, followed by a vertical 2. Blank spaces represent zero.

Positive	0	1	2	3	4	5	6	7	8	9
Vertical		𝟙	𝟙𝟙	𝟙𝟙𝟙	𝟙𝟙𝟙𝟙	𝟙𝟙𝟙𝟙𝟙	⊤	⊤	⊤	⊤
Horizontal		—	=	☰	☰	☰	⊥	⊥	⊥	⊥

Negative	0	−1	−2	−3	−4	−5	−6	−7	−8	−9
Vertical		𝟙	𝟙𝟙	𝟙𝟙𝟙	𝟙𝟙𝟙𝟙	𝟙𝟙𝟙𝟙𝟙	⊤	⊤	⊤	⊤
Horizontal		—	=	☰	☰	☰	⊥	⊥	⊥	⊥

resulting in a deficit (a negative quantity). For their commercial arithmetic, the ancient Chinese used small bamboo rods, laid out on a large board. Positive and negative quantities were represented by rods of different colors and could be added together. The Chinese military strategist Sun Tzu, who lived around 500 BCE, used such rods to make calculations before battles.

By 150 BCE, the rod system had developed into alternating horizontal and vertical rods in sets of up to five. Later, during the Sui dynasty (581–618 CE), the Chinese also used triangular rods for positive quantities and rectangular rods for negative quantities. The system was employed for trading and tax calculations: amounts received were represented by red rods, and debts by black rods.

When rods of different colors were added together, they canceled each other out—like income erasing a debt. The polarized nature of positive numbers (red rods) and negative numbers (black rods) was also in tune with the Chinese concept that opposing but complementary forces—yin and yang—governed the Universe.

Fluctuating fortunes

Over a period of several centuries, starting around 200 BCE, the ancient Chinese produced a book of collected scholarship called *The Nine Chapters on the Mathematical Art* (see box). This work, which encapsulated the essence of their mathematical knowledge, included algorithms that assumed negative quantities were possible—for example, as solutions to problems on profit and loss. »

Mathematics in ancient China

Jiuzhang suanshu, or *The Nine Chapters on the Mathematical Art*, reveals the mathematical methods known to the ancient Chinese. It is written as a collection of 246 practical problems and their solutions.

The first five chapters are mostly about geometry (areas, lengths, and volumes) and arithmetic (ratios, and square and cube roots). Chapter six covers taxes, and includes the ideas of direct, inverse, and compound proportions, most of which did not appear in Europe until around the 1500s. Chapters seven and eight deal with solutions to linear equations, including the rule of "double false position," whereby two test (or "false") values for the solution to a linear equation are used in repeated steps to yield the actual solution. The final chapter concerns applications of the "Gougu" (equivalent to Pythagoras's theorem), and the solving of quadratic equations.

Temperature readings on the Celsius scale display negative numbers to show when something such as an ice crystal is colder than 0°C—the point at which water freezes.

×	−4	−3	−2	−1	0	1	2	3	4
−4	16	12	8	4	0	−4	−8	−12	−16
−3	12	9	6	3	0	−3	−6	−9	−12
−2	8	6	4	2	0	−2	−4	−6	−8
−1	4	3	2	1	0	−1	−2	−3	−4
0	0	0	0	0	0	0	0	0	0
1	−4	−3	−2	−1	0	1	2	3	4
2	−8	−6	−4	−2	0	2	4	6	8
3	−12	−9	−6	−3	0	3	6	9	12
4	−16	−12	−8	−4	0	4	8	12	16

A negative multiplied by a negative makes a positive.
This is why all positive numbers have two square roots
(a positive and a negative) and negative numbers have no
real square roots—because a positive number squared is
positive, and a negative number squared is also positive.

■ Positive number
□ Negative number

The Persian mathematican and poet al-Khwarizmi (c. 780–c. 850)—whose theories, particularly on algebra, influenced later European mathematicians—was familiar with the rules of Brahmagupta and understood the use of negative numbers for dealing with debts. However, he could not accept the use of negative numbers in algebra, believing them to be meaningless. Instead, al-Khwarizmi followed geometric methods to solve linear or quadratic equations.

Accepting the negative

Throughout the Middle Ages, European mathematicians remained unsure of negative quantities as numbers. This was still the case in 1545 when Italian polymath Gerolamo Cardano published his *Ars Magna* (*The Great Art*), in which he explained how to solve linear, quadratic, and cubic equations. He could not exclude negative solutions to his equations and even used a sign, "m," to denote a negative number. He could not, however, accept the value of negative numbers, calling them "fictitious." René Descartes (1596–1650) also accepted negative quantities as solutions to equations but referred to them as "false roots" rather than true numbers.

In contrast, the mathematics of ancient Greece was based on geometry and geometrical magnitudes, or their ratios. As these quantities—actual lengths, areas, and volumes—can only be positive, the idea of a negative number did not make sense to Greek mathematicians.

By the time of Diophantus, around 250 CE, linear and quadratic equations were used to solve problems, but any unknown quantity was still represented geometrically—by a length. So the idea of negative numbers as solutions to these equations was still seen as an absurdity.

An important advance in the arithmetical use of negative numbers came around 400 years later from India, in the work of the mathematician Brahmagupta (c. 598–668). He set out arithmetic rules for negative quantities, and even used a symbol to indicate negative numbers. Like the ancient Chinese, Brahmagupta looked at numbers in financial terms, as "fortunes" (positive) and "debts" (negative), and stated the following rules for multiplying with positive and negative quantities:

The product of two fortunes is a fortune. The product of two debts is a fortune. The product of a debt and a fortune is a debt. The product of a fortune and a debt is a debt.

It makes no sense to find the product of two piles of coins, as only the actual quantities can be multiplied, not the money itself (just as you cannot multiply apples by apples). Brahmagupta was therefore performing arithmetic with positive and negative quantities, while using fortunes and debts as a way to try to understand what negative numbers represented.

Negative numbers are evidence of inconsistency or absurdity.
Augustus De Morgan
British mathematician

English mathematician John Wallis (1616–1703) gave some meaning to negative numbers by extending the number line below zero. This way of seeing numbers as points on a line finally led to the acceptance of negative numbers on equal terms with positive numbers, and by the end of the 1800s, they had been formally defined within mathematics, separate from notions of quantities. Today, negative numbers are used in many areas, ranging from banking and temperature scales to the charge on subatomic particles. Any ambiguity about their status in mathematics is long gone. ■

Investors rush to withdraw their money from the Seamen's Savings Bank in New York in 1857. The panic was caused by American banks loaning out many millions of dollars (a negative quantity) without the reserves (a positive quantity) to back this up.

In **15th-century Europe**, the letters p and m are used for plus and minus.

The + and − signs are introduced in the **16th century**.

But **negative numbers** are seen as absurd and viewed with **hostility** and **suspicion**.

It is not until the **1600s** that negative numbers are **accepted in Europe**, when they are placed on a **number line** for the first time.

THE VERY FLOWER OF ARITHMETIC

DIOPHANTINE EQUATIONS

IN CONTEXT

KEY FIGURE
Diophantus (c. 200–c. 284 CE)

FIELD
Algebra

BEFORE
c. 800 BCE The Indian scholar Baudhayana finds solutions to some "Diophantine" equations.

AFTER
c. 1600 François Viète lays the foundations for solutions of Diophantine equations.

1657 Pierre de Fermat writes his last theorem (about a Diophantine equation) in his copy of *Arithmetica*.

1900 The 10th problem on David Hilbert's list of unsolved research problems is the quest to find an algorithm to solve all Diophantine equations.

1970 Mathematicians in Russia show that there is no algorithm that can solve all Diophantine equations.

Diophantus tried to solve equations with **more than two unknown quantities** and only **integer** or **rational number** solutions.

Equations like this are now known as **Diophantine equations**.

While some have a simple solution, **most have many solutions**—or none at all.

Diophantine equations have proved **endlessly fascinating for mathematicians**.

I n the 3rd century CE, the Greek mathematician Diophantus, a pioneer of number theory and arithmetic, created a prodigious work called *Arithmetica*. In 13 volumes, only six of which have survived, he explored 130 problems involving equations and was the first person to use a symbol for an unknown quantity—a cornerstone of algebra. It is only in the past 100 years that mathematicians have fully explored what are now known as Diophantine equations. Today, the equations are considered to be one of the most interesting areas of number theory.

Diophantine equations are a type of polynomial—an equation in which the powers of the variables (unknown quantities) are integers, such as $x^3 + y^4 = z^5$. The aim of Diophantine equations is to find all the variables, but solutions must be integers or rational numbers (those that can be written as one integer

See also: The Rhind papyrus 32–33 ▪ Pythagoras 36–43 ▪ Hypatia 82 ▪ The equals sign and other symbology 126–27
▪ 23 problems for the 20th century 266–67 ▪ The Turing machine 284–89 ▪ Proving Fermat's last theorem 320–23

> The symbolism that Diophantus introduced for the first time… provided a short and readily comprehensible means of expressing an equation.
> **Kurt Vogel**
> **German mathematical historian**

divided by another, such as $^8/_3$). In Diophantine equations, the coefficients—integers such as the 4 in $4x$, that multiply a variable—are also rational numbers. Diophantus only used positive numbers, but mathematicians now look for negative solutions as well.

The quest for solutions

Many of the problems now called Diophantine equations were known well before Diophantus's time. In India, mathematicians explored some of them from around 800 BCE, as the ancient *Shulba Sutras* texts reveal. In the 6th century BCE, Pythagoras created his quadratic equation for calculating the sides of a right-angled triangle; its $x^2 + y^2 = z^2$ form is a Diophantine equation.

Diophantine equations of the kind $x^n + y^n = z^n$ may look simple to calculate, but only those with squares are solvable. If the power (n in the equation) is greater than 2, the equation has no integer solutions for x, y, and z—as Fermat

The *Arithmetica* of Diophantus strongly influenced 17th-century mathematicians as the study of modern algebra developed. This volume of the book was published in Latin in 1621.

asserted in a marginal note in 1657 and British mathematician Andrew Wiles finally proved in 1994.

A source of fascination

Diophantine equations are vast in number and form, and mostly very difficult to solve. In 1900, David Hilbert suggested that the question of whether or not they could all be solved was one of the greatest challenges facing mathematicians.

The equations are now grouped in three classes: those with no solution, those with a finite number of solutions, and those with an infinite number of solutions. Rather than finding solutions, however, mathematicians are often more interested in discovering whether solutions exist at all. In 1970, Russian mathematician Yuri Matiyasevich settled Hilbert's query, which he and three others

had studied for years, concluding that no general algorithm to solve a Diophantine equation exists. Yet studies continue, as the fascination of these equations is largely theoretical. Mathematicians, who are driven by curiosity, believe there is still more to discover. ▪

Diophantus

Little is known about the life of the Greek mathematician and philosopher Diophantus, but he was probably born in Alexandria, Egypt, in c. 200 CE. His 13-volume *Arithmetica* was well-received—the Alexandrian mathematician Hypatia wrote about the first six volumes—but fell into relative obscurity until the 1500s, when interest in his ideas was revived.

The *Greek Anthology*, a compilation of mathematical games and verses published around 500 CE, contains one number problem purporting to be an epitaph to Diophantus that appeared on his tombstone. Written as a puzzle, it suggests he married at the age of 35, and five years later had a son, who died at the age of 40 when he was half his father's age. Diophantus is then said to have lived a further four years, dying at the age of 84.

Key work

c. 250 CE *Arithmetica*

AN INCOMPARABLE STAR IN THE FIRMAMENT OF WISDOM
HYPATIA

IN CONTEXT

KEY FIGURE
Hypatia of Alexandria
(c. 355–415 CE)

FIELDS
Arithmetic, geometry

BEFORE
6th century BCE Pythagoras's wife Theano and other women actively participate in the Pythagorean community.

c. 100 BCE Mathematician and astronomer Aglaonike of Thessaly wins renown for her ability to predict lunar eclipses.

AFTER
1748 Italian mathematician Maria Agnesi writes the first textbook to explain differential and integral calculus.

1874 Russian mathematician Sofia Kovalevskaya is the first woman to be awarded a doctorate in mathematics.

2014 Iranian mathematician Maryam Mirzakhani is the first woman to win the Fields Medal.

History mentions only a few pioneering female mathematicians in the ancient world, among them Hypatia of Alexandria. An inspirational teacher, she was appointed head of the city's Platonist school in 400 CE.

Hypatia is not known to have contributed any original research, but she is credited with editing and writing commentaries on several classic mathematical, astronomical, and philosophical texts. It is likely that she helped her father, Theon, a respected Alexandrian scholar, to produce his definitive edition of Euclid's *Elements*, and his *Almagest* and *Handy Tables* of Ptolemy. She also continued his project of preserving and expanding the classic texts, in particular providing commentaries on Diophantus's 13-volume *Arithmetica*, and Apollonius's work on conic sections. Hypatia may have intended these editions to serve as textbooks for students, as she offered commentaries providing clarification, and developed some of the concepts further.

The Alexandrian scholar Hypatia, depicted here in an 1889 painting by Julius Kronberg, was revered as a heroic martyr after her murder. She later became a symbol for feminists.

Hypatia won great renown for her teaching, scientific knowledge, and wisdom, but in 415 she was killed by Christian zealots for her "pagan" philosophy. As attitudes toward women in academia became less tolerant, mathematics and astronomy would be almost exclusively male preserves until the Enlightenment opened up new opportunities for women in the 1700s. ∎

See also: Euclid's *Elements* 52–57 ▪ Conic sections 68–69 ▪ Diophantine equations 80–81 ▪ Emmy Noether and abstract algebra 280–81

THE CLOSEST APPROXIMATION OF PI FOR A MILLENNIUM
ZU CHONGZHI

IN CONTEXT

KEY FIGURE
Zu Chongzhi (429–501 CE)

FIELD
Geometry

BEFORE
c. 1650 BCE The area of a circle is calculated using π as $(^{16}/_9)^2 \approx$ 3.1605 in the Rhind papyrus.

c. 250 BCE Archimedes finds an approximate value for π using a polygon algorithm method.

AFTER
c. 1500 Indian astronomer Nilakantha Somayaji uses an infinite series (the sum of terms of an infinite sequence, such as $\frac{1}{2} + \frac{1}{4} + \frac{1}{8} + \frac{1}{16}$) to compute π.

1665–66 Isaac Newton calculates π to 15 digits.

1975–76 Iterative algorithms allow computer calculations of π to millions of digits.

L
ike their counterparts in Greece, mathematicians in ancient China realized the importance of π (pi)—the ratio of a circle's circumference to its diameter—in geometric and other calculations. Various values for π were suggested from the 1st century CE onward. Some were sufficiently accurate for practical purposes, but several Chinese mathematicians sought more precise methods for determining π. In the 3rd century, Liu Hui approached the task using the same method as Archimedes—drawing regular polygons with increasing numbers of sides inside and outside a circle. He found that a 96-sided polygon allowed a calculation of π as 3.14, but by repeatedly doubling the number of sides up to 3,072, he reached a value of 3.1416.

More precision
In the 5th century, astronomer and mathematician Zu Chongzhi, who was renowned for his meticulous calculations, set about obtaining an even more accurate value for π. Using a 12,288-sided polygon, he calculated that π is between 3.1415926 and 3.1415927, and suggested two fractions to express the ratio: the *Yuelü*, or approximate ratio, of $^{22}/_7$, which had been in use for some time; and his own calculation, the *Milü*, or close ratio, of $^{355}/_{113}$. This later became known as "Zu's ratio." Zu's calculations of π were not bettered until European mathematicians set about the task during the Renaissance, almost a millennium later. ∎

I cannot help thinking that Zu Chongzhi was a genius of Antiquity.
Takebe Katahiro
Japanese mathematician

THE MID
AGES
500–1500

The Indian mathematician Brahmagupta **establishes the role and use of zero** and uses "debts" to designate negative quantities.

The **House of Wisdom** is established in Baghdad, facilitating the **exchange and development of ideas** within the Islamic/Arabic world.

Al-Khwarizmi and al-Kindi explain the **use of Hindu numerals**, the precursors of our **modern "Arabic" numerals**.

c. 628 CE   LATE 8TH CENTURY   c. 825–830

8TH CENTURY   c. 820   c. 930

The spread of Islam into some parts of India leads to **Indian mathematicians sharing their knowledge** with Arab scholars.

Al-Khwarizmi writes his **book on algebra**, introducing many methods for solving equations that are still important today.

Death of Abu Kamil, writer of *The Book of Algebra*, **a key influence for Fibonacci** three centuries later.

As the Roman Empire collapsed and Europe entered the Middle Ages, the center of scientific and mathematical scholarship shifted from the eastern Mediterranean to China and India. From about the 5th century CE, India began a "Golden Age" of mathematics, building on its own long tradition of scholarship, but also on ideas brought in by the Greeks. Indian mathematicians made significant advances in the fields of geometry and trigonometry, which had practical applications in astronomy, navigation, and engineering, but the most far-reaching innovation was the development of a character to represent the number zero.

The use of a specific symbol—a simple circle, rather than a blank space or placeholder—to denote zero is attributed to the brilliant mathematician Brahmagupta, who described the rules of its use in calculation. In fact, the character may already have been in use for some time. It would have fitted well with India's numeral system, which is the prototype of our modern Hindu–Arabic numerals. Yet it is thanks to Islam that these and other ideas from India's Golden Age (which continued until the 12th century) went on to influence the history of mathematics.

Persian powerhouse

After the death of the Prophet Mohammed in 632, Islam rapidly became a major political as well as religious power in the Middle East and beyond, spreading from Arabia across Persia and into Asia as far as the Indian subcontinent. The new religion had a high regard for philosophy and scientific enquiry, and the "House of Wisdom," a center of learning and research established in Baghdad, attracted scholars from all over the expanding Islamic Empire.

This thirst for knowledge prompted the study of ancient texts, especially those of the great Greek philosophers and mathematicians. Islamic scholars not only preserved and translated the ancient Greek texts, but provided commentaries on them and developed their own original concepts. Open to new ideas, they also adopted many of the Indian innovations, in particular their numeral system. The Islamic world, like India, entered a "Golden Age" of learning that lasted until the 1300s, and

The **binomial theorem**, which makes it possible to **solve equations without relying on geometric diagrams**, is laid out by al-Karaji.

Robert of Chester translates **al-Khwarizmi's work** into Latin.

Historian Ibn Khallikan makes the **first written mention** of the **"wheat on a chessboard"** problem.

The anonymously published *Treviso Arithmetic* becomes the first **printed mathematics textbook** in Europe.

c. 1020   **1145**   **1256**   **1478**

c. 1070   **1202**   **14TH CENTURY**

Omar Khayyam invents a method for **classifying and solving cubic equations**.

Fibonacci's *Liber Abaci* (*The Book of Calculation*) introduces many ideas from the Arabic world, including the **Hindu-Arabic numeral system** and his famous sequence.

The **Oxford Calculators** at Merton College give the University of Oxford a **prominent position in Western mathematics**.

---

produced a succession of influential mathematicians—such as al-Khwarizmi, a key figure in the development of algebra (the word "algebra" derives from the Arabic term for rejoining), and other scholars whose contributions to the binomial theorem and the treatment of quadratic and cubic equations were groundbreaking.

From East to West

In Europe, mathematical study was under the control of the Church, and was confined to a few early translations of some of Euclid's work. Progress was hindered by the continued use of the cumbersome Roman system of numerals, necessitating the use of the abacus for calculation. However, from the 12th century onward, during the Crusades,

contact with the Islamic world increased, and some recognized the wealth of scientific knowledge Islamic scholars had amassed. Christian scholars now gained access to Greek and Indian philosophical and mathematical texts, and to the work of the Islamic scholars. Al-Khwarizmi's treatise on algebra was translated into Latin in the 12th century by Robert of Chester, and soon after, complete translations of Euclid's *Elements* and other important texts began to appear in Europe.

Mathematical renaissance

City-states in Italy were quick to trade with the Islamic Empire, and it was an Italian, Leonardo of Pisa, nicknamed Fibonacci, who spearheaded the revival of mathematics in the West. He

adopted the Hindu-Arabic numeral system, and the use of symbols in algebra, and contributed many original ideas, including the Fibonacci arithmetical sequence.

With the growth in trade in the later Middle Ages, mathematics—especially the fields of arithmetic and algebra—became increasingly important. Advances in astronomy also demanded sophisticated calculations. Mathematical education was now taken more seriously. With the invention of the movable-type printing press in the 1400s, books of all sorts, including the *Treviso Arithmetic*, became widely available, spreading the newfound knowledge across Europe. These books inspired a "scientific revolution" that would accompany the cultural rebirth known as the Renaissance. ∎

A FORTUNE SUBTRACTED FROM ZERO IS A DEBT

ZERO

IN CONTEXT

KEY FIGURE
Brahmagupta (c. 598–668 CE)

FIELD
Number theory

BEFORE
c. 700 BCE On a clay tablet, a Babylonian scribe indicates a placeholder zero with three hooks; it is later written as two slanted wedge marks.

36 BCE A shell-shaped zero is recorded on a Mayan stela (stone slab) in Central America.

c. 300 CE Parts of the Indian *Bakshali* text reveal many circular placeholder zeros.

AFTER
1202 In his book *Liber Abaci*, Leonardo of Pisa (Fibonacci) introduces zero to Europeans.

17th century Zero is finally established as a number and is in widespread use.

A number that represents the absence of something is a difficult concept, which may be why zero took so long to become widely accepted. Several ancient civilizations, including the Babylonians and the Sumerians, could claim to have invented zero, but its use as a number was pioneered in the 7th century CE, by Brahmagupta, an Indian mathematician.

The development of zero

Any system for recording numbers eventually reaches a point at which it becomes positional; that is to say, digits are ordered according to their value to cope with increasingly large

See also: Positional numbers 22–27 ▪ Negative numbers 76–79 ▪ Binary numbers 176–77 ▪ The law of large numbers 184–85 ▪ The complex plane 214–15

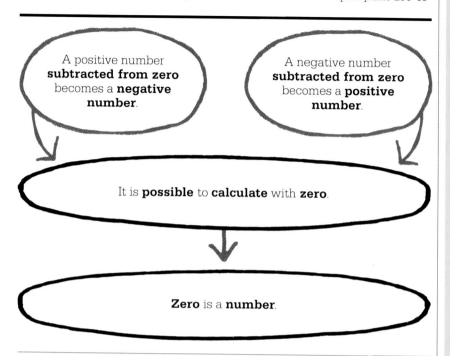

> A positive number **subtracted from zero** becomes a **negative number**.

> A negative number **subtracted from zero** becomes a **positive number**.

> It is **possible** to **calculate** with **zero**.

> **Zero** is a **number**.

numbers. All place value (positional) systems require a way of denoting "there is nothing here." The Babylonians (1894–539 BCE), for example, who at first used context to differentiate between, say, 35 and 305, eventually used a double wedge mark rather like inverted commas to indicate the empty value. In this way, zero entered the world as a form of punctuation.

The problem for historians has been finding evidence for early civilizations using zero and recognizing it as such, which has been made more difficult by the fact that zero fell in and out of use over time. In about 300 BCE, for example, the Greeks were starting to develop a more sophisticated form of mathematics based on geometry, with quantities being represented by the lengths of lines. There was no need for zero, or indeed negative numbers (numbers less than 0), as the Greeks did not have a positional number system (lengths cannot be nonexistent or negative).

As the Greeks developed the use of mathematics in astronomy, they began to use an "O" to represent zero, although it is not clear why. In his astronomical manual *Almagest*, written in the 2nd century CE, the Greco-Roman scholar Ptolemy used a circular symbol positionally between digits and at the end of a number, but did not consider it a number in its own right.

In Central America, during the 1st millenium CE, the Mayans used a place value system, which included zero as a numeral, denoted by a shell shape. It was one of three symbols used by the Mayans for arithmetic; the other two were a dot representing 1 and a bar for 5. While the Mayans could calculate up to hundreds of millions, their geographical isolation meant that their mathematics never spread to other cultures. »

An abax, a table or board covered in sand, was used by the Greeks to count. Some scholars have suggested that "O" was used because it was the shape left when a counter was removed.

Brahmagupta

Born in 598 CE, astronomer and mathematician Brahmagupta lived in Bhillamala, northwest India—a center of learning in those fields. He became head of the leading astronomical observatory at Ujjain, and incorporated new work on number theory and algebra into his studies on astronomy.

Brahmagupta's use of the decimal number system and the algorithms he devised spread throughout the world and informed the work of later mathematicians. His rules for calculating with positive and negative numbers, which he called "fortunes" and "debts," are still cited today. Brahmagupta died in 668, only a few years after completing his second book.

Key works

628 *Brahmasphutasiddhanta (The Correctly Established Doctrine of Brahma)*
665 *Khandakhadyaka (Morsel of Food)*

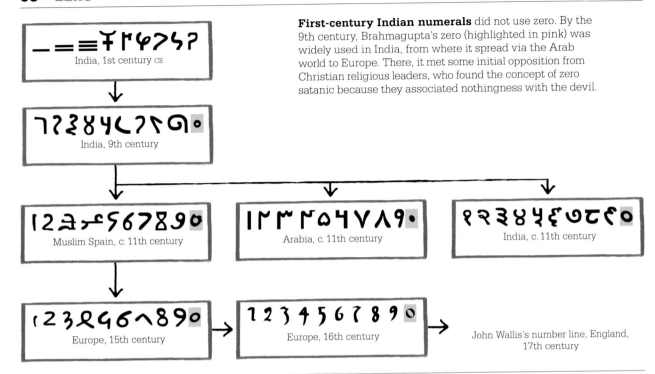

First-century Indian numerals did not use zero. By the 9th century, Brahmagupta's zero (highlighted in pink) was widely used in India, from where it spread via the Arab world to Europe. There, it met some initial opposition from Christian religious leaders, who found the concept of zero satanic because they associated nothingness with the devil.

India, 1st century CE

India, 9th century

Muslim Spain, c. 11th century

Arabia, c. 11th century

India, c. 11th century

Europe, 15th century

Europe, 16th century

John Wallis's number line, England, 17th century

The Nadi Yali yantra is part of an 18th-century observatory in Ujjain, India. A center of mathematics and astronomy since Brahmagupta worked there in the 7th century, it lies on the intersection of a former zero meridian of longitude and the Tropic of Cancer.

In India, mathematics advanced rapidly in the early centuries of the 1st millennium CE. By the 3rd and 4th centuries, a place value system had long been in use, and by the 7th century—the time of Brahmagupta—the use of a circular symbol as a placeholder was already well established there.

Zero as a number

Brahmagupta established rules for calculating with zero. He began by defining it as the result of subtracting a number from itself— for example, $3 - 3 = 0$. That established zero as a number in its own right as opposed to simply a figurative notation or placeholder. He then explored the effect of calculating with zero. Brahmagupta showed that if he added zero to a negative number, the result was equal to that negative number. Similarly, adding zero to a positive number produced the same positive number. Brahmagupta

also described subtracting zero from both a negative and a positive number, and noted again that it left the numbers unchanged.

Brahmagupta went on to describe the effect of subtracting numbers from zero. He calculated that a positive number subtracted from zero becomes a negative number and a negative number subtracted from zero becomes a

Black holes are where God divided by zero.
Steven Wright
American comedian

positive number. This calculation brought negative numbers into the same number system as positive numbers. Like zero, negative numbers were an abstract concept rather than positive values such as lengths or quantities.

Multiplying and dividing

Brahmagupta went on to examine zero in relation to multiplication and described how the product of multiplying any number with zero is zero, including zero multiplied by zero. The next step was to explain division by zero, which was more problematic. Recording the result of dividing a number, *n*, by zero as ⁿ/₀, Brahmagupta suggested that a number is unchanged when it is divided by zero. However, this was later found to be impossible, as is demonstrated by multiplying any number by zero (division being defined as finding the missing number in a multiplication). The result cannot be the original number, as any number multiplied by zero equals zero.

Mathematicians now describe division by zero as "undefined." Some have suggested that the

Zero is the most magical number we know. It is the number we're striving toward every day.
Bill Gates

required answer to ⁿ/₀ is "infinity," but infinity is not a number and cannot be used in calculations. Dividing zero itself by zero has proved even trickier. The result could be zero, if zero divided by any number is thought to be zero. It could also be 1, as any number divided by itself is 1.

The spread of Islam through parts of India in the 8th century led to Indian mathematicians sharing their knowledge, including the concept of zero, with scholars in the Arab world. In the 9th century, the Islamic mathematician al-Khwarizmi wrote a treatise on

Hindu–Arabic numbers, which described the place value system including zero. Yet 300 years later when Leonardo of Pisa (better known as Fibonacci) introduced Hindu–Arabic numerals to Europe, he was still wary of zero and treated it as an operator like + and − rather than a number. Even in the 1500s, Italian polymath Gerolamo Cardano solved quadratic and cubic equations without zero. Europeans finally accepted zero in the 1600s, when English mathematician John Wallis incorporated zero in his number line.

A vital concept

Mathematics without zero would mean many of the articles in this book could not have been written: there would be no negative numbers, no coordinate systems, no binary systems (and hence no computers), no decimals, and no calculus, because it would not be possible to describe infinitesimally small quantities. Advances in engineering would have been severely restricted. Zero is perhaps the most important number of all. ■

The *Treviso Arithmetic*

The figure zero first became known in Italy from the *Arte dell' Abbaco* (*Art of Calculation*, also known as *The Treviso Arithmetic*), published anonymously in 1478 and the first printed mathematics textbook in Europe. It was revolutionary because it was written in everyday Venetian for merchants and anyone else who wanted to solve calculation problems. It outlined the Hindu–Arabic decimal place value system and described how

the number system worked. The unknown author makes 0 the 10th number and calls it a "cipher" or "nulla"—something that has no value unless it is written to the right of other numbers to increase their value.

In the Treviso description, zero is just a placeholder number, which itself was still a new notion. The idea of zero as a number was not accepted for centuries. It was also of little interest to the readers of the *Arte dell' Abbaco*, most of whom wanted to learn how to use numbers in practical business calculations in everyday trading.

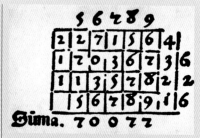

This grid method of multiplication from the *Treviso Arithmetic* multiplies the number 56,289 by 1,234. Zero is used as a placeholder in the calculation and in the final solution—70,072,626. The book also illustrated other methods of multiplication.

ALGEBRA IS
A SCIENTIFIC ART

ALGEBRA

IN CONTEXT

KEY FIGURE
Al-Khwarizmi (c. 780–c. 850)

FIELD
Algebra

BEFORE
1650 BCE The Egyptian Rhind papyrus includes solutions to linear equations.

300 BCE Euclid's *Elements* lays the foundations of geometry.

3rd century CE Greek mathematician Diophantus uses symbols to represent unknown quantities.

7th century CE Brahmagupta solves the quadratic equation.

AFTER
1202 Leonardo of Pisa's *Liber Abaci* uses the Hindu-Arabic number system.

1591 François Viète introduces symbolic algebra, in which letters are used to abbreviate terms in equations.

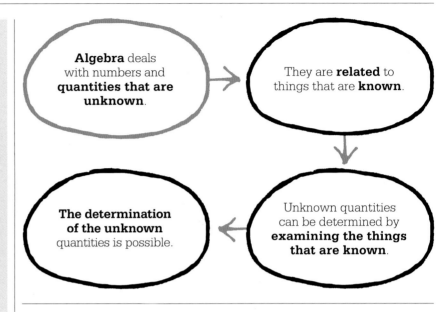

Algebra deals with numbers and **quantities that are unknown**.

They are **related** to things that are **known**.

Unknown quantities can be determined by **examining the things that are known**.

The determination of the unknown quantities is possible.

The origins of algebra—a mathematical method for calculating unknown quantities—can be traced back to ancient Babylonians and Egyptians, as equations on cuneiform tablets and papyri reveal. Algebra evolved from the need to solve practical problems, often of a geometrical nature, requiring the determination of a length, area, or volume. Mathematicians gradually developed rules to handle a wider range of general problems. To work out lengths and areas, equations involving variables (unknown quantities) and squared terms were devised. Using tables, the Babylonians could also calculate volumes, such as the space within a grain store.

A search for new methods
Over the centuries, as mathematics developed, problems became longer and more complex, and scholars

Al-Khwarizmi

Born in c. 780 CE near what is now Khiva, Uzbekistan, Muhammad Ibn Musa al-Khwarizmi moved to Baghdad, where he became a scholar at the House of Wisdom.

Al-Khwarizmi is regarded as the "father of algebra" for his systematic rules for solving linear and quadratic equations. These were outlined in his major work on calculation by "completion and balancing"—methods he devised that are still used today. Other achievements include his text on Hindu numerals, which, in its Latin translation, introduced Europe to Hindu-Arabic numerals.

He wrote a book on geography, helped construct a world map, took part in a project to determine the circumference of Earth, developed the astrolabe (an earlier Greek tool for navigation), and compiled a set of astronomical tables. Al-Khwarizmi died around 850.

Key works

c. 820 *On the Calculation with Hindu Numerals*
c. 830 *The Compendious Book on Calculation by Completion and Balancing*

See also: Quadratic equations 28–31 ▪ The Rhind papyrus 32–33 ▪ Diophantine equations 80–81 ▪ Cubic equations 102–05 ▪ The algebraic resolution of equations 200–01 ▪ The fundamental theorem of algebra 204–09

sought new ways to shorten and simplify them. Although early Greek mathematics was largely geometry-based, Diophantus developed new algebraic methods in the 3rd century CE, and was the first to use symbols for unknown quantities. However, it would be more than a thousand years before standard algebraic notation was accepted.

After the fall of the Roman Empire, mathematics in the Mediterranean area declined, but the spread of Islam from the 7th century had a revolutionary impact on algebra. In 762 CE, Caliph al-Mansur established a capital in Baghdad, which swiftly became a major center of culture, learning, and commerce. Its status was enhanced by the acquisition and translation of manuscripts from earlier cultures, including works by the Greek mathematicians Euclid, Apollonius, and Diophantus, as well as Indian scholars such as Brahmagupta. They were housed in a great library, the House of Wisdom, which became a center for research and the dissemination of knowledge.

The early algebraists

Scholars at the House of Wisdom produced their own research, and in 830, Muhammad Ibn Musa al-Khwarizmi presented his work to the library—*The Compendious Book on Calculation by Completion and Balancing*. It revolutionized ways of calculating algebraic problems, introducing principles that are the foundation of modern algebra. As in earlier periods, the types of problems discussed were largely geometrical. The study of geometry was important in the

Islamic world, partly because the human form was forbidden in religious art and architecture, so many Islamic designs were based on geometric patterns.

Al-Khwarizmi introduced some fundamental algebraic operations, which he described as reduction, rejoining, and balancing. The process of reduction (simplifying

an equation) could be done by rejoining (*al-jabr*)—moving subtracted terms to the other side of an equation—and then balancing the two sides of the equation. The word "algebra" comes from *al-jabr*.

Al-Khwarizmi was not working in a total vacuum, as he had the translated works of earlier Greek »

Key texts in the House of Wisdom

Treatise on Demonstration of Problems of Algebra, Omar Khayyam (1070 CE)

Book of Rare Things in the Art of Calculation, Abu Kamil (c. 850–950 CE)

Glorious on Algebra, Al-Karaji (980–1030 CE)

The Compendious Book on Calculation by Completion and Balancing, Al-Khwarizmi (830 CE)

Book of Algebra, Abu Kamil (850–930 CE)

Arithmetica, Diophantus (3rd century CE)

Elements, Euclid (c. 300 BCE)

The Correctly Established Doctrine of Brahma, Brahmagupta (628 CE)

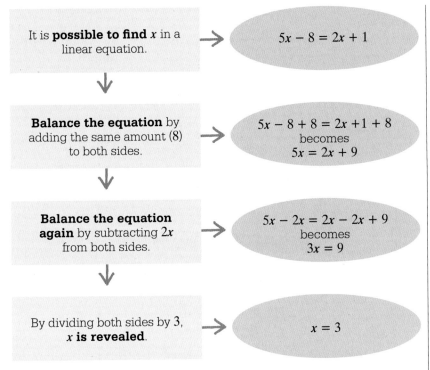

It is **possible to find** x in a linear equation.

$$5x - 8 = 2x + 1$$

Balance the equation by adding the same amount (8) to both sides.

$$5x - 8 + 8 = 2x + 1 + 8$$
becomes
$$5x = 2x + 9$$

Balance the equation again by subtracting $2x$ from both sides.

$$5x - 2x = 2x - 2x + 9$$
becomes
$$3x = 9$$

By dividing both sides by 3, x **is revealed**.

$$x = 3$$

---

and Indian mathematicians at his disposal. He introduced the Indian decimal place-value system to the Islamic world, which later led to the adoption of the Hindu-Arabic numeral system widely used today.

Al-Khwarizmi began by studying linear equations, so-called because they create a straight line when plotted on a graph. Linear equations involve only one variable, which is expressed only to the power of 1, rather than squared or to any higher power.

Quadratic equations

Al-Khwarizmi did not employ symbols; he wrote his equations in words, supported by diagrams. For example, he wrote out the equation $(x/3 + 1)(x/4 + 1) = 20$ as: "A quantity: I multiplied a third of it and a dirham by a fourth of it and a dirham; it becomes twenty," a dirham being a single coin, used by al-Khwarizmi to signify a single

unit. According to al-Khwarizmi, by using his completion and balancing methods, all quadratic equations—those in which the highest power of x is x^2—can be simplified to one of six basic forms. In modern notation, these would be: $ax^2 = bx$; $ax^2 = c$; $ax^2 + bx = c$; $ax^2 + c = bx$; $ax^2 = bx + c$; and $bx = c$. In these six types, the letters a, b, and c all represent known numbers, and x represents the unknown quantity.

Al-Khwarizmi approached more complex problems too, producing a geometrical method for solving quadratic equations that used the technique known as "completing the square" (right). He went on to search for a general solution to cubic equations—in which the highest power of x is x^3—but was unable to find one. However, his pursuit of this goal showed how mathematics had progressed since the time of the ancient Greeks.

For centuries, algebra had just been a tool to solve geometric problems, but now became a discipline in its own right, where calculating increasingly difficult equations was the end goal.

Rational answers

Many of the equations that al-Khwarizmi was dealing with had solutions that could not be expressed rationally and completely using the Hindu-Arabic decimal system. Although numbers such as $\sqrt{2}$—the square root of 2—had been known since ancient Greek times and from even earlier Babylonian clay tablets, in 825 CE, al-Khwarizmi was the first to make the distinction between rational numbers—which can be made into fractions—and irrational numbers, which have an indefinite string of decimals with no recurring pattern. Al-Khwarizmi described rational numbers as "audible" and irrational numbers as "inaudible."

Al-Khwarizmi's work was developed further by Egyptian mathematician Abu Kamil Shuja ibn Aslam (c. 850–930 CE), whose *Book of Algebra* was designed to

The principal object of Algebra... is to determine the value of quantities which were before unknown... by considering attentively the conditions given... expressed in known numbers.
Leonhard Euler

> Algebra is but written geometry and geometry is but figured algebra.
> **Sophie Germain**
> **French mathematician**

be an academic treatise for other mathematicians, rather than for educated people who had a more amateur interest. Abu Kamil embraced irrational numbers as possible solutions to quadratic equations, rather than rejecting them as awkward anomalies. In his *Book of Rare Things in the Art of Calculation*, Abu Kamil attempted to solve indeterminate equations (those with more than one solution). He further explored this topic in his *Book of Birds*, in which he posed a miscellany of bird-related algebra problems, including: "How many ways can one buy 100 birds in the market with 100 dirhams?"

Geometric solutions

Up until the era of the Arab "algebraists"—from al-Khwarizmi in the 9th century to the death of the Moorish mathematician al-Qalasadi in 1486—the key developments within algebra were underpinned by geometrical representations. For example, al-Khwarizmi's method of "completing the square" in order to solve quadratic equations relies on consideration of the properties of a real square; later scholars worked in a similar way. »

Al-Khwarizmi showed how to solve quadratic equations by a method known as "completing the square." This example shows how to find x in the equation $x^2 + 10x = 39$.

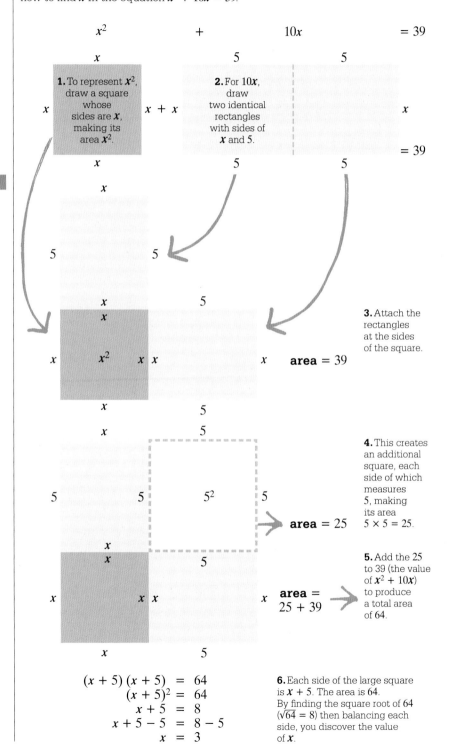

$$x^2 \qquad + \qquad 10x \qquad = 39$$

1. To represent x^2, draw a square whose sides are x, making its area x^2.

2. For $10x$, draw two identical rectangles with sides of x and 5.

area = 39

3. Attach the rectangles at the sides of the square.

5^2 area = 25

4. This creates an additional square, each side of which measures 5, making its area $5 \times 5 = 25$.

area = 25 + 39

5. Add the 25 to 39 (the value of $x^2 + 10x$) to produce a total area of 64.

$$(x + 5)\,(x + 5) = 64$$
$$(x + 5)^2 = 64$$
$$x + 5 = 8$$
$$x + 5 - 5 = 8 - 5$$
$$x = 3$$

6. Each side of the large square is $x + 5$. The area is 64. By finding the square root of 64 ($\sqrt{64} = 8$) then balancing each side, you discover the value of x.

Mathematician and poet Omar Khayyam, for example, was interested in solving problems using the relatively new discipline of algebra, but employed both geometrical and algebraic methods. His *Treatise on Demonstration of Problems of Algebra* (1070) notably includes a fresh perspective on the difficulties within Euclid's postulates, a set of geometric rules that are assumed to be true without requiring a proof. Picking up on earlier work by al-Karaji, Khayyam also develops ideas about binomial coefficients, which determine how many ways there are to select a number of items from a larger set. He solved cubic equations, too, inspired by al-Khwarizmi's use of Euclid's geometrical constructions for working out quadratic equations.

Polynomials

During the 10th and early 11th centuries, a more abstract theory of algebra was developed, which was not reliant on geometry—an important factor in establishing its academic status. Al-Karaji was instrumental in this development. He established a set of procedures for performing arithmetic on polynomials—expressions that contain a mixture of algebraic terms. He created rules for calculating with polynomials, in much the same way that there were rules for adding, subtracting, or multiplying numbers. This allowed mathematicians to work on increasingly complex algebraic expressions in a more uniform

An ounce of algebra is worth a ton of verbal argument.
John B. S. Haldane
British mathematical biologist

Islamic mathematicians gather in the library of a mosque in an illustration from a manuscript by the 12th-century poet and scholar Al-Hariri of Basra.

way, and reinforced algebra's essential links with arithmetic.

Mathematical proof is a vital part of modern algebra and one of the tools of proof is called mathematical induction. Al-Karaji used a basic form of this principle, whereby he would show an algebraic statement to be true for the simplest case (say $n = 1$), then use that fact to show that it must also be true for $n = 2$ and so on, with the inevitable conclusion that the statement must hold true for all possible values of n.

One of al-Karaji's successors was the 12th-century scholar Ibn Yahya al-Maghribi al-Samaw'al. He noted that the new way of thinking of algebra as a kind of arithmetic with generalized rules involved the algebraist "operating on the unknown using all the

> As the sun eclipses the stars by its brilliancy, so the man of knowledge will eclipse the fame of others in assemblies of the people if he proposes algebraic problems, and still more if he solves them.
> **Brahmagupta**

arithmetical tools, in the same way as the arithmetician operates on the known." Al-Samaw'al continued al-Karaji's work on polynomials, but also developed the laws of indices, which led to much later work on logarithms and exponentials, and was a significant step forward in mathematics.

Plotting equations

Cubic equations had challenged mathematicians since the time of Diophantus of Alexandria. Al-Khwarizmi and Khayyam had made significant progress in understanding them—work further developed by Sharaf al-Din al-Tusi, a 12th-century scholar, probably born in Iran, whose mathematics appears to have been inspired by the work of earlier Greek scholars, especially Archimedes. Al-Tusi was more interested in determining types of cubic equation than al-Khwarizmi and Khayyam had been. He also developed an early understanding of graphical curves, articulating the significance of maximum and minimum values. His work strengthened the connection between algebraic

equations and graphs—between mathematical symbols and visual representations.

A new algebra

The discoveries and rules set down by medieval Arab scholars still form the basis of algebra today. The works of al-Khwarizmi and his successors were key to establishing algebra as a discipline in its own right. It was not until the 1500s, however, that mathematicians began to abbreviate equations by using letters to stand for known and unknown variables. French mathematician François Viète was key to this development. In his works, he pioneered the move away from the Arabic algebra of procedures toward what is known as symbolic algebra.

In his *Introduction to the Analytic Arts* (1591), Viète suggested that mathematicians should use letters to symbolize the variables in an equation: vowels to

represent unknown quantities and consonants to represent the known. Although this convention was eventually replaced by René Descartes—in which letters at the beginning of the alphabet represent known numbers and letters at the end represent the unknown—Viète nonetheless was responsible for simplifying algebraic language far beyond what the Arab scholars had imagined. The innovation allowed mathematicians to write out increasingly complex and detailed abstract equations, without using geometry. Without symbolic algebra, it would be difficult to imagine how modern mathematics would have ever developed. ∎

Islamic algebraists wrote equations as text with accompanying diagrams, as in the 14th-century *Treatise on the Question of Arithmetic Code* by Master Ala-El-Din Muhammed El Ferjumedhi.

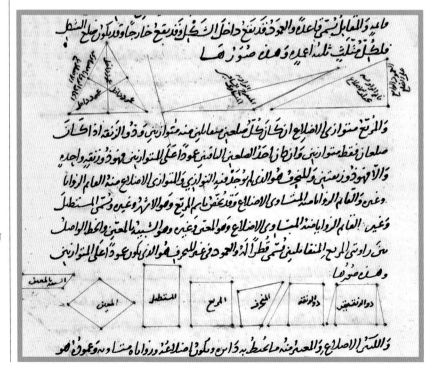

FREEING ALGEBRA FROM THE CONSTRAINTS OF GEOMETRY

THE BINOMIAL THEOREM

IN CONTEXT

KEY FIGURE
Al-Karaji (c. 980–c. 1030)

FIELD
Number theory

BEFORE
c. 250 CE In *Arithmetica*, Diophantus lays down ideas about algebra later taken up by al-Karaji.

c. 825 CE The Persian astronomer and mathematician al-Khwarizmi develops algebra.

AFTER
1653 In *Traité du triangle arithmétique* (*Treatise on the Arithmetical Triangle*), Blaise Pascal reveals the triangular pattern of coefficients in the bionomial theorem in what is later called Pascal's triangle.

1665 Isaac Newton develops the general binomial series from the binomial theorem, forming part of the basis for his work on calculus.

In ancient Greece, mathematics was almost entirely based on **geometric** arguments.

⬇

Al-Karaji **broke from** this **tradition** and treated the solution of equations in purely **numerical** terms.

⬇

He created a set of **algebraic rules**, including the binomial theorem.

⬇

Algebraic solutions no longer had to rely on geometric diagrams.

At the heart of many mathematical operations lies an important basic theorem—the binomial theorem. It provides a shorthand summary of what happens when you multiply out a binomial, which is a simple algebraic expression consisting of two known or unknown terms added together or subtracted. Without the binomial theorem, many mathematical operations would be almost impossible to achieve. The theorem shows that when binomials are multiplied out, the results follow a predictable pattern that can be written as an algebraic expression or displayed on a triangular grid (known as Pascal's triangle after Blaise Pascal, who explored the pattern in the 1600s).

Making sense of binomials

The binomial pattern was first observed by mathematicians in ancient Greece and India, but the man credited with its discovery is the Persian mathematician al-Karaji, one of many scholars who flourished in Baghdad from the 8th to the 14th century. Al-Karaji explored the multiplication of algebraic terms. He defined

See also: Positional numbers 22–27 ▪ Diophantine equations 80–81 ▪ Zero 88–91 ▪ Algebra 92–99 ▪ Pascal's triangle 156–61 ▪ Probability 162–65 ▪ Calculus 168–75 ▪ The fundamental theorem of algebra 204–09

Al-Karaji created a table to work out the coefficients of binomial equations. The first five lines of it are shown here. The top line is for powers, with the coefficients for each power listed in the column below. The first and final numbers are always 1. Each other number is the sum of its adjacent number in the preceding column and the number above that adjacent number.

The expansion of $(a + b)^3$ can be found by looking in the column headed by 3. ⟶

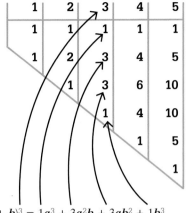

$$(a + b)^3 = 1a^3 + 3a^2b + 3ab^2 + 1b^3$$

> The binomial theorem and a Bach fugue are, in the long run, more important than all the battles of history.
> **James Hilton**
> **British novelist**

single terms called monomials"—x, x^2, x^3, and so on—and showed how they can be multiplied or divided. He also looked at "polynomials" (expressions with multiple terms), such as $6y^2 + x^3 - x + 17$. But it was his discovery of the formula for multiplying out binomials that had the most impact.

The binomial theorem concerns powers of binomials. For example, multiplying out the binomial $(a + b)^2$ by converting it to $(a + b)(a + b)$ and multiplying each term in the first parentheses by each term in the second parentheses results in $(a + b)^2 = a^2 + 2ab + b^2$. The calculation for the power 2 is manageable, but for greater powers, the resulting expression becomes increasingly complicated. The binomial theorem simplifies the problem by unlocking the pattern in the coefficients—numbers, such as 2 in $2ab$, by which the unknown terms are multiplied. As al-Karaji discovered, the coefficients can be laid out in a grid, with the columns showing the coefficients needed for multiplying out each power. The coefficients in a column are

calculated by adding together pairs of numbers in the preceding column. To determine the powers in the expansion, you take the degree of the binomial as n. In $(a + b)^2$, $n = 2$.

Algebra breaks free

Al-Karaji's discovery of the binomial theorem helped to open the way for the full development of algebra, by allowing mathematicians to manipulate complicated algebraic

expressions. The algebra developed by al-Khwarizmi 150 years or so previously had used a system of symbols to work out unknown quantities and was limited in scope. It was tied to the rules of geometry, and the solutions were geometric dimensions, such as angles and side lengths. Al-Karaji's work showed how algebra could instead be based entirely on numbers, liberating it from geometry. ∎

Al-Karaji

Born around 980 CE, Abu Bakr ibn Muhammad ibn al-Husayn al-Karaji most likely got his name from the city of Karaj, near Tehran, but he lived most of his life in Baghdad, at the court of the caliph. It was here around 1015 that he probably wrote his three key mathematics texts. The work in which al-Karaji developed the binomial theorem is now lost, but later commentators preserved his ideas. Al-Karaji was also an engineer, and his book

Extraction of Hidden Waters is the first known manual on hydrology.

Later in life, al-Karaji moved to "mountain countries" (possibly the Elburz mountains near Karaj), where he spent his time working on practical projects for drilling wells and building aqueducts. He died around 1030 CE.

Key works

Glorious on algebra
Wonderful on calculation
Sufficient on calculation

FOURTEEN FORMS WITH ALL THEIR BRANCHES AND CASES

CUBIC EQUATIONS

IN CONTEXT

KEY FIGURE
Omar Khayyam (1048–1131)

FIELD
Algebra

BEFORE
3rd century BCE Archimedes solves cubic equations using the intersection of two conics.

7th century CE Chinese scholar Wang Xiaotong solves a range of cubic equations numerically.

AFTER
16th century Mathematicians in Italy create jealously guarded methods to solve cubic equations in the fastest time.

1799–1824 Italian scholar Paolo Ruffini and Norwegian mathematician Niels Henrik Abel show that no algebraic formulas exist for equations involving terms to the power of 5 and higher.

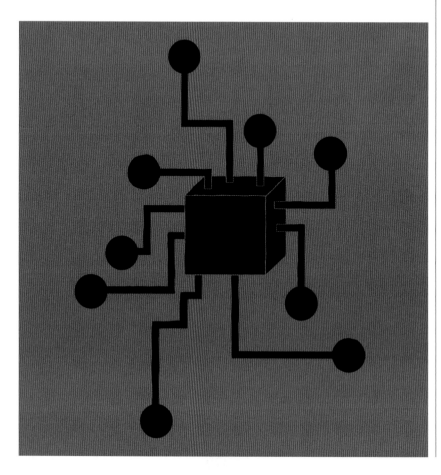

In the ancient world, scholars considered problems in a geometric way. Simple linear equations (which describe a line), such as $4x + 8 = 12$, where x is to the power of 1, could be used to find a length, while a squared variable (x^2) in a quadratic equation could represent an unknown area—a two-dimensional space. The next step up is the cubic equation, where the x^3 term is an unknown volume—a three-dimensional space.

The Babylonians could solve quadratic equations in 1800 BCE, but it took another 3,000 years until Persian poet-scientist Omar Khayyam found an accurate method

See also: Quadratic equations 28–31 ▪ Euclid's *Elements* 52–57 ▪ Conic sections 68–69 ▪ Imaginary and complex numbers 128–31 ▪ The complex plane 214–15

Cubic equations involve a **variable to the power of 3** (x^3).

↓

The ancient Greeks attempted to **solve cubic equations** using only a **ruler** and **compasses**.

↓

Omar Khayyam devised **more accurate methods** for solving cubic equations by:

↓

breaking the equation down into a simpler equation with **squares** (powers of 2) and **lengths** (powers of 1).

drawing geometric diagrams to explore where **shapes intersect**.

Omar Khayyam

Born in Nishapur, Persia (now Iran), in 1048, Omar Khayyam was educated in philosophy and the sciences. Although he won renown as an astronomer and mathematician, when his patron Sultan Malik Shah died in 1092, he was forced into hiding. Finally rehabilitated 20 years later, he lived quietly and died in 1131.

In mathematics, Khayyam is best remembered for his work on cubic equations, but he also produced an important commentary on Euclid's fifth postulate, known as the parallel postulate. As an astronomer, he helped to construct a highly accurate calendar that was used until the 1900s. Ironically, Khayyam is now best known for a work of poetry for which he may not have been the sole author—the *Rubaiyat*, which was translated into English by Edward Fitzgerald in 1859.

Key works

c. 1070 *Treatise on Demonstration of Problems of Algebra*
1077 *Commentaries on the difficult postulates of Euclid's book*

for solving cubic equations, using curves called conic sections—such as circles, ellipses, hyperbolas, or parabolas—formed by the intersection of a plane and a cone.

Problems with cubes

The ancient Greeks, who used geometry to work out complex problems, puzzled over cubes. A classic conundrum was how to produce a cube that was twice the volume of another cube. For example, if the sides of a cube are each equal to 1 in length, what length sides do you need for a cube twice the volume? In modern terms, if a cube with side length 1 has a volume of 1^3, what side length cubed (x^3) produces twice that volume; that is, since $1^3 = 1$, what

is x if $x^3 = 2$? The ancient Greeks used a ruler and compasses to attempt constructing a solution to this cubic equation but they never succeeded. Khayyam saw that such tools were not enough to solve all cubic equations, and set out his use of conic sections and other methods in his treatise on algebra.

Using modern conventions, cubic equations can be expressed simply, such as $x^3 + bx = c$. Without the economy of modern notation, Khayyam expressed his equations in words, describing x^3 as "cubes", x^2 as "squares," x as "lengths," and numbers as "amounts." For example, he described $x^3 + 200x = 20x^2 + 2{,}000$ as a problem of finding a cube that "with two hundred times its side" » is equal to "twenty squares of its side and two thousand." For a simpler equation, such as $x^3 + 36x = 144$, Khayyam's method was to draw a geometric diagram. He found that he could break down the cubic equation into two simpler equations: one for a circle, and the other for a parabola. By working out the value of x for which both these simpler equations are true simultaneously, he could solve the original cubic equation. This is shown in the graph below. At the time, mathematicians did not have these graphical methods and Khayyam would have constructed the circle and parabola geometrically.

Khayyam had also explored the properties of conic sections, and had deduced that a solution to the cubic equation could be found by giving the circle in the diagram a diameter of 4. This measure was arrived at by dividing c by b, or $144/36$ in the example below. The circle passed through the origin (0,0) and its center was on the x axis at (2,0). Using this diagram, Khayyam drew a perpendicular line from the point where the circle and parabola intersected down to the x axis. The point where the line crossed the x axis (where $y = 0$) gives the value for x in the cubic equation. In the case of $x^3 + 36x = 144$, the answer is $x = 3.14$ (rounded to two decimal places).

Khayyam did not use coordinates and axes (which were invented about 600 years later). Instead, he would have drawn the shapes as accurately as possible and carefully measured the lengths on their diagrams. He would then have found an approximate numerical solution using trigonometric tables, which were common in astronomy.

For Khayyam, the solution would always have been a positive number. There is an equally valid negative answer, as shown by the minus numbers in the graph below, but although the concept of negative numbers was recognized in Indian mathematics, it was not generally accepted until the 1600s.

Khayyam's contribution

While Archimedes, working in the 3rd century BCE, may well have examined the intersection of conic sections in a bid to solve cubic equations, what marks Khayyam out is his systematic approach. This enabled him to produce a general theory. He extended his mix of geometry and algebra to solve cubic equations using circles, hyperbolas, and ellipses, but never explained how he constructed them, simply saying he "used instruments."

Khayyam was among the first to realize that a cubic equation could have more than one root, and therefore more than one solution. As can be shown on a modern graph that plots a cubic equation as a curve snaking above and below the x axis, a cubic equation has up to three roots. Khayyam suspected two, but would not have considered

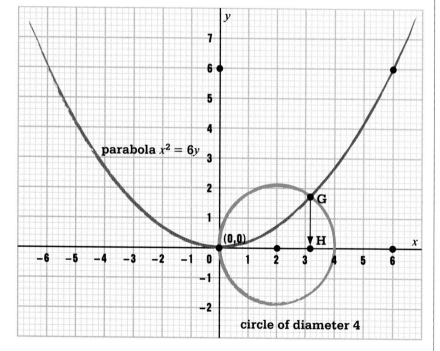

I have shown how to find the sides of the square-square, quatro-cube, cubo-cube… to any length, which has not been [done] before now.
Omar Khayyam

A parabola (pink) for the equation $x^2 = 6y$ intersects the circle (blue) $(x-2)^2 + y^2 = 4$. A line from G, the point of intersection, to H on the x axis, gives the value for x (3.14) in the cubic equation $x^3 + 36x = 144$.

Graph labels: parabola $x^2 = 6y$; (0,0); G; H; circle of diameter 4

> Algebras are
> geometric facts
> which are proved by
> propositions.
> **Omar Khayyam**

negative values. He did not like having to use geometry as well as algebra to find a solution, and hoped that his geometrical efforts would one day be replaced by arithmetic.

Khayyam anticipated the work of 16th-century Italian mathematicians, who solved cubic equations without direct recourse to geometry. Scipione del Ferro produced the first algebraic solution to cubic equations, discovered in his notebook after his death. He and successors Niccolò Tartaglia, Lodovico Ferrari, and Gerolamo Cardano all worked on algebraic

formulae to solve cubic equations. Cardano published Ferro's solution in his book *Ars Magna* in 1545. Their solutions were algebraic but differed from those of today, partly because zero and negative numbers were little used at the time.

Toward modern algebra

Mathematicians who continued the quest for cubic equation solutions included Rafael Bombelli. He was among the first to state that a cubic root could be a complex number, that is, a number that makes use

A passion for geometric forms is evident in Islamic architecture, seen here in the tile patterns, curved arches, and domes of the Masjid-i Kabud, the "Blue Mosque," in Tabriz, Iran.

of an "imaginary" unit derived from the square root of a negative number, something not possible with "real" numbers. In the late 1500s, Frenchman François Viète created more modern algebraic notation, using substitution and simplifying to reach his solutions. By 1637, René Descartes had published a solution to the quartic equation (involving x^4), reducing it to a cubic equation and then to two quadratic equations to solve it. Today, a cubic equation can be written in the form $ax^3 + bx^2 + cx + d = 0$, provided a itself is not 0. Where the coefficients (a, b, and c, which multiply the variable x) are real numbers, rather than complex numbers, the equation will have at least one real root and up to three roots in total.

Khayyam's method is still taught today. His painstaking work advanced early algebra, while later mathematicians have continued to refine its expression and scope. ∎

The length of the year

In 1074, the ruling sultan of Persia, Jalal al-Din Malik Shah I, commissioned Omar Khayyam to reform the lunar calendar used since the 7th century, replacing it with a solar calendar. A new observatory was built in the capital Isfahan, and Khayyam assembled a team of eight astronomers to assist him with the work.

The year—computed to a highly accurate 365.24 days—began at the vernal equinox in March, when the center of the

visible Sun is directly above the equator. Each month was worked out by the passage of the sun into the corresponding zodiac region, which required both computations and actual observations. Because solar transit times could vary by 24 hours, months were between 29 and 32 days long, but their length could differ from year to year. The new Jalali calendar, named after the sultan, was adopted on March 15, 1079 and was only modified in 1925.

THE UBIQUITOUS MUSIC OF THE SPHERES

THE FIBONACCI SEQUENCE

IN CONTEXT

KEY FIGURE
Leonardo of Pisa, also known as Fibonacci
(1170–c. 1250)

FIELD
Number theory

BEFORE
200 BCE The number sequence later known as the Fibonacci sequence is cited by the Indian mathematician Pingala in relation to Sanskrit poetic meters.

700 CE The Indian poet and mathematician Virahanka writes about the sequence.

AFTER
17th century In Germany, Johannes Kepler notices that the ratio of successive terms in the sequence converges.

1891 Édouard Lucas coins the name Fibonacci sequence in *Théorie des Nombres* (*Number Theory*).

Number sequences are lists of numbers **linked by a rule**.

In the Fibonacci sequence, starting with 0 and 1, the next number is the sum of the previous two.

The **sequence continues infinitely**.

0 + 1 = 1; 1 + 1 = 2;
1 + 2 = 3; 2 + 3 = 5;
3 + 5 = 8; 8 + 5 = 13...

O ne sequence of numbers occurs time and again in the natural world. In this sequence, every number is the sum of the previous two (0, 1, 1, 2, 3, 5, 8, 13, 21, 34, and so on). Originally referred to by the Indian scholar Pingala in around 200 BCE, it was later called the Fibonacci sequence after Leonardo Pisano (Leonardo of Pisa), an Italian mathematician known as Fibonacci. Fibonacci explored the sequence in his 1202 book *Liber Abaci* (*The Book of Calculation*). The sequence has important forecasting applications in nature, geometry, and business.

A problem with rabbits

One of the problems Fibonacci raised in *Liber Abaci* concerned the growth of rabbit populations. Starting with a single pair of rabbits, he asked his readers to work out how many pairs there would be in each successive month. Fibonacci made several assumptions: no rabbit ever died;

Fibonacci

Born Leonardo Pisano, probably in Pisa, Italy, in 1170, Fibonacci did not become known as Fibonacci ("son of Bonacci") until long after his death. Leonardo traveled widely with his diplomat father and studied at a school of accounting in Bugia, North Africa. There he came across the Hindu–Arabic symbols used to represent the numbers 1 to 9. Impressed by these numerals' simplicity compared with the lengthy Roman numerals used in Europe, he discussed them in *Liber Abaci* (*The Book of Calculation*), which he wrote in 1202.

Leonardo also traveled to Egypt, Syria, Greece, Sicily, and Provence, exploring different number systems. His work was widely read and came to the attention of the Holy Roman Emperor, Frederick II. Fibonacci died *c.* 1240–50.

Key works

1202 *Liber Abaci* (*The Book of Calculation*)
1220 *Practica Geometriae* (*Practical Geometry*)
1225 *Liber Quadratorum* (*The Book of Squares*)

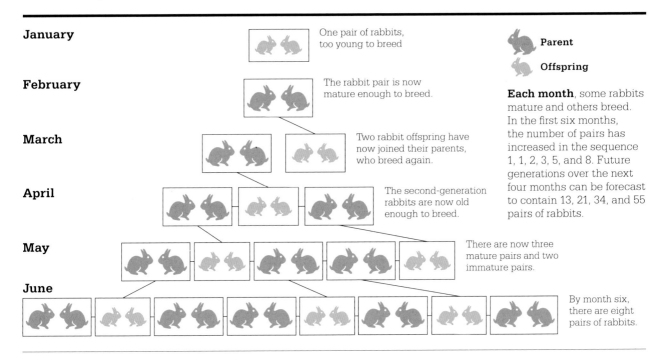

January — One pair of rabbits, too young to breed

February — The rabbit pair is now mature enough to breed.

March — Two rabbit offspring have now joined their parents, who breed again.

April — The second-generation rabbits are now old enough to breed.

May — There are now three mature pairs and two immature pairs.

June — By month six, there are eight pairs of rabbits.

 Parent

 Offspring

Each month, some rabbits mature and others breed. In the first six months, the number of pairs has increased in the sequence 1, 1, 2, 3, 5, and 8. Future generations over the next four months can be forecast to contain 13, 21, 34, and 55 pairs of rabbits.

rabbit pairs mated every month, but only after they were two months old, the age of maturity; and each pair produced one male and one female offspring every month. For the first two months, he said, there would only be the original pair: by the end of three months, there would be a total of two pairs; and at the end of four months, there would be three pairs, as only the original pair was old enough to breed.

Thereafter, the population grows more quickly. In the fifth month, both the original pair and their first offspring produce baby rabbits, although the second pair of offspring is still too young. This results in a total of five pairs of rabbits. The process continues in successive months, resulting in a number sequence in which each number is the sum of the previous two: 1, 1, 2, 3, 5, 8, 13, 21, 34, 55, 89, 144, and so on – a sequence that became known as the Fibonacci

sequence. As with many mathematical problems, it is based on a hypothetical situation: Fibonacci's assumptions about how the rabbits behave are unrealistic.

Generations of bees

An example of the Fibonacci sequence cropping up in nature concerns bees in a beehive. A male bee, or drone, develops from the unfertilized egg of a queen bee. Since the egg is unfertilized, the drone has only one parent, its "mother." Drones have different roles in the beehive, one of which is to mate with the queen and fertilize her eggs. Fertilized eggs develop into female bees, which can either be queens or workers. This means that one generation back the drone has only one ancestor, its mother; two generations back it has two ancestors, or "grandparents"—the mother and father of its mother; and three generations back, it has

three "great grandparents"—its grandmother's two parents and its grandfather's mother. Further back, there are five members of the previous generation, eight of the one before that, and so on. The pattern is clear: the number of members in each generation of ancestors forms the Fibonacci sequence. The sum of the »

> The Fibonacci sequence turns out to be the key to understanding how nature designs.
> **Guy Murchie**
> **American writer**

If a **number** in a sequence is **divided by** the **previous one**, it creates **a ratio**.

→

The ratios of any two consecutive **Fibonacci numbers** get closer and closer to **1.618**.

↓

Like the Fibonacci sequence, the **golden ratio** often occurs in the **natural world**.

←

1.618 is an approximation of the **"golden ratio,"** which is actually $(1 + \sqrt{5}) \div 2$.

six petals, so while numbers from the sequence are common, other patterns are also found.

Each Fibonacci number is the sum of the previous two, so the first two have to be stated before the third can be calculated. The Fibonacci sequence can be defined by a recurrence relation—an equation that defines a number in a sequence in terms of its previous numbers. The first Fibonacci number is written as f_1, the second as f_2, and so on. The equation is $f_n = f_{(n-1)} + f_{(n-2)}$, where n is greater than 1. If you are trying to find the fifth Fibonacci number (f_5), for example, you must add together f_4 and f_3.

Fibonacci ratios
Calculating the ratios of successive terms in the Fibonacci sequence is particularly interesting. Dividing

number of parents of a male and a female from the same generation of bees is three. Their parents total five grandparents, whose own parents add up to eight great-grandparents. When the pattern is traced back to earlier generations, the Fibonacci sequence continues, with 13, 21, 34, 55 ancestors, and so on.

Plant life
The Fibonacci sequence can also be seen in the arrangement of leaves and seeds in some plants. Pine cones and pineapples, for example, display Fibonacci numbers in the spiral formation of their exterior scales. Many flowers

have three, five, or eight petals—numbers that belong to the Fibonacci sequence. Ragwort flowers have 13 petals, chicory often has 21, and different types of daisy have 34 or 55. However, many other flowers have four or

[If] a spider climbs so many feet up a wall each day and slips back a fixed number each night, how many days does it take him to climb the wall?
Fibonacci

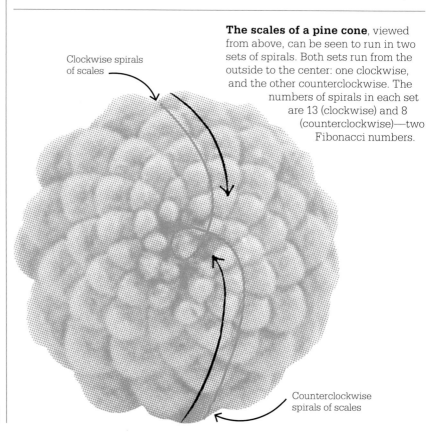

Clockwise spirals of scales

The scales of a pine cone, viewed from above, can be seen to run in two sets of spirals. Both sets run from the outside to the center: one clockwise, and the other counterclockwise. The numbers of spirals in each set are 13 (clockwise) and 8 (counterclockwise)—two Fibonacci numbers.

Counterclockwise spirals of scales

A piano keyboard scale from C to C spans 13 keys, eight white and five black. The black keys are in groups of two and three. These numbers all form part of the Fibonacci sequence.

each number by the previous number in the sequence produces the following: $^1/_1 = 1$, $^2/_1 = 2$, $^3/_2 = 1.5$, $^5/_3 = 1.666...$, $^8/_5 = 1.6$, $^{13}/_8 = 1.625$, $^{21}/_{13} = 1.61538...$, $^{34}/_{21} = 1.61904...$ By continuing this process indefinitely, it can be shown that the numbers approach 1.618, approximately. This is referred to as the golden ratio or the golden mean. The same number is also significant in a curve called the golden spiral, which gets wider by a factor of 1.618 for every quarter turn it makes. This spiral crops up commonly in nature: for example, the seeds of pine cones, sunflowers, and coneflowers tend to grow in golden spirals.

Arts and analysis

The Fibonacci sequence can also be found in poetry, art, and music. A pleasing rhythm in poetry, for example, is produced when successive lines have 1, 1, 2, 3, 5, and 8 syllables, and there is a long tradition of 6-line, 20-syllable poetry structured in this way. Around 200 BCE, Pingala was aware of this pattern in Sanskrit poetry, and the Roman poet Virgil used it in the 1st century BCE.

The sequence has also been used in music. French composer Claude Debussy (1862–1918) employed Fibonacci numbers in several compositions. In the dramatic climax of his *Cloches à travers les feuilles* (*Bells Through the Leaves*), the ratio of total bars in the piece to climax bars is approximately 1.618.

Although it is often associated with the arts, the Fibonacci sequence has also proved a useful tool in finance. Today, ratios derived from the sequence are used as an analytical tool to forecast the point at which stock market prices will stop rising or falling. ■

A page from the original manuscript of *Liber Abaci* shows the Fibonacci sequence listed on the right.

Practical solutions

Fibonacci's work was intended to have a useful purpose. In *Liber Abaci* (1202), for example, he described how to solve many of the problems encountered in commerce, including calculating profit margins and converting currencies. In *Practica Geometriae* (1220), he solved problems associated with surveying, such as finding the height of a tall object using similar triangles (triangles that have identical angles, but different sizes). In his *Liber Quadratorum* (1225), he tackled several topics in number theory, including finding Pythagorean triples—groups of three integers that represent the lengths of the sides of right-angled triangles. In these triangles, the square of the length of the longest side (the hypotenuse) equals the sum of the squares of the lengths of the two shorter sides. Fibonacci found that, starting with 5, every second number in his sequence (13, 34, 89, 233, 610, and so on) is the length of the hypotenuse of a right-angled triangle when the lengths of the two shorter sides are integers.

THE POWER OF DOUBLING
WHEAT ON A CHESSBOARD

IN CONTEXT

KEY FIGURE
Sissa ben Dahir
(3rd or 4th century CE)

FIELD
Number theory

BEFORE
c. 300 BCE Euclid introduces the concept of a power to describe squares

c. 250 BCE Archimedes uses the law of exponents, which states that multiplying exponents can be achieved by adding the powers.

AFTER
1798 British economist Thomas Malthus predicts that the human population will grow exponentially while the food supply will increase more slowly, causing a catastrophe.

1965 American co-founder of Intel Gordon Moore observes how the number of transistors on a microchip doubles roughly every 18 months.

The first written record of the wheat on a chessboard problem was made in 1256 by Muslim historian Ibn Khallikan, though it is probably a retelling of an earlier version that arose in India in the 5th century. According to the story, the inventor of chess, Sissa ben Dahir, was summoned to an audience with his ruler, King Sharim. The king was so delighted with the game of chess that he offered to grant Sissa any reward that he wanted. Sissa asked for some

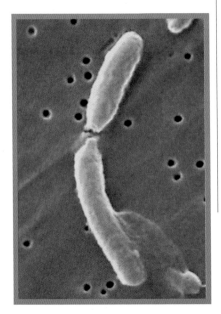

grains and explained the quantity he desired using the squares on the 8 × 8 chessboard. One grain of wheat (or rice, in some versions of the story) was to be placed on the bottom left square of the chessboard. Moving right, the number of grains would then be doubled, so the second square had two grains, the third had four, and so on, moving left to right along each row to the 64th square at the top right.

Puzzled by what seemed to be a paltry reward, the king ordered that the grains be counted out. The 8th square had 128 grains, the 24th had more than 8 million, and the 32nd, the last square on the chessboard's first half, had over 2 billion. By then, the king's granary was running low, and he realized that the next square alone, number 33, would need 4 billion grains, or one large field's worth. His advisers calculated that the final square would need 9.2 million trillion grains, and the total number of

Bacteria dividing is an example of exponential growth; when a single cell divides, it creates two cells that divide to make four, and so on. This allows bacteria to spread very quickly.

Sissa's concept of wheat on a chessboard is an early example of how quickly numbers can increase with exponential growth. (Numbers from 1 million onward are approximate.) The wheat on this chessboard would total over 18 million trillion grains.

72 thousand trillion	144 thousand trillion	288 thousand trillion	0.6 million trillion	1.2 million trillion	2.3 million trillion	4.6 million trillion	9.2 million trillion
281 trillion	562 trillion	1,123 trillion	2,252 trillion	4,504 trillion	9,007 trillion	18 thousand trillion	36 thousand trillion
1 trillion	2 trillion	4 trillion	8 trillion	17 trillion	35 trillion	70 trillion	140 trillion
4 billion	8 billion	16 billion	33 billion	66 billion	131 billion	262 billion	524 billion
16 million	32 million	64 million	128 million	256 million	512 million	1 billion	2 billion
65,536	131,072	262,144	524,288	1 million	2 million	4 million	8 million
256	512	1,024	2,048	4,096	8,192	16,384	32,768
1	2	4	8	16	32	64	128

grains on the chessboard would be 18,446,744,073,709,551,615 ($2^{64} - 1$). The story has two alternative endings: in one, the king made Sissa his chief adviser; in the other, Sissa was executed for making the king look foolish.

Sissa's concept is an example of what is known as a geometric series, in which every successive term is the previous one multiplied by two: $1 + 2 + 4 + 8 + 16$, and so on. From 2 onward, these numbers are all powers of 2: $1 + 2 + 2^2 + 2^3 + 2^4$, and so on. The superscript number, the exponent, shows how many times the other number, in this case 2, is multiplied by itself. The last term in the series, 2^{63}, is 2 multiplied by itself 63 times.

Power of exponents

The growth of the values in this series is described as exponential. Exponents can be viewed as instructions for how many times 1 should be multiplied by a given number. For example, 2^3 means that 1 will be multiplied by 2 three times: $1 \times 2 \times 2 \times 2 = 8$, while 2^1

The second half of the chessboard

Recent thinkers have used the chessboard problem as a metaphor for the rate of change in technology over recent years. In 2001, computer scientist Ray Kurzweil wrote an influential essay describing the exponential growth in technology over previous years. He predicted that, like the wheat on the second half of the chessboard, the rate of technological development would rapidly grow out of control, following the model of doubling its previous growth with every leap forward.

Kurzweil argued that this rate of growth in technology would eventually lead to the singularity, which is defined in physics as a point at which a function takes an infinite value. When applied to technology, the singularity marks the point at which the cognitive ability of artificial intelligence will surpass that of humans.

means that 1 will be multiplied by 2 just once: $1 \times 2 = 2$. The first square of the chessboard contains 1 grain, so 1 is the first term of this series. The number 1 can be written as 2^0, because it is equivalent to 1 multiplied by 2 zero times, leaving 1 unaffected. In fact, any number to the power of 0 equals 1 for this reason.

Exponential growth and decay relate to many aspects of everyday life. For example, a radioactive isotope decays into another atomic form at an exponential rate, and that results in a half-life, where half the material takes the same amount of time to decay, irrespective of the starting quantity. ▪

THE RENAISS

1500–1680

ANCE

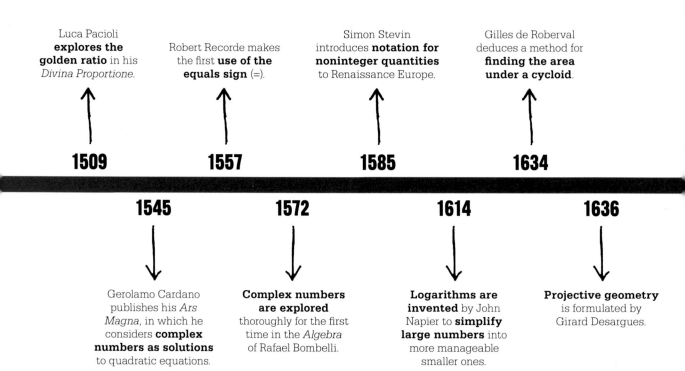

Luca Pacioli **explores the golden ratio** in his *Divina Proportione*.

Robert Recorde makes the first **use of the equals sign** (=).

Simon Stevin introduces **notation for noninteger quantities** to Renaissance Europe.

Gilles de Roberval deduces a method for **finding the area under a cycloid**.

1509   **1557**   **1585**   **1634**

1545   **1572**   **1614**   **1636**

Gerolamo Cardano publishes his *Ars Magna*, in which he considers **complex numbers as solutions** to quadratic equations.

Complex numbers are explored thoroughly for the first time in the *Algebra* of Rafael Bombelli.

Logarithms are invented by John Napier to **simplify large numbers** into more manageable smaller ones.

Projective geometry is formulated by Girard Desargues.

Throughout the Middle Ages, the Catholic Church wielded considerable political power across Europe, and had a virtual monopoly of learning, but in the 1400s, its authority was being challenged. A new cultural movement, known as the Renaissance ("rebirth"), was inspired by renewed interest in the arts and philosophy of the Graeco-Roman Classical period.

The Renaissance thirst for discovery also accelerated a "Scientific Revolution"—classic texts of mathematics, philosophy, and science had become widely available, and inspired a new generation of thinkers. So too did the Protestant Reformation that challenged the hegemony of the Catholic Church in the 1500s.

Renaissance art also influenced mathematics. Luca Pacioli, an early Renaissance mathematician, investigated the mathematics of the golden ratio that was so important in Classical art, and the innovative use of perspective in painting inspired Girard Desargues to explore the mathematics behind it and develop the field of projective geometry. Practical considerations also prompted progress: commerce required more sophisticated means of accounting, and international trade drove advances in navigation, which demanded a deeper understanding of trigonometry.

Mathematical innovation

A major advance in the business of calculation came with the adoption of the Hindu-Arabic number system and an increase in the use of symbols to represent functions such as equals, multiplication, and division. Another significant development was the formalization of a number system of base-10, and Simon Stevin's introduction of the decimal point in 1585.

To meet the era's practical needs, mathematicians devised tables of relevant calculations, and John Napier developed a means of calculating with logarithms in the 1600s. The first mechanical aids to calculation were invented during this period, such as William Oughtred's slide rule, and Gottfried Leibniz's mechanical calculating device, which was a first step toward true computing devices.

Other mathematicians took a more theoretical path, inspired by the ideas in the newly available

The Cartesian system of **coordinates and axes** still in use today is formalized by René Descartes.

Blaise Pascal publishes his **study on the triangle** that bears his name.

Christiaan Huygens's **solution to the tautochrone problem** leads to more accurate clocks.

Leibniz proposes **a machine that calculates using binary principles**, laying the foundations for future computer coding.

1637   **1653**   **1656**   **1679**

1644   **1654**   **1665–75**

The monk Marin Mersenne describes the **method for finding the primes** named after him.

The correspondence between Pascal and Pierre de Fermat lays the **groundwork for probability theory**.

Calculus is developed by Gottfried Leibniz and Isaac Newton, **probably independently** of each other.

---

texts. In the 1500s, the solution of cubic and quartic equations occupied Italian mathematicians such as Gerolamo Cardano, while Marin Mersenne devised a method of finding prime numbers, and Rafael Bombelli laid down rules for using imaginary numbers. In the 1600s, the pace of mathematical discovery accelerated as never before, and several pioneering modern mathematicians emerged. Among these was philosopher, scientist, and mathematician René Descartes, whose methodical approach to problem-solving set the scene for the modern scientific era. His major contribution to mathematics was the invention of a system of coordinates to specify the position of a point in relation to axes, establishing the new field of analytic geometry, in which lines and shapes are described in terms of algebraic equations.

Another late-Renaissance mathematician who has become almost a household name is Pierre de Fermat, whose claim to fame rests largely on his enigmatic last theorem, which remained unsolved until 1994. Less well known are his contributions to the development of calculus, number theory, and analytic geometry. He and fellow mathematician Blaise Pascal corresponded about gambling and games of chance, laying the foundations for the field of probability.

The birth of calculus

One of the key mathematical concepts of the 1600s was developed independently by two scientific giants of the time, Gottfried Leibniz and Isaac Newton. Following on from the work of Gilles de Roberval in finding the area under a cycloid, Leibniz and Newton worked on the problems of calculation of such things as continuous change and acceleration, which had puzzled mathematicians ever since Zeno of Elea had presented his famous paradoxes of motion in ancient Greece. Their solution to the problem was the theorem of calculus, a set of rules for calculating using infinitesimals. For Newton, calculus was a practical tool for his work in physics and especially on the motion of planets, but Leibniz recognized its theoretical importance and refined the rules of differentiation and integration. ∎

THE GEOMETRY OF ART AND LIFE

THE GOLDEN RATIO

IN CONTEXT

KEY FIGURE
Luca Pacioli (1445–1517)

FIELD
Applied geometry

BEFORE
447–432 BCE Designed by the Greek sculptor Phidias, the Parthenon is later said to approximate the golden ratio.

c. 300 BCE Euclid makes the first known written reference to the golden ratio in his *Elements*.

1202 CE Fibonacci introduces his famous sequence.

AFTER
1619 Johannes Kepler proves that the numbers in the Fibonacci sequence approach the golden ratio.

1914 Mark Barr, an American mathematician, is credited with using the Greek letter phi (ϕ) for the golden ratio.

[The golden proportion] is a scale of proportions which makes the bad difficult [to produce] and the good easy.
Albert Einstein

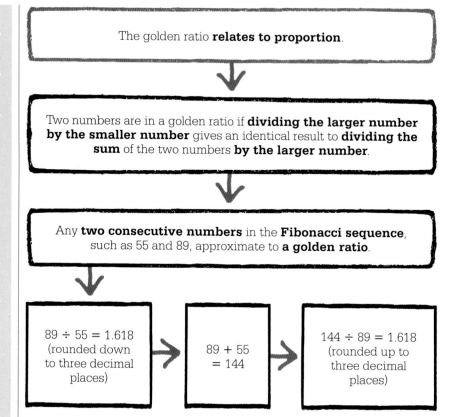

The golden ratio **relates to proportion**.

Two numbers are in a golden ratio if **dividing the larger number by the smaller number** gives an identical result to **dividing the sum** of the two numbers **by the larger number**.

Any **two consecutive numbers** in the **Fibonacci sequence**, such as 55 and 89, approximate to **a golden ratio**.

89 ÷ 55 = 1.618 (rounded down to three decimal places)

89 + 55 = 144

144 ÷ 89 = 1.618 (rounded up to three decimal places)

The Renaissance was a time of intellectual creativity, in which disciplines such as art, philosophy, religion, science, and mathematics were considered to be much closer to each other than they are today. One area of interest was in the relationship between mathematics, proportion, and beauty. In 1509, Italian priest and mathematician Luca Pacioli wrote *Divina Proportione* (*The Divine Proportion*), which discussed the mathematical and geometric underpinnings of perspective in architecture and the visual arts. The book was illustrated by Pacioli's friend and colleague Leonardo da Vinci, a leading artist and polymath of the Renaissance.

Since the Renaissance, the mathematical analysis of art by means of the "golden ratio," "golden mean"—or, as Pacioli called it, the Divine Proportion—has come to symbolize geometrical perfection. The ratio can be found by dividing a straight line into two parts, so that the ratio of the longer length (*a*) to the smaller length (*b*) is the same as the ratio of the whole line (*a* + *b*) divided by the longer length (*a*). So: $(a + b) \div a = a \div b$. The value of this ratio is a mathematical constant denoted by the Greek letter ϕ ("phi"). The name ϕ comes from the ancient Greek sculptor Phidias (500–432 BCE), who is believed to have been one of the first to recognize the aesthetic possibilities of the golden ratio. He allegedly used the ratio in the design of the Parthenon in Athens.

Like π (3.1415...), ϕ is an irrational number (a number that cannot be expressed as a fraction) and can therefore be expanded to an infinite number of decimals in a nonrepeating random pattern. Its approximate value is 1.618. It is one of the wonders of mathematics that this seemingly unremarkable number should produce such aesthetically pleasing proportions in art, architecture, and nature.

Discovering phi

Some believe that proportions related to ϕ can be found in ancient Greek architecture—and even earlier in ancient Egyptian culture, with the Great Pyramid built at Giza in c. 2560 BCE, which has a base to height ratio of 1.5717. Yet there is no evidence that ancient architects were conscious of this ideal ratio. Approximations to the golden ratio may have been the result of an unconscious tendency rather than any deliberate mathematical intention.

The Pythagoreans, a semi-mystical group of mathematicians and philosophers associated with Pythagoras of Samos (570–495 BCE) had the pentagram, or five-pointed

star, as their symbol. Where one side of the pentagram crosses another, it divides each side into two parts, the ratio of which is ϕ. The Pythagoreans were convinced that the Universe was based on numbers; they also believed that all numbers could be described as the ratio of two integers. According to Pythagorean doctrine, any two lengths are both integer multiples of some fixed smaller length. In other words, their ratio is a rational number, so it can be expressed as the ratio of integers. Supposedly, when one of Pythagoras's followers, Hippasus, discovered that this was not true, his fellow Pythagoreans drowned him in disgust.

Written records

The earliest written references to the golden ratio are found in the work of the Alexandrian mathematician Euclid, c. 300 BCE. Euclid's *Elements* discussed the Platonic solids described earlier by Plato (such as the tetrahedron), and demonstrated the golden ratio (which Euclid called the

The good, of course, is always beautiful, and the beautiful never lacks proportion.
Plato

"extreme and mean ratio") in their proportions. Euclid showed how to construct the golden ratio using a ruler and compass.

Phi and Fibonacci

The golden ratio is also closely related to another well-known mathematical phenomenon—the set of numbers known as the Fibonacci sequence. It was introduced by Leonardo of Pisa, or Fibonacci, in his 1202 book *Liber Abaci* (*The Book of Calculation*). »

Luca Pacioli

Luca Pacioli was born in 1445 in Tuscany. After moving to Rome in his youth, he received training from the artist–mathematician Piero della Francesca as well as the renowned architect Leon Battista Alberti, and gained knowledge of geometry, artistic perspective, and architecture. He became a teacher and traveled throughout Italy. He also took his vows as a Franciscan friar, combining monastic pursuits with teaching. In 1496, Pacioli moved to Milan to work as a payroll clerk. While there, he also gave mathematics tuition, one of his

students being Leonardo da Vinci, who illustrated Pacioli's *Divina Proportione*. Pacioli also devised a method of accounting that is still in use today. He died in 1517, in Sansepolcro, Tuscany.

Key works

1494 *Summa de arithmetica, geometria, proportioni et proportionalita* (*Summary of arithmetic, geometry, proportions, and proportionality*)
1509 *Divina Proportione* (*The Divine Proportion*)

Leonardo da Vinci supposedly used golden rectangles in his composition of *The Last Supper* (1494–98). Other Renaissance artists—such as Raphael and Michelangelo—also used the ratio.

Subsequent numbers in the Fibonacci sequence are found by adding the previous two together: 1, 1, 2, 3, 5, 8, 13, 21, 34, 55, 89….

It took until 1619 for German mathematician and astronomer Johannes Kepler to show that the golden ratio is revealed if a number in the Fibonacci sequence is divided by the one that precedes it. The further along the sequence this calculation is attempted, the closer the answer is to ϕ. For example, $6{,}765 \div 4{,}181 = 1.61803$. Both Fibonacci's sequence and the golden ratio appear to exist widely in nature. For example, many species of flower have a Fibonacci number of petals, and the scales of a pine cone, viewed from below, are arranged in 8 clockwise spirals and 13 counterclockwise spirals.

Another golden ratio approximated in nature is the golden spiral, which gets wider by a factor of ϕ for every quarter turn it makes. The golden spiral can be drawn by splitting a golden rectangle (a rectangle with side lengths in the golden ratio) into successively smaller squares and golden rectangles, and inscribing quarter circles inside the squares (see opposite). Natural spiral shapes, such as the nautilus shell, have a resemblance to the golden spiral, but do not strictly fit the proportions.

The golden spiral was first described by French philosopher, mathematician, and polymath René Descartes in 1638 and was studied by Swiss mathematician Jacob Bernoulli. It was classified as a type of "logarithmic spiral" by French mathematician Pierre Varignon because the spiral can be generated by a logarithmic curve.

Art and architecture

While the golden ratio can be found in music and poetry, it is more often associated with the art of the Renaissance in the 15th and 16th centuries. Da Vinci's painting *The Last Supper* (1494–98) is said to incorporate the golden ratio. His famous drawing of the "Vitruvian Man"—a "perfectly proportioned" man inscribed in a circle and square—for *Divina Proportione* is also said to contain many instances of the golden ratio in the proportions of the ideal

The problem with using the golden ratio to define human beauty is that if you're looking hard enough for a pattern, you'll almost certainly find one.
Hannah Fry
British mathematician

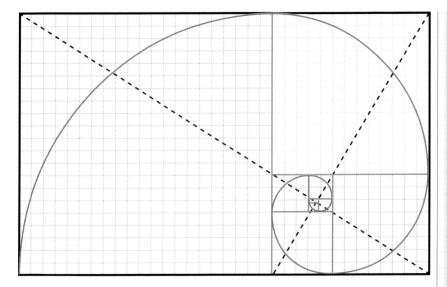

A golden spiral can be inscribed within a golden rectangle. It is created by splitting the rectangle into squares and a smaller golden rectangle, then repeating the process in the smaller rectangle. If quarter circles are then inscribed in the squares, it creates a golden spiral.

human body. In reality, the Vitruvian Man, which illustrated the theories of ancient Roman architect Vitruvius, does not quite align with golden proportions. Despite this, many people have subsequently attempted to relate the golden ratio to the notion of attractiveness in people (see box, right).

Against the golden ratio
In the 1800s, German psychologist Adolf Zeising argued that the perfect human body aligned with the golden ratio; it could be found by measuring the person's total height and dividing this by the height from their feet to their navel. In 2015, Stanford mathematics professor Keith Devlin argued that the golden ratio is a "150-year scam." He blamed Zeising's work for the idea that the golden ratio has historically had a relationship to aesthetics. Devlin argues that Zeising's ideas have led people to look back at historical art

and architecture and retrospectively apply the golden ratio. Similarly, in 1992, American mathematician George Markowsky suggested that supposed discoveries of the golden ratio in the human body were a result of imprecise measurements.

Modern uses
Although ϕ's historical use is debated, the golden ratio can still be traced in modern works, such as Salvador Dalí's *Sacrament of the Last Supper* (1955), in which the shape of the painting itself is a golden rectangle. Beyond the arts, the golden ratio has also appeared in modern geometry, particularly in the work of British mathematician Roger Penrose, whose Fibonacci tiles incorporate the golden ratio in their structure. Standard aspect ratios for television and computer monitor screens, such as the 16:9 display, also come close to ϕ, as do modern bank cards, which are almost perfect golden rectangles. ∎

The ratio of beauty
Studies indicate that facial symmetry plays a major role in determining a person's perceived attractiveness. However, the proportions defined by the golden ratio appear to play an even greater role. People whose faces have proportions that approximate to the golden ratio (the ratio of the length of the head to its width, for instance) are often cited as being more attractive than those whose faces do not. Studies to date, however, are inconclusive and often contradictory; there is little scientific basis for believing that the golden ratio makes a face more attractive.

Stephen Marquardt, an American plastic surgeon, created a "mask" (see below) based on applying the golden ratio to the human face. The more closely a face aligns with the mask, the more beautiful it supposedly is. Some, however, see the mask—used as a template for plastic surgery—as an unethical, unfounded use of mathematics.

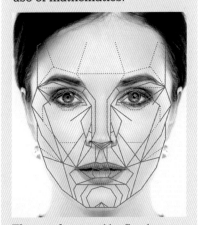

The mask created by Stephen Marquardt has been criticized for defining beauty based on white, Western models.

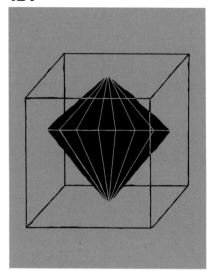

LIKE A LARGE DIAMOND

MERSENNE PRIMES

IN CONTEXT

KEY FIGURES
Hudalrichus Regius (early 1500s), **Marin Mersenne** (1588–1648)

FIELD
Number theory

BEFORE
c. 300 BCE Euclid proves the fundamental theorem of arithmetic that every integer greater than 1 can be expressed as a product of primes in only one way.

c. 200 BCE Eratosthenes devises a method for calculating prime numbers.

AFTER
1750 Leonhard Euler confirms that the Mersenne number $2^{31} - 1$ is prime.

1876 French mathematician Édouard Lucas verifies that $2^{127} - 1$ is a Mersenne prime.

2018 The largest known prime to date is found to be $2^{82,589,933} - 1$.

Prime numbers—numbers that can only be divided by themselves or 1—have fascinated scholars since the ancient Greeks of Pythagoras's school first studied them, not least because they can be thought of as the building blocks of all natural numbers (positive integers). Until 1536, mathematicians believed that all prime numbers for n, when employed in the equation $2^n - 1$, would lead to another prime as the solution. However, in his *Utriusque Arithmetices Epitome* (*Epitome of Both Arithmetics*), published in 1536, a scholar known to us only as Hudalrichus Regius pointed out that $2^{11} - 1 = 2,047$. This is not a prime number, as $2,047 = 23 \times 89$.

Mersenne's influence

Regius's work on primes was continued by others who proposed new hypotheses with $2^n - 1$. The most significant was that of French monk Marin Mersenne in 1644). He stated that $2^n - 1$ was valid when $n = 2, 3, 5, 7, 13, 17, 19, 31, 67, 127,$ and 257. Mersenne's work rekindled interest in the topic, and primes generated by $2^n - 1$ are now known as Mersenne primes (M_n).

The use of computers has made it possible to find more Mersenne primes. Two of Mersenne's n values (67 and 257) were proved incorrect, but in 1947, three new primes were found: $n = 61, 89,$ and 107 (M_{61}, M_{89}, M_{107}), and in 2018, the Great Internet Mersenne Prime Search uncovered the 51st known Mersenne prime. ∎

The beauty of number theory [is] related to the contradiction between the simplicity of the integers and the complicated structure of the primes.
Andreas Knauf
German mathematician

See also: Euclid's *Elements* 52–57 ▪ Eratosthenes' sieve 66–67 ▪ The Riemann hypothesis 250–51 ▪ The prime number theorem 260–61

SAILING ON A RHUMB
RHUMB LINES

From around 1500, as ships began to cross the world's oceans, navigators met a problem—plotting a course across the world that took account of the Earth's curved surface. The problem was solved by the introduction of the rhumb line by Portuguese mathematician Pedro Nunes in his *Treatise on the Sphere* (1537).

The rhumb spiral
A rhumb line cuts across every meridian (line of longitude) at the same angle. Because meridians get closer toward the poles, rhumb lines bend around into a spiral. Such spirals were called loxodromes by Dutch mathematician Willebrord Snell in 1617; they became a key concept in the geometry of space.

The rhumb line helps navigators because it gives a single compass bearing for a voyage. In 1569, Mercator maps—on which lines of longitude are drawn parallel, so that all rhumb lines are straight—were introduced. This further enabled people to plot a course just by drawing a straight line on the

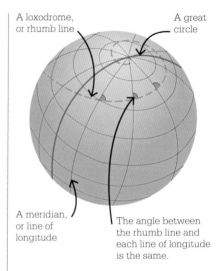

A loxodrome, or rhumb line

A great circle

A meridian, or line of longitude

The angle between the rhumb line and each line of longitude is the same.

A loxodrome starts at the North or South Pole, and spirals around the globe, crossing each meridian at the same angle. A rhumb line is all or part of this spiral.

map. The shortest distance across the globe is not a rhumb, however, but a "great circle"—any circle that centers on the center of the Earth. It only became practical to follow a great circle course with the invention of GPS. ∎

A PAIR OF EQUAL-LENGTH LINES

THE EQUALS SIGN AND OTHER SYMBOLOGY

IN CONTEXT

KEY FIGURE
Robert Recorde (c. 1510–58)

FIELD
Number systems

BEFORE
250 CE Greek mathematician Diophantus uses symbols to represent variables (unknown quantities) in *Arithmetica*.

1478 The *Treviso Arithmetic* explains in simple language how to perform addition, subtraction, multiplication, and division calculations.

AFTER
1665 In England, Isaac Newton develops infinitesimal calculus, which introduces ideas such as limits, functions, and derivatives. These processes require new symbols for abbreviation.

1801 Carl Friedrich Gauss introduces the symbol for congruence—equal size and shape.

I n the 16th century, when Welsh doctor and mathematician Robert Recorde began his work, there was little consensus on the notation used in arithmetic. Hindu–Arabic numerals, including zero, were already established, but there was little to represent calculations.

In 1543, Recorde's *The Grounde of Artes* introduced the symbols for addition (+) and subtraction (−) to mathematics in England. These signs had first appeared in print in *Mercantile Arithmetic* (1489), by German mathematician Johannes

Widman, but were probably already used by German merchants before Widman's book was published. These symbols slowly replaced the letters "p" for plus and "m" for minus as they were taken up by scholars, first in Italy, then in England.

In 1557, Recorde went on to recommend a new symbol of his own. In *The Whetstone of Witte*, he used a pair of identical parallel lines (=) to represent "equals," claiming that "no two things can be more equal" than these. Recorde suggested that symbols would save mathematicians from having to write out calculations in words. The equals sign was widely adopted, and the 17th century also saw the creation of many of the other symbols used today, such as those for multiplication (×) and division (÷).

Notating algebra

While the earliest algebraic techniques date back more than two millennia to the Babylonians, most calculations before the

Robert Recorde tested the equals sign (=) in his own calculations, as seen here in one of his exercise books. Recorde's sign was noticeably longer than the modern form.

See also: Positional numbers 22–27 ▪ Negative numbers 76–79 ▪ Algebra 92–99
▪ Decimals 132–37 ▪ Logarithms 138–41 ▪ Calculus 168–75

The creation of symbols			
Symbol	**Meaning**	**Inventor**	**Date**
−	Subtraction	Johannes Widman	1489
+	Addition	Johannes Widman	1489
=	Equals	Robert Recorde	1557
×	Multiplication	William Oughtred	1631
<	Less than	Thomas Harriot	1631
>	Greater than	Thomas Harriot	1631
÷	Division	Johann Rahn	1659

16th century were recorded in words—sometimes abbreviated, but not in a uniform way. English mathematician Thomas Harriot and French mathematician François Viète, who each made important contributions to developments in algebra, used letters to produce consistent symbolic notation. In their system, the most noticeable difference from today's notation is the use of a repeated letter to indicate a power. For example, a^3 was aaa and x^4 was $xxxx$.

A modern system

French mathematician Nicholas Chuquet used superscripts in 1484 to represent exponents ("to the power of"), but did not record them as such; for example, $6x^2$ was 6.2. It took more than 150 years for superscripts to become common; René Descartes used recognizable examples in 1637 when writing $3x + 5x^3$, yet continued to write x^2 as xx. Only in the early 1800s, when the influential German mathematician Carl Gauss favored using x^2, did superscript notation begin to stick. Descartes also made a contribution with his use of x, y, and z for the unknowns in equations, and a, b, and c for known figures.

Algebraic notation may have taken a long time to catch on, but when a symbol made sense and helped mathematicians work through problems, it became the norm. Improved contact between mathematicians in different parts of the world in the 1600s also led to such notations being adopted much more swiftly. ▪

To avoid the tedious repetition of these words, is equal to, I will set, as I do often in work use, a pair of parallels.
Robert Recorde

Robert Recorde

Born in Tenby, Wales, around 1510, Recorde grew up to study medicine first at Oxford University, then at Cambridge, where he qualified as a physician in 1545. He taught mathematics at both universities and wrote the first English book on algebra in 1543. In 1549, after a period practicing medicine in London, Recorde was made controller of the Bristol mint. However, after he refused to issue funds to William Herbert, the future Earl of Pembroke, for his army, the mint was closed.

In 1551, Recorde was given charge of the Dublin mint, which included silver mines in Germany. When he failed to show a profit, the mines were also closed. Recorde later tried to sue Pembroke for misconduct, but was instead countersued for libel. Sent to a London prison in 1557 for failure to pay the fine, Recorde died there in 1558.

Key works

1543 *Arithmetic: or the Grounde of Artes*
1551 *The Pathway to Knowledge*
1557 *The Whetstone of Witte*

PLUS OF MINUS TIMES PLUS OF MINUS MAKES MINUS

IMAGINARY AND COMPLEX NUMBERS

IN CONTEXT

KEY FIGURE
Rafael Bombelli (1526–72)

FIELD
Algebra

BEFORE
1500s In Italy, Scipione del Ferro, Tartaglia, Antonio Fior, and Ludovico Ferrari compete publicly to solve cubic equations.

1545 Gerolamo Cardano's *Ars Magna*, a book of algebra, includes the first published calculation involving complex numbers.

AFTER
1777 Leonhard Euler introduces the notation i for $\sqrt{-1}$.

1806 Jean-Robert Argand publishes a geometrical interpretation of complex numbers, leading to the Argand diagram.

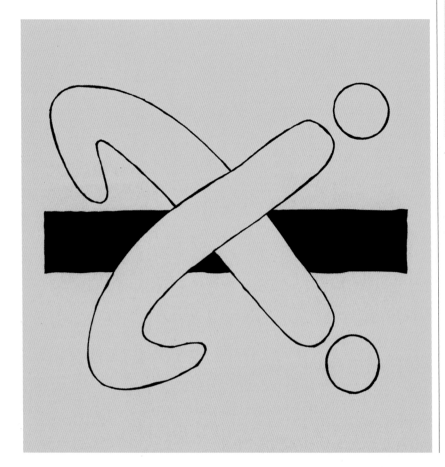

I n the late 1500s, Italian mathematician Rafael Bombelli broke new ground when he laid down the rules for using imaginary and complex numbers in his book *Algebra*. An imaginary number, when squared, produces a negative result, defying the usual rules that any number (positive or negative) results in a positive number when squared. A complex number is the sum of any real number (on the number line) and an imaginary number. Complex numbers take the form $a + bi$, where a and b are real and $i = \sqrt{-1}$.

Over the centuries, scholars have needed to extend the concept of the number in order to solve

See also: Quadratic equations 28–31 ▪ Irrational numbers 44–45 ▪ Negative numbers 76–79 ▪ Cubic equations 102–05 ▪ The algebraic resolution of equations 200–01 ▪ The fundamental theorem of algebra 204–09 ▪ The complex plane 214–15

A **real number** gives a **positive result** when it is squared.

An **imaginary number** gives a **negative result** when it is squared.

A complex number is the sum of a real number and an imaginary number.

Complex numbers enable us to solve polynomial equations (those that place a sum of powers of x equal to zero, such as $3x^3 - 2x^2 + x - 5 = 0$).

Some people believe in imaginary friends. I believe in imaginary numbers.
R. M. ArceJaeger
American author

different problems. Imaginary and complex numbers were new tools in this endeavor, and Bombelli's *Algebra* advanced understanding of how these and other numbers work. To solve the simplest equations, such as $x + 1 = 2$, only natural numbers (positive integers) are needed. To solve $x + 2 = 1$, however, x must be a negative integer, while solving $x^2 + 2 = 1$ requires the square root of a negative number. This did not exist with the numbers at Bombelli's disposal, so had to be invented—leading to the concept of the imaginary unit ($\sqrt{-1}$). Negative numbers were still mistrusted in the 1500s; imaginary and complex numbers were not widely accepted for many decades.

Fierce rivalry

The idea of complex numbers first emerged early in Bombelli's lifetime as Italian mathematicians

sought to find solutions to cubic equations as efficiently as possible, without relying on the geometrical methods devised by Persian polymath Omar Khayyam in the 12th century. As most quadratic equations could be solved with an algebraic formula, the search was on for a similar formula that worked for cubic equations. Scipione del Ferro, a mathematics professor at Bologna University, took a major step forward when he discovered an algebraic method for solving some cubic equations, but the quest for a comprehensive formula continued.

Italian mathematicians of this era would publicly challenge one another to solve cubic equations and other problems in the least possible time. Achieving fame in such contests became essential for any scholar who wanted to gain a post as a mathematics professor at

a prestigious university. As a result, many mathematicians kept their methods secret rather than sharing them for the common good. Del Ferro tackled equations of the form $x^3 + cx = d$. He passed his technique on to only two people, Antonio Fior and Annibale della Nave, swearing them to secrecy. Del Ferro soon had competition from Niccolò Fontana (known as Tartaglia, or "the stutterer"). An itinerant teacher of considerable mathematical ability, but with few financial resources, Tartaglia discovered a general method »

I shall call [the imaginary unit] 'plus of minus' when added and when subtracted, 'minus of minus.'
Rafael Bombelli

Bombelli's rules for the combination of imaginary numbers

Rafael Bombelli set out the rules for operations on complex numbers. He used the term "plus of minus" to describe a positive imaginary unit and "minus of minus" to describe a negative imaginary unit. Multiplying a positive imaginary unit by a negative imaginary unit, for example, equals a positive integer; while multiplying a negative imaginary unit by a negative imaginary unit equals a negative integer.

Plus of minus	×	Plus of minus	=	Minus
Plus of minus	×	Minus of minus	=	Plus
Minus of minus	×	Minus of minus	=	Minus
Minus of minus	×	Plus of minus	=	Plus

for solving cubic equations independently of del Ferro. When del Ferro died in 1526, Fior decided the time had come for him to unleash del Ferro's formula upon the world. He challenged Tartaglia to a cubic duel, but was beaten by Tartaglia's superior methods. Gerolamo Cardano heard of this and persuaded Tartaglia to share his methods with him. As with del Ferro, the condition was that the method should never be published.

Beyond positive numbers

At this time all equations were solved using positive numbers. Working with Tartaglia's method, Cardano had to grapple with the notion that using the square roots of negative numbers might help solve cubic equations. He was evidently prepared to experiment with the method, but appears not to have been convinced. He called such negative solutions "fictitious" and "false" and described the intellectual effort involved in finding them as "mental torture." His *Ars Magna* shows his use of the negative square root. He wrote: "Multiply $5 + \sqrt{-15}$ by $5 - \sqrt{-15}$, making $25 - (-15)$, which is $+15$. Hence this product is 40." This is the first recorded calculation involving complex numbers, but the significance of this breakthrough escaped Cardano; he branded his work "subtle" and "useless."

Explaining the numbers

Rafael Bombelli assimilated the tussles between the various mathematicians solving cubic equations. He read Cardano's *Ars Magna* with great admiration. His own work, *Algebra*, was a more accessible version, and was a thorough and innovative survey of the subject. It investigated the arithmetic of negative numbers, and included some economical notation that represented a major advance on what had gone before.

The work outlines the basic rules for calculating with positive and negative quantities, such as: "Plus times plus makes plus; Minus times minus makes plus." It then sets out new rules for adding, subtracting, and multiplying imaginary numbers in terminology that differs from that used by mathematicians today. For example, he stated that "Plus of minus multiplied by plus of minus makes minus"—meaning a positive imaginary number multiplied by a positive imaginary number equals a negative number: $\sqrt{-n} \times \sqrt{-n} = -n$. Bombelli also gave practical examples of how to apply his rules

Rafael Bombelli

Born in Bologna, Italy, in 1526, Rafael Bombelli was the eldest of six children; his father was a wool merchant. Although Bombelli did not receive a college education, he was taught by an engineer–architect and became an engineer himself, specializing in hydraulics. He also developed an interest in mathematics, studying the work of ancient and contemporary mathematicians. While waiting for a drainage project to recommence, he embarked on his major work, *Algebra*, which laid out a primitive but thorough arithmetic of complex numbers for the first time.

Greatly impressed by a copy of Diophantus's *Arithmetica* found in the Vatican library, Bombelli helped to translate it into Italian – work that led him to revise *Algebra*. Three volumes were published in 1572, the year he died; the last two incomplete volumes were published in 1929.

Key work

1572 *Algebra*

> The shortest route between two truths in the real domain passes through the complex domain.
> **Jacques Hadamard**
> **French mathematician**

for complex numbers to cubic equations, where solutions require finding the square root of some negative number. Although Bombelli's notation was advanced for his time, the use of algebraic symbols was still in its infancy. Two centuries later, Swiss mathematician Leonhard Euler introduced the symbol *i* to denote the imaginary unit.

Applying complex numbers

Imaginary and complex numbers joined the ranks of other sets, such as natural numbers, real numbers, rational numbers, and irrational numbers, that were used to solve equations and perform a range of other increasingly sophisticated mathematical tasks.

Over the decades, sets of such numbers acquired their own universal symbols that could be used in formulae. For instance, the bold capital N is used for natural numbers from the set $\{0, 1, 2, 3, 4…\}$, enclosed in curly brackets to denote a set. In 1939, American mathematician Nathan Jacobson established the bold capital C to signify the set of complex numbers, $\{a + bi\}$, where a and b are real and $i = \sqrt{-1}$.

Complex numbers enable all polynomial equations to be solved completely, but have also proved immensely useful in many other branches of mathematics—even in number theory (the study of integers, especially positive numbers). By treating the integers as complex numbers (the sum of a real value and an imaginary value), number theorists can use powerful techniques of complex analysis (a study of functions with complex numbers) to investigate the integers. The Riemann zeta

> There is an ancient and innate sense in people that numbers ought not to misbehave.
> **Douglas Hofstadter**
> **Cognitive scientist**

function, for example, is a function of complex numbers that provides information about primes. In other practical areas, physicists use complex numbers in the study of electromagnetism, fluid dynamics, and quantum mechanics, while engineers need them for designing electronic circuits, and for studying audio signals. ∎

A series of cups shows blue food dye being dripped over an ice cube (left). As the ice cube melts, the heavier blue dye sinks. Complex numbers are used to model the velocity (speed and direction) of such fluids.

THE ART OF TENTHS

DECIMALS

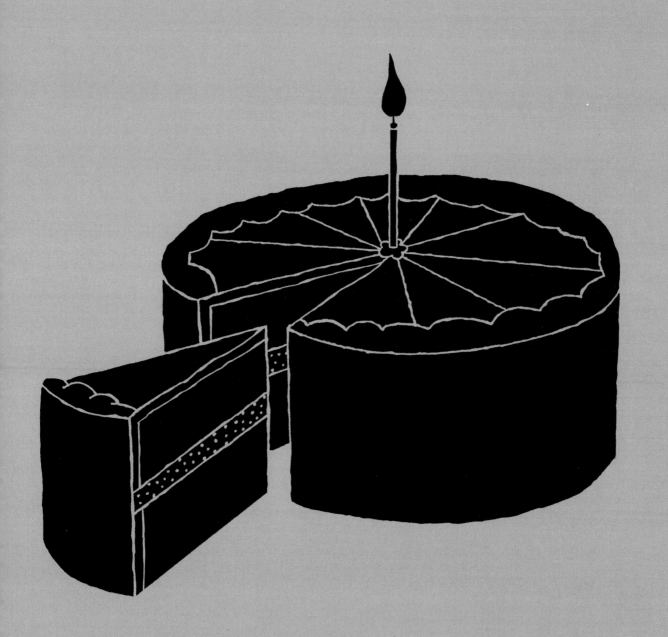

IN CONTEXT

KEY FIGURE
Simon Stevin (1548–1620)

FIELD
Number systems

BEFORE
830 CE Al-Kindi's four-volume *On the use of Indian numerals* spreads the place value system based on the Hindu numerals throughout the Arab world.

1202 Leonardo of Pisa's *Liber Abaci* (*The Book of Calculation*) brings the Arabic number system to Europe.

AFTER
1799 The metric system is introduced for French currency and measures during the French Revolution.

1971 Britain introduces decimalization, dispensing with pounds, shillings, and pence, which stemmed from the Latin system.

F ractions—so named for the Latin word *fractio*, meaning "break"—were used from around 1800 BCE in Egypt to express parts of a whole. At first they were limited to unit fractions, which are those with a 1 as the numerator (top number). The ancient Egyptians had symbols for $\frac{2}{3}$ and $\frac{3}{4}$, but other fractions were expressed as the sum of unit fractions, for example as $\frac{1}{3} + \frac{1}{13} + \frac{1}{17}$. This system worked well for recording amounts but not for doing calculations. It was not until after Simon Stevin's *De Thiende* (*The Art of Tenths*) was published in 1585 that a decimal system became commonplace.

The importance of 10

Simon Stevin, a Flemish engineer and mathematician in the late 16th and early 17th century, used many calculations in his work. He simplified these by using fractions with a base system of tenth powers. Stevin correctly predicted that a decimal system would eventually be universal.

Cultures throughout history had used many different bases for expressing parts of a whole. In ancient Rome, fractions were based

By relieving the brain of all unnecessary work, a good notation sets it free to concentrate on more advanced problems.
Alfred North Whitehead
British mathematician

on a system of twelfths, and written out in words: $\frac{1}{12}$ was called *uncia*, $\frac{6}{12}$ was *semis*, and $\frac{1}{24}$ was *semiuncia*, but this cumbersome system made it difficult for people to do any calculations. In Babylon, fractions were expressed using their base-60 number system, but in writing, it was difficult to distinguish which numbers represented integers and which were part of the whole.

For many centuries, Europeans used Roman numerals to record numbers and to do calculations.

Simon Stevin

Born in 1548 in Bruges, now in Belgium, Simon Stevin worked as a bookkeeper, cashier, and clerk before entering the University of Leiden in 1583. There he met Prince Maurice, the heir of William of Orange, and they became friends. Stevin tutored the prince in mathematics and also advised him on military strategy, leading to some significant victories over the Spanish. In 1600, Prince Maurice asked Stevin, who was also an outstanding engineer, to found a School of Engineering at the University in 1600. As quarter-master general from 1604, Stevin

was responsible for several innovative military and engineering ideas that were adopted across Europe. He authored many books on a variety of subjects, including mathematics. He died in 1620.

Key works

1583 *Problemata geometrica* (*Geometric Problems*)
1585 *De Thiende* (*The Art of Tenths*)
1585 *De Beghinselen der Weeghconst* (*Principles of the Art of Weighing*)

See also: Positional numbers 22–27 ▪ Irrational numbers 44–45 ▪ Negative numbers 76–79 ▪ The Fibonacci sequence 106–11 ▪ Binary numbers 176–77

Medieval Italian mathematician Leonardo of Pisa (also known as Fibonacci) came across the Indian place-value number system while he was traveling in the Arab world. He quickly realized its usefulness and efficiency for both recording and calculating with whole numbers. His *Liber Abaci* (1202), which brought many useful Arabic ideas to the west, also introduced a new notation for fractions to Europe that would form the basis of the notation used today. Fibonacci employed a horizontal bar to divide the numerator and denominator (bottom number), but followed the Arabic practice of writing the fraction to the left of the integer, rather than to the right.

Introducing decimals

Finding that conventional fractions were both time-consuming and prone to errors, Stevin began using a decimal system. The idea of "decimal fractions"—which have powers of 10 as the denominator—had been used five centuries before Stevin, in the Middle East, but it

> To **decimalize**, fractions are converted to **decimal fractions**, where the **denominator** (bottom number) is a **power of 10**.

> The **numerator** (top number) of the converted fraction is used to write the fraction as a **decimal**—for example, $^{25}/_{100}$ becomes 0.25.

> The numerator is placed after a **decimal separator**, such as a **decimal point**, to show that it is **not a whole number**.

> The **decimal system** makes it easy to **add** and **subtract** amounts that are not whole numbers.

> Decimals [are] a kind of arithmetic invented by the tenth progression, consisting in characters of cyphers.
> **Simon Stevin**

was Stevin who made decimals commonplace in Europe, both for recording and calculating with parts of a whole. He suggested a notation system for decimal fractions, replicating the advantages of the Indian place-value system for whole numbers.

In Stevin's new notation, numbers that would previously have been written as the sum of fractions—for example, $32 + ^5/_{10} +$ $^6/_{100} + ^7/_{1,000}$—could now be written as a single number. Stevin placed circles after each number; these were shorthand for the denominator of the original decimal fraction. The whole 32 would be followed by a 0, because 32 is an integer, whereas the $^6/_{100}$, for example, was expressed as 6 and a 2 inside a circle. This 2 denoted the power of 10 of the original denominator, as 100 is 10^2. In the same vein, »

Stevin's notation used circles to indicate the power of ten of the denominator of the converted fraction. This represents how Stevin would have written the number now expressed as 32.567.

32①0⓪5①6②7③

The decimal system makes it easier to divide and multiply fractions, especially by 10. Shown here with the example of 32.567 (or $32 + \frac{5}{10} + \frac{6}{100} + \frac{7}{1,000}$), numbers shift one column to the left or right, crossing over the decimal separator.

	Hundreds 100	Tens 10	Units 1	Tenths $\frac{1}{10}$	Hundredths $\frac{1}{100}$	Thousandths $\frac{1}{1,000}$	Ten-thousandths $\frac{1}{10,000}$
× 1		3	2	5	6	7	
× 10	3	2	5	6	7		
÷ 10			3	2	5	6	7

the $\frac{7}{1,000}$ became a 7 followed by a 3 inside a circle. The entire sum could be written out following this pattern (see p.135, bottom right). The symbol that is placed between the whole-number part and the fractional part of a number is called the decimal separator. Stevin's zero inside a circle later evolved into a dot, now called the decimal point. The dot was positioned on the midline (at a middle height) in Stevin's notation but has now moved to be on the baseline to avoid confusion with the dot notation sometimes used for multiplication. Stevin's circled numbers for tenth powers were also done away with, meaning that $32 + \frac{5}{10} + \frac{6}{100} + \frac{7}{1,000}$ could now be written as 32.567.

Different systems

The decimal point has never become universally accepted. Many countries use a comma as the decimal separator instead of a point. There would be no problem with the two common notations if not for the use of delimiters—symbols that separate groups of three digits in the whole-number section of a very large or sometimes very small number. For example,

in the UK, the commas in the number 2,500,000 are delimiters and are used to make it easier both to read the number and to recognize its size. The UK uses a point for the decimal separator and a comma as a delimiter. Elsewhere in the world, if a comma is used for the decimal separator, a point is then used as the delimiter. In Vietnam, for example, a price of two hundred thousand Vietnamese dong is often written as 200.000.

Usually, the context is sufficient for people to interpret the notation correctly, but this can go badly

wrong. In an attempt to solve this problem, the 22nd General Conference on weights and measures—a meeting of delegates from 60 nations of the International Bureau of Weights and Measures—decided in 2003 that, although either a point or comma on the line could be used as the decimal separator, the delimiter was to be

In Spain, the decimal separator is a comma, as seen in the prices at this market stall in Catalonia. In handwritten Spanish, an upper comma (similar to an apostrophe) is also common.

a space rather than either of the previous symbols. This notation is yet to become universal.

Benefits of decimals

The same processes of addition, subtraction, multiplication, and division of whole numbers can be used with decimal numbers, resulting in a far simpler way of performing basic arithmetic than the previous method, which relied on learning a different set of rules for calculations with fractions. When multiplying fractions, for example, the numerators would be multiplied separately from the denominators, and the resulting fraction would then be reduced. With decimal fractions, multiplying and dividing by powers of 10 is straightforward— as in the example of 32.567 (see top left), the decimal separator can be simply moved left or right.

Stevin believed that the universal introduction of decimal coinage, weights, and measures would only be a matter of time. The introduction of decimal measures for length and weight (using meters and kilograms) arrived in Europe some 200 years later, during the French Revolution. When it introduced the metric system,

This marble plaque on the rue de Vaugirard, Paris, is one of 16 original meter markers installed in 1791, after the French Académie des Sciences defined the meter for the first time.

France also tried to introduce a decimal system for time; there would be 10 hours in a day, 100 minutes in each hour, and 100 seconds in each minute. The attempt was so unpopular that it was dropped after just one year. The Chinese had introduced various forms of decimal time over some 3,000 years, but finally abandoned it in 1645 CE.

In the US, the use of a decimal system for measurement and coinage was championed by Thomas Jefferson. His 1784 paper persuaded Congress to introduce a decimal system for money using dollars, dimes, and cents. In fact, the name "dime" originates from *Disme*, the French title of *The Art of Tenths*. Yet Jefferson's view did not hold sway for measurement, and inches, feet, and yards are still used today. While many European currencies were decimalized in the 1800s, it was not until 1971 that decimal currency was introduced in the UK. ∎

Terminating and recurring decimals

Fractions are converted to decimals by dividing the numerator by the denominator. If the denominator is only divisible by 2 or 5 and no other prime numbers–as is the case for 10—then the decimal will terminate. For example, $3/40$ can be expressed as 0.075, and this value is exact because 40 is only divisible by the primes 2 and 5.

Other fractions become recurring decimals—meaning that they do not end. For example, $2/11$ is decimalized as 0.18181818..., denoted as $0.1\dot{8}$ to show that both the 1 and 8 recur. The length of the recurring cycle (two numbers in the case of $0.1\dot{8}$) can be predicted as it will be a factor of the denominator minus 1 (so if the denominator of the fraction is 11, the number of digits in the cycle is a factor of 10). These differ from irrational numbers, which do not terminate and have no pattern of recurrence. Irrational numbers cannot be expressed as a fraction of two integers.

Perhaps the most important event in the history of science... [is] the invention of the decimal system...
Henri Lebesgue
French mathematician

TRANSFORMING MULTIPLICATION INTO ADDITION

LOGARITHMS

IN CONTEXT

KEY FIGURE
John Napier (1550–1617)

FIELD
Number systems

BEFORE
14th century The Indian mathematician Madhava of Kerala constructs an accurate table of trigonometric sines to aid calculation of angles in right-angled triangles.

1484 In France, Nicolas Chuquet writes an article about calculation using geometric series.

AFTER
1622 English mathematician and clergyman William Oughtred invents the slide rule using logarithmic scales.

1668 In *Logarithmo-technia*, German mathematician Nicholas Mercator first uses the term "natural logarithms."

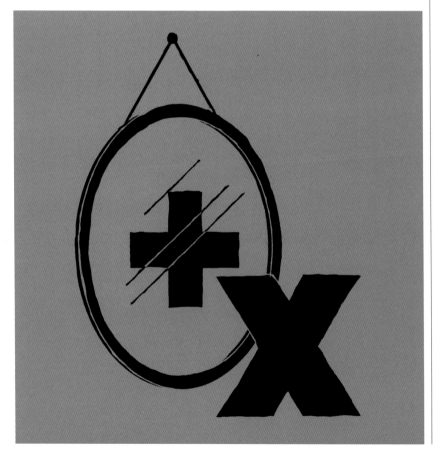

For thousands of years, most calculations were carried out by hand, using devices such as counting boards or the abacus. Multiplication was especially long-winded and much more difficult than addition. In the scientific revolution of the 16th and 17th centuries, the lack of a reliable calculating tool hampered progress in areas such as navigation and astronomy, where the potential for error was greater because of the lengthy calculations involved.

Solving by series

In the 1400s, French mathematician Nicolas Chuquet investigated how the relationships between

See also: Wheat on a chessboard 112–13 ▪ The problem of maxima 142–43 ▪ Euler's number 186–91 ▪ The prime number theorem 260–61

In the 1500s, **multiplying large numbers** was a **long and laborious** process.

John Napier **simplified** the process by developing **tables of numbers**.

In these tables, each number had its equivalent "**artificial number**," or logarithm.

Two numbers could be multiplied by **adding their logarithms together** and **converting** the result back to an ordinary number.

Logarithms enable mathematicians to complete **complex multiplications through addition**.

John Napier

Born into a wealthy family in 1550 at Merchiston Castle, near Edinburgh, John Napier would later become 8th Laird of Merchiston. Aged just 13, he entered St. Andrews University and became passionately interested in theology. Before graduating, however, he left to study in Europe, although few details of this time are known.

Napier returned to Scotland in 1571 and devoted much time to his estates, where he devised new methods of agriculture to improve his land and livestock. A fervent Protestant, he also wrote a prominent book attacking Catholicism. His keen interest in astronomy, and a desire to find simpler ways to perform the calculations that it required, led to his invention of logarithms. He also created Napier's Bones, a calculation device using numbered rods. Napier died at Merchiston Castle in 1617.

Key works

1614 *Mirifici Logarithmorum Canonis Descriptio* (*A Description of the Marvellous Rule of Logarithms*)
1617 *Rabdologiae*

arithmetic and geometric sequences could aid calculation. In an arithmetic sequence, each number differs from the one preceding it by a constant quantity, such as 1, 2, 3, 4, 5, 6… (going up by 1), or 3, 6, 9, 12… (going up by 3). In a geometric sequence, each number after the first term is determined by multiplying the previous number by a fixed amount, called the "common ratio." For example, the sequence 1, 2, 4, 8, 16 has a common ratio of 2. Setting down a geometric sequence (such as 1, 2, 4, 8…) and above it

an arithmetic sequence (such as 1, 2, 3, 4…), it can be seen that the top numbers are the exponents to which 2 is raised to arrive at the series below. It was a much more sophisticated version of this scheme that lay at the heart of the tables of logarithms developed by Scottish landowner John Napier.

Generating logarithms

Napier was fascinated by numbers and spent much of his time finding ways of making calculations easier. In 1614, he published the first description »

0	1	2	3	4	5	6	7	8	9	10
1	2	4	8	16	32	64	128	256	512	1024

The lower row of this table is a geometric sequence (progressing powers of 2), while the top row is an arithmetic sequence that reveals the exponents (powers) by which 2 is raised to arrive at the numbers in the lower row. (Anything to the power of 0 is 1.) To multiply the numbers 16 and 32 in the lower row, their exponents (4 + 5) can be added together to produce 2^9 (= 512).

and table of logarithms; a logarithm of a given number is the exponent or power to which another fixed number (the base) is raised to produce that given number. The use of such tables facilitated complex calculations and advanced the development of trigonometry. As Napier recognized, the basic principle of calculating was simple enough: he could replace the tedious task of multiplication by the simpler operation of addition. Each number would have its equivalent "artificial number" as he initially termed it. (Napier later settled on the name "logarithm," derived by combining the Greek words *logos*, meaning proportion, and *arithmos*, meaning number.) Adding the two logarithms, and then converting the answer back to an ordinary number, produces the result of multiplying the original numbers. For division, one logarithm is subtracted from another and the result is converted back.

To generate his logarithms, Napier imagined two particles traveling along two parallel lines. The first line was of infinite length, while the second was of fixed length. Each particle left the same

I found at length some excellent brief rules.
John Napier

starting position at the same time and at the same velocity. The particle on the infinite line traveled with uniform motion, so it covered equal distances in equal times. The velocity of the second particle was proportional to the distance remaining to the end of the line. Halfway between the starting point and the end of the line, the second particle is traveling at half the velocity it started with; at the three-quarter point, it is traveling with a quarter of its initial velocity; and so on. This means that the second particle is never going to reach the end of the line, and equally, the first particle, on its

infinite line, will never arrive at the end of its journey. At any instant there is a unique correspondence between the positions of the two particles. The distance the first particle has traveled is the logarithm of the distance the second particle has yet to go. The first particle's progress can be viewed as an arithmetic progression, while that of the second particle is geometric.

Improving the method
It took Napier 20 years to complete his calculations and to publish his first logarithm tables as *Mirifici Logarithmorum Canonis Descriptio* (*A Description of the Marvellous Rule of Logarithms*). Henry Briggs, professor of mathematics at the University of Oxford, recognized the significance of Napier's tables but thought they were unwieldy.

Briggs visited Napier in 1616 and again in 1617. Following their discussions, the two agreed that the logarithm of 1 should be redefined as 0 and the logarithm of 10 as 1. This approach made logarithms much easier to use. Briggs also helped with the calculation of logarithms of

The pH logarithmic scale measures alkalinity and acidity. A pH of 2 is 10 times more acidic than a pH of 3 and 100 times more acidic than pH 4.

Logarithmic scales

When measuring physical variables, such as sound, flow, or pressure, where values may change exponentially, rather than by regular increments, a logarithmic scale is often used. Such scales use the logarithm of a value instead of the actual value of whatever is being measured. Each step on a logarithmic scale is a multiple of the preceding step. For example, on a $\log_{10}$ scale, every unit up the scale represents a 10-fold increase in whatever is being measured.

In acoustics, sound intensity is measured in decibels. The decibel scale takes the hearing threshold, defined as 0 dB, as its reference level. A sound 10 times louder is assigned a decibel value of 10; a sound 100 times louder has a decibel value of 20; a sound 1,000 times louder a value of 30, and so on. This logarithmic scale fits well with the way we hear things, as a sound must become 10 times more intense to sound twice as loud to the human ear.

Napier's book describing logarithms was published in 1614, as its title page shows. The principles behind his logarithm tables were published in 1619, two years after his death.

ordinary numbers based on the logarithm of 10 being 1 and spent several years recalculating the tables. The results were published in 1624 with the logarithms calculated to 14 decimal places. The base-10 logarithms calculated by Briggs are known as $\log_{10}$ or common logarithms. The earlier table to the power of 2 (see p.139) can be thought of as a simple base-2, or $\log_2$ table.

The impact of logarithms

Logarithms had an immediate impact on science, and on astronomy in particular. German astronomer Johannes Kepler had published his first two laws of planetary motion in 1605, but only after the invention of log tables was he able to make the breakthrough to discover his third law. This describes how the time it takes

for a planet to complete one orbit of the Sun is related to its average orbital distance. When he published this finding in 1620 in his book *Ephemerides novae motuum coelestium*, Kepler dedicated it to Napier.

The exponential function

Later in the 1600s, logarithms revealed something of further significance. While studying number series, Italian mathematician Pietro Mengoli showed that the alternating series $1 - ½ + ⅓ - ¼ + ⅕ - \ldots$ has a value of around 0.693147, which he demonstrated to be the natural logarithm of 2. A natural logarithm (*ln*)—so-called because it occurs naturally, revealing the time required to reach a certain level of growth—has a special base, later known as *e*, with an approximate value of 2.71828. This number is hugely significant in mathematics due to its links with natural growth and decay.

It was through work such as that of Mengoli that the important concept of the exponential function came to light. This function is used to represent exponential growth—where the rate of growth of a quantity is proportional to its size at any particular moment, so the bigger it is, the faster it grows—which is relevant to fields

By shortening the labors, [Napier] doubled the life of the astronomer.
Pierre-Simon Laplace

such as finance and statistics, and most areas of science. The exponential function is given in the form $f(x) = b^x$, where *b* is greater than 0, but does not equal 1, and *x* can be any real number. In mathematical terms, logarithms are the inverse of exponentials (powers of a number) and can be to any base.

A basis for Euler's work

The push for accurate log tables spurred mathematicians such as Nicholas Mercator to pursue further research in this area. In *Logarithmotechnica*, published in 1668, he set out a series formula for the natural logarithm $ln(1 + x) = x - x^2/2 + x^3/3 - x^4/4 + \ldots$ This was an extension of Mengoli's formulation, in which the value of *x* was 1. In 1744, more than 130 years after Napier produced his first logarithm table, Swiss mathematician Leonhard Euler published a full treatment of e^x and its relationship to the natural logarithm. ∎

The slide rule, used here in 1941 by a member of the Women's Auxiliary Air Force, is marked with logarithmic scales that facilitate multiplication, division, and other functions. Invented in 1622, it was a vital mathematical tool before the advent of pocket calculators.

NATURE USES AS LITTLE AS POSSIBLE OF ANYTHING
THE PROBLEM OF MAXIMA

IN CONTEXT

KEY FIGURE
Johannes Kepler (1571–1630)

FIELD
Geometry

BEFORE
c.240 BCE In *Method of Mechanical Theorems*, Archimedes uses indivisibles to estimate the areas and volumes of curvilinear shapes.

AFTER
1638 Pierre de Fermat circulates his *Method for determining Maxima and Minima and Tangents for Curved Lines*.

1671 In *Treatise on the Method of Series and Fluxions*, Isaac Newton produces new analytical methods for solving problems such as the maxima and minima of functions.

1684 Gottfried Leibniz publishes *New Method for Maximums and Minimums*, his first work on calculus.

Astronomer Johannes Kepler is best known for his discovery of the elliptical shape of the planets' orbits and his three laws of planetary motion, but he also made a major contribution to mathematics. In 1615, he devised a way of working out the maximum volumes of solids with curved shapes, such as barrels.

Kepler's interest in this field began in 1613, when he married his second wife. He was intrigued when the wine merchant at the wedding feast measured the wine in the barrel by sticking a rod diagonally through a hole in the top and checking how far up the stick the wine went. Kepler wondered whether this worked equally well for all shapes of barrel and, concerned that he may have been cheated, decided to analyze the issue of volumes. In 1615, he published his results in *Nova stereometria doliorum vinariorum* (*New solid geometry of wine barrels*).

Kepler looked at ways of calculating the areas and volumes of curved shapes. Since ancient times, mathematicians had discussed using "indivisibles"—elements so tiny they cannot be divided. In theory these can be

Kepler felt **cheated by wine merchants** and wanted an accurate way to **measure a barrel's contents**.

Inspired by Archimedes, he used the **method of infinitesimals** to divide the barrel into thin sections and find the **exact volume of wine**.

The method Kepler used was a key step in **the development of calculus**.

fitted into any shape and added up. The area of a circle could be determined, for example, by using slender pie-slice triangles.

To find the volume of a barrel or any other 3-D shape, Kepler imagined it as a stack of thin

Barrel 1

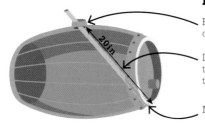

The merchant's rod is submerged to an equal extent when pushed at a diagonal into these two barrels, so he charges the same price for both. However, the elongated shape of the second barrel means it has a smaller volume, containing less wine but for the same price as the first.

Bunghole in the center of the barrel

Distance between the bung hole and the opposite edge

Merchant's rod

Barrel 2

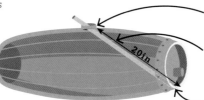

Bunghole in the center of the barrel

Distance between the bunghole and the opposite edge

Merchant's rod

Johannes Kepler

Born near Stuttgart, Germany, in 1571, Johannes Kepler witnessed the "Great Comet" of 1577 and a lunar eclipse, and remained interested in astronomy throughout his life.

Kepler taught at the Protestant seminary in Graz, Austria. In 1600, non-Catholics were expelled from Graz and Kepler moved to Prague, where his friend Tycho Brahe lived. Following the death of his first wife and son, he moved to Linz in Austria, where his main job as imperial mathematician was to make astronomical tables.

Kepler was convinced that God had made the Universe according to a mathematical plan. He is best known for his work in astronomy, especially his laws of planetary motion and his astronomical tables. A year after his death in 1630, the transit of Mercury was observed as he had predicted.

Key works

1609 *New Astronomy*
1615 *New Solid Geometry of Wine Barrels*
1619 *Harmonies of the World*
1621 *Epitome of Copernican Astronomy*

layers. The total volume is the sum of the volumes of the layers. In a barrel, for example, each layer is a shallow cylinder.

Infinitesimals

The problem with cylinders is that if they have thickness, their straight sides will not fit into the curve of a barrel, while cylinders without thickness have no volume. Kepler's solution was to accept the notion of "infinitesimals"—the thinnest slices that can exist without vanishing. This idea had already been mooted by ancient Greeks such as Archimedes. Infinitesimals bridge the gap between continuous things and things broken into discrete units.

Kepler then used his cylinder method to find the barrel shapes with the maximum volume. He worked with triangles defined by the cylinders' height, diameter, and a diagonal from top to bottom. He investigated how, if the diagonal was fixed, like the merchant's rod, changing the barrel height would change its volume. It turned out that the maximum volume is held in short, squat barrels with a height just under 1.5 times the diameter—like the barrels at his wedding. In contrast, the tall barrels from Kepler's homeland on the Rhine River held much less wine.

Kepler also noticed that the closer to the maximum the shape gets, the less the rate at which the volume increases: an observation that contributed to the birth of calculus, opening up the exploration into maxima and minima. Calculus is the mathematics of continuous change, and maxima and minima are the turning points, or limits in any change—the peak and trough of any graph.

Pierre de Fermat's analysis of maxima and minima, which quickly followed Kepler's, opened the way for the development of calculus by Isaac Newton and Gottfried Leibniz later in the 17th century. ▪

THE FLY
ON THE CEILING
COORDINATES

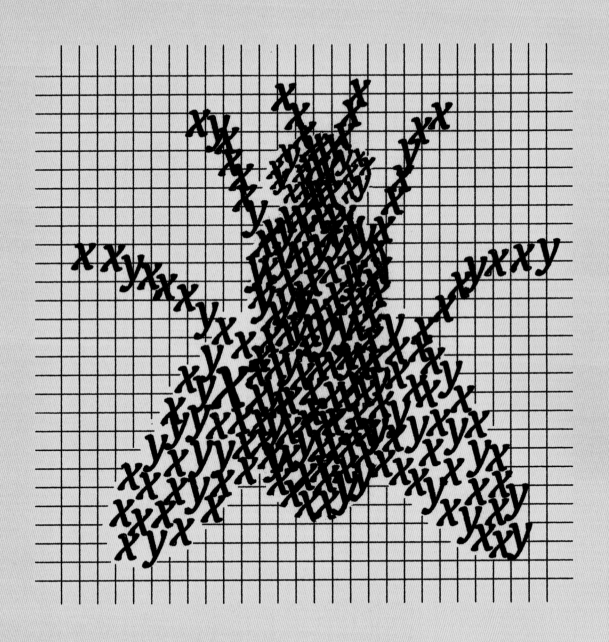

IN CONTEXT

KEY FIGURE
René Descartes (1596–1650)

FIELD
Geometry

BEFORE
2nd century BCE Apollonius of Perga explores positions of points within lines and curves.

c. 1370 French philosopher Nicole Oresme represents qualities and quantities as lines defined by coordinates.

1591 French mathematician François Viète introduces symbols for variables in algebraic notation.

AFTER
1806 Jean-Robert Argand uses a coordinate plane to represent complex numbers.

1843 Irish mathematician William Hamilton adds two new imaginary units, creating quaternions, which are plotted in four-dimensional space.

In geometry (the study of shapes and measurements), coordinates are employed to define a single point—an exact position—using numbers. Several different systems of coordinates are in use, but the dominant one is the Cartesian system, named after Renatus Cartesius, the Latinized name of French philosopher René Descartes. Descartes presented his coordinate geometry in *La Géométrie* (*Geometry*, 1637), one of three appendices to his philosophical work *Discours de la Méthode* (*Discourse on the Method*), in which he proposed methods for arriving at truth in the sciences. The other two appendices were on light and the weather.

Building blocks

Coordinate geometry transformed the study of geometry, which had barely evolved since Euclid had written *Elements* in ancient Greece some 2,000 years earlier. It also revolutionized algebra by turning equations into lines (and lines into equations). By using Cartesian coordinates, scholars could visualize mathematical relationships. Lines, surfaces, and shapes could also be

Problems which can be constructed by means of circles and straight lines only.
René Descartes
describing geometry

interpreted as a series of defined points, which changed the way people thought about natural phenomena. In the case of events such as volcanic eruptions or droughts, plotting elements such as intensity, duration, and frequency could help to identify trends.

Finding a new method

There are two accounts of how Descartes came to develop the coordinate system. One suggests that the idea dawned on him as he watched a fly moving over the ceiling of his bedroom. He realized he could plot its position, using

René Descartes

The son of a minor noble, René Descartes was born in Touraine, France, in 1596. His mother died shortly after his birth, and he was sent to live with his grandmother. He later attended a Jesuit college, then went to study law in Poitiers. In 1618, he left France for the Netherlands and joined the Dutch States Army as a mercenary.

Around this time, Descartes began to formulate philosophical ideas and mathematical theorems. Returning to France in 1623, he sold his property there in order to secure a lifelong income, then moved back to the Netherlands to study. In 1649, he was invited by Christina, Queen of Sweden, to tutor her and to launch a new academy. His weak constitution could not resist the cold winter. In February 1650, Descartes caught pneumonia and died.

Key works

1630–33 *Le Monde* (*The World*)
1630–33 *L'Homme* (*Man*)
1637 *Discours de la Méthode* (*Discourse on the Method*)
1637 *La Géométrie* (*Geometry*)
1644 *Principia philosophia* (*Principles of Philosophy*)

See also: Pythagoras 36–43 ▪ Conic sections 68–69 ▪ Trigonometry 70–75 ▪ Rhumb lines 125 ▪ Viviani's triangle theorem 166 ▪ The complex plane 214–15 ▪ Quaternions 234–35

A rectangular ceiling has a **length** and a **width**.

⬇

Two-dimensional **coordinates** are pinpointed using **horizontal (*x*)** and **vertical (*y*)** measurements.

⬇

The location of a fly on the ceiling can therefore be expressed in mathematical terms.

---

numbers to describe where it was in relation to the two adjacent walls. Another account relates that the idea came to him in dreams in 1619, when he was serving as a mercenary in southern Germany. It was at this time, too, that he is thought to have figured out the relationship between geometry and algebra that is the basis of the coordinate system.

I realized that it was necessary… to start again right from the foundations if I wanted to establish anything in the sciences that was stable and likely to last.
René Descartes

The simplest Cartesian coordinate system is one-dimensional; it indicates positions along a straight line. One endpoint of the line is set as the zero point, and all other points on the line are counted from there in equal lengths, or fractions of a length. Just a single coordinate number is needed to describe an exact point on the line—as when measuring a distance with a ruler from zero to a unit of length. More commonly, coordinates are used to describe points on two-dimensional surfaces that have a length and width, or within a three-dimensional space, which also has depth. To achieve this, more than one number line is needed—each starting at the same zero point, or origin. For a point on a plane (a flat two-dimensional surface), two number

This edition of *La Géometrie* (in Latin because that was the language of scholars) was printed in 1639. Descartes originally published the book in French so it could be read by less well-educated people.

lines are needed. The horizontal line, called the *x*-axis, and the vertical *y*-axis are always perpendicular to each other; the origin is the only place they will ever meet. The term for the *x*-axis is abscissa, while the *y*-axis is the ordinate. Two numbers, one from each axis, "coordinate" to pinpoint an exact position.

When taking a graph reading, these two numbers are now presented as a tuple—a strictly ordered sequence listed inside brackets. The abscissa (value of *x*) always precedes the ordinate (value of *y*) to create the tuple (*x*,*y*). Although they were conceived before negative numbers were fully accepted, coordinates now often include both negative and positive values—negative values below and to the left of the origin; positive values above and to the right of the origin. Together, the two axes »

create a field of points called a coordinate plane, which extends outward in two dimensions with the origin (0,0) at the center. Any point on that plane, which could stretch to infinity, can be described exactly using a pair of numbers.

Plotting 3-D space

For a three-dimensional space, the coordinates require a third number, ordered in the tuple (x, y, z). The z refers to a third axis, which is perpendicular to the plane formed by the x and y axes (see p.151). Each pair of axes creates its own coordinate plane; these intersect at right angles to each other, thus dividing the space into eight zones called octants. The coordinates within each octant follow one of eight sequences of values for $x, y,$ and z, ranging from all negative values to all positive values, with six possible negative and positive combinations in between.

Curved lines

La Géométrie sets out what soon became the foundation of the coordinate system. Descartes, however, was primarily interested in finding out how coordinates could help him use algebra to better understand lines, especially curved lines. In so doing he created a new field of mathematics, called analytic geometry, where shapes are described in terms of their coordinates and the relationships between a pair of variables, x and y. This was very different from Euclid's "synthetic geometry," in which shapes are defined by the

> Each problem that I solved became a rule which served afterwards to solve other problems.
> **René Descartes**

way they are constructed using a ruler and pair of compasses. The ancient method was limiting; Descartes' new method opened up all sorts of new possibilities.

La Géométrie contains much discussion about curves, which were the subject of renewed interest in the 1600s—partly because treatises by ancient Greek mathematicians had been newly translated, but also because curves featured prominently in fields of scientific exploration such as astronomy and mechanics.

Coordinates make it possible to convert curves and shapes into algebraic equations, which can be shown visually. A straight line that runs diagonally from the origin,

---

Any **point on a line** can be defined **using a number** x.

↓

Any **point on a plane** (flat surface) can be defined **using two numbers** x and y.

→

The **points in a straight line** all share the **same relationship** between x and y.

↓

All lines **can be described** in terms of **an algebraic equation**.

←

All **equations** can be **plotted as a line**.

↓

Coordinates **allow curves and shapes** to be **converted into equations** and **equations to be plotted**.

> With me, everything turns into mathematics.
> **René Descartes**

A geometric shape such as the curve of a roller-coaster can be mapped on to a graph and described in relation to the x and y axes. The straight section of the curve has the equation $y = x$.

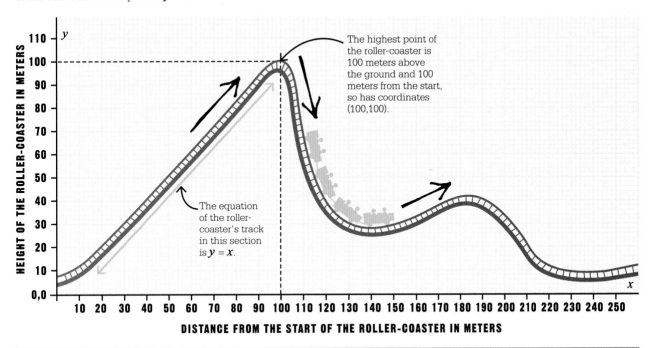

HEIGHT OF THE ROLLER-COASTER IN METERS

The highest point of the roller-coaster is 100 meters above the ground and 100 meters from the start, so has coordinates (100,100).

The equation of the roller-coaster's track in this section is $y = x$.

DISTANCE FROM THE START OF THE ROLLER-COASTER IN METERS

equidistant from both axes, can be described using algebra as $y = x$, and has coordinates (0,0); (1,1); (2,2), and so on. The line $y = 2x$ would follow a steeper path along a line including the coordinates (0,0); (1,2); (2,4), for instance. A line running parallel to $y = 2x$ would pass through the y axis at a point other than the origin, such as at (0,2). The formula for this particular line is $y = 2x + 2$ and that includes the points (0,2); (1,4); (2,6).

Cartesian coordinates help to reveal the great power of algebra to generalize relationships. All the straight lines described above have the same general equation: $y = mx + c$, where the coefficient m is the slope of the line, indicating how much bigger (or smaller) y is compared to x. The constant c, meanwhile, shows where the line meets the y axis when x is equal to zero.

The circle equation

In analytic geometry, all circles centered on the origin can be defined as $r = \sqrt{(x^2 + y^2)}$, known as the circle equation. This is because a circle can be thought of as all the points that lie at an equal distance from a central point (that distance being the radius of the circle). If

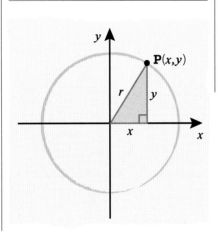

Any point P, with coordinates (x, y), on the circumference of a circle can be connected to the center of the circle (0, 0) by a straight line (the circle's radius) that forms the hypotenuse of a right-angled triangle with sides of length x and y. The equation of the circle is $r^2 = x^2 + y^2$.

that central point is (0,0) on an x, y graph, the circle equation emerges, by drawing on Pythagoras's theorem. The circle's radius can be conceived as the hypotenuse of a right-angled triangle with short sides x and y, so $r^2 = x^2 + y^2$, which can be rewritten as $r = \sqrt{(x^2 + y^2)}$. The circle can then be plotted on axes using different values of x and y that give the same value of r. For example, if r is 2, then the circle crosses the x axis at (2,0) and (−2,0), and it crosses the y axis at (0,2) and (0,−2). All the other points on the circle can be seen as one »

Polar coordinates

In mathematics, polar coordinates, which define points on a plane using two numbers, are the closest rivals to Descartes' system. The first number, the radial coordinate r, is the distance from the central point—called the pole, not the origin. The second number, the angular coordinate (θ), is the angle that is defined as 0° from a single polar axis. To compare it with the Cartesian system, the polar axis would be the Cartesian x axis, and the polar coordinates (1,0°) would replace the Cartesian coordinates (1,0). The polar version of the Cartesian point (0,1) is (1,90°).

Polar coordinates are used to help manipulate complex numbers plotted on a plane, especially for multiplication. Multiplying complex numbers is simplified when they are treated as polar coordinates, a process that involves multiplying the radial coordinates and adding the angular ones.

Coordinates of A are r, θ

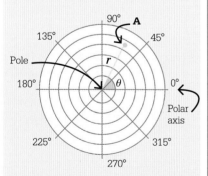

The polar coordinate system is often used to calculate the movement of objects around, or in relation to, a central point.

corner of a right-angled triangle moving around in a circle. As the corner moves around the circle, the short sides of the triangle vary in length, but the hypoteneuse does not because it is always the radius of the circle. The line formed by a point moving in this defined way is called a locus. This idea was developed by the Greek geometer Apollonius of Perga about 1,750 years before Descartes' birth.

Exchange of ideas

In addition to drawing on theorems formulated by the ancient Greeks, Descartes exchanged ideas with other French mathematicians, among them Pierre de Fermat, with whom he frequently corresponded. Descartes and Fermat both made use of algebraic notation, the x and y system that François Viète had introduced at the end of the 1500s. Fermat also independently developed a coordinate system, but he did not publish it. Descartes was aware of Fermat's ideas, no doubt using them to improve his own. Fermat also helped Dutch mathematician Frans van Schooten to understand Descartes' ideas. Van Schooten translated *La*

A modified form of polar coordinates that gives an aircraft's destination in terms of angle and distance can be used as an alternative to GPS.

Géométrie into Latin and also popularized the use of coordinates as a mathematical technique.

New dimensions

Van Schooten and Fermat had both suggested extending Cartesian coordinates into the third dimension. Today, mathematicians and physicists use coordinates to go much further than that and to imagine a space with any number of dimensions. Although it is almost impossible to visualize such a space, mathematicians can use these tools to describe lines moving in four, five, or as many spatial dimensions as they desire.

Coordinates can also be used to examine the relationship between two quantities. This idea was pioneered as long ago as the 1370s, when a French monk called Nicole Oresme used rectangular coordinates and the geometric forms created by his results to understand, for instance,

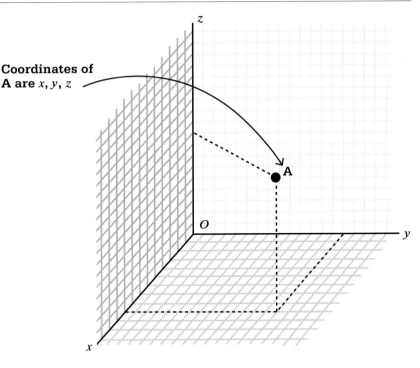

Coordinates of
A are x, y, z

3-D Cartesian coordinates can be used to plot an object that has, for instance, width, depth, and height. Three axes (x, y, z) are set at right angles to each other. Where they meet is the origin (O).

the relationship of elements such as speed and time, or the links between heat intensity and the degree of expansion due to heat.

Some quantities can be represented using coordinates known as vectors, and exist in a purely mathematical "vector space." Vectors are quantities with two values, which can be plotted as a magnitude (the length of a line) and a direction. Velocity is a vector as it has exactly those values (a quantity of speed and a direction of motion), while other vectors, such as Oresme's heat and expansion, are

visualized in this way to make it easier to add and subtract different sets of values or to manipulate them in another way.

Mathematicians in the 1800s also found new purposes for Cartesian coordinates. They used them to represent complex numbers (sums of imaginary numbers, such as $\sqrt{-1}$, and real numbers) or quaternions (the system that extends complex numbers) as vectors plotted in two, three, or more dimensions.

The key coordinates

The Cartesian coordinate system is by no means the only one. Geographic coordinates plot points on the globe as angles from preset great circles—the Equator and the Greenwich Meridian. A similar system, using celestial coordinates,

describes the location of stars in an imaginary sphere centered on Earth and extended infinitely into space. Polar coordinates, determined by distance and angles from the center of Earth, are also useful for certain types of calculation.

Cartesian coordinates remain an ubiquitous tool, however, able to plot anything from simple survey data to the movements of atoms. Without them, breakthroughs such as analytical calculus (which divides quantities into infinitesimally small amounts) and advances in space-time and non-Euclidean geometries could not have happened. Cartesian coordinates have had an immense impact in mathematics, and in many fields of science and the arts, from engineering and economics to robotics and computer animation. ■

A DEVICE OF MARVELOUS INVENTION

THE AREA UNDER A CYCLOID

IN CONTEXT

KEY FIGURES
Bonaventura Cavalieri
(1598–1647), **Gilles Personne
de Roberval** (1602–75)

FIELD
Applied geometry

BEFORE
c.240 BCE Archimedes
investigates the volume
and surface area of spheres
in his *Method Concerning
Mechanical Theorems.*

1503 French mathematician
Charles de Bovelles gives the
first description of a cycloid
in *Introductio in geometriam*
(*Introduction to Geometry*).

AFTER
1656 Dutch mathematician
Christiaan Huygens bases
his invention of the pendulum
clock on the curve of a cycloid.

1693 De Roberval's solution
to the area of a cycloid is
published more than 60 years
after its discovery and 18
years after his death.

The ancient Greeks puzzled
over problems relating to
areas and volumes of
figures bounded by curves. They
compared the areas of shapes
by transforming each one into a
square with the same area as the
original shape, then compared the
sizes of the squares. This was easy
for shapes with straight edges, but
curvilinear shapes caused problems.

These problems remained
unresolved until 1629, when Italian
mathematician and Jesuate priest
Bonaventura Cavalieri found a
method for calculating the areas
and volumes of curved shapes by
slicing them into parallel pieces
(Cavalieri's principle, see above
right), although he did not publish
his results until six years later. In
1634, Gilles Personne de Roberval
used this method to work out that
the area beneath a cycloid (the arc
traced by the rim of a rolling wheel)
is three times the area of the circle
used to generate the cycloid.

Squaring the circle
The ancient Greek mathematician
Archimedes had used an ingenious
method of exhaustion to determine

This wheel has rolled over a piece of gum. The graph shows the path
of the gum as the wheel rotates, creating a cycloid shape, which, as
de Roberval discovered, has an area three times that of the wheel.

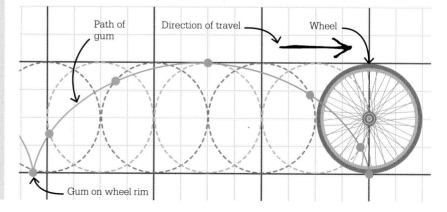

Path of gum

Direction of travel

Wheel

Gum on wheel rim

> A pretty result
> which I had
> not noticed before.
> **René Descartes**
> on de Roberval's method for finding
> the area under a cycloid

Since this shark-fin shape (left) and triangle (right) are the same height and the same width at equivalent points along their height, Cavalieri's principle states that they can be sliced into parallel pieces that have similar area.

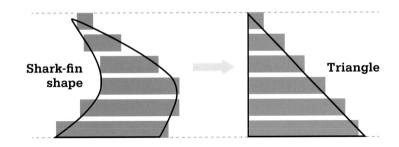

the area between a parabola and a straight line. It entailed inscribing a triangle of known area to fit inside the parabola, then inscribing ever smaller triangles in all the gaps that remained. By adding together the areas of the triangles, Archimedes obtained a close approximation of the area he sought. The straight-edge-and-compass methods of his day, however, had their limitations. When he tried to calculate the surface area of a 3-D sphere using quadrature, a process which involves constructing a square of an area equal to a circle, he failed. He knew the surface area of the sphere was four times that of a circle of the same radius, but could not find a square that would give the surface area.

New spins on the problem

The first description of a cycloid was published by Charles de Bovelles in 1503. Italian polymath Galileo gave the cycloid its name (from the Greek for "circular") and tried to calculate its area by cutting up models of a cycloid and a circle, weighing the pieces, and comparing the results. Around 1628, Frenchman Marin Mersenne challenged his fellow mathematicians, including de Roberval, René Descartes, and Pierre de Fermat, to find both the area under the arch of a cycloid and a tangent to a point on the curve. When de Roberval told Descartes of his success, the latter dismissed it as "so small a result." Descartes, in turn, discovered the tangent to a cycloid in 1638, and challenged de Roberval and Fermat to do the same. Only Fermat succeeded.

In 1658, English architect Christopher Wren calculated the length of an arc of a cycloid as four times the diameter of the generating circle. The same year, Blaise Pascal calculated the area of any vertical slice of a cycloid. He also imagined rotating these vertical slices about a horizontal axis, and worked out the surface area and volume of the disks swept out by this rotation. Pascal's use of infinitely small slices of shapes to solve the properties of cycloids would lead to the "fluxions" introduced by Isaac Newton as he developed early calculus. ▪

Gilles Personne de Roberval

Born in 1602, in a field near Roberval in northern France, where his mother was bringing in the harvest, Gilles Personne de Roberval was tutored in classics and mathematics by the local priest. In 1628, he moved to Paris, where he joined Marin Mersenne's circle of intellectuals.

In 1632, de Roberval became professor of mathematics at the Collège Gervais, and two years later he won a competition for a highly prestigious post at the Collège Royale. He lived frugally, but managed to buy a farm for his extended family and leased out plots to generate income. He continued to practice mathematics all his life. In 1669, he invented a set of scales known as the Roberval balance. He died in 1675.

Key work

1693 *Traité des Indivisibles* (*Treaty on Indivisibles*)

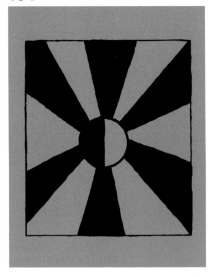

THREE DIMENSIONS MADE BY TWO

PROJECTIVE GEOMETRY

U nlike traditional Euclidean geometry, where all 2-D figures and objects belong in the same plane, projective geometry is concerned with how the apparent shape of an object is altered by the perspective from which that object is viewed. The 17th-century French mathematician Girard Desargues was a founder of such geometry.

The idea of perspective had been addressed two centuries earlier by Renaissance artists and

Linear perspective and geometry

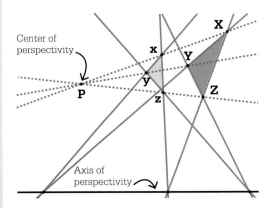

These two triangles are in perspective from a viewpoint called the center of perspectivity (P). Lines connecting the corresponding vertices of the triangles (X to x; Y to y, and Z to z) will always meet at P. If XYZ were a real triangular object, it would appear as the triangle xyz when viewed from P. Desargues' theorem states that lines extending from the corresponding sides of each triangle will always meet on a line known as the axis of perspectivity.

Perspective makes the parallel lines on sides of this flat-roofed building appear as though they will eventually meet. This meeting point is called a vanishing point.

> Good architecture
> should be a projection
> of life itself.
> **Walter Gropius**
> **German architect**

architects. Fillipo Brunelleschi had rediscovered the principles of linear perspective known to the ancient Greeks and Romans, and explored them in his architectural plans, sculptures, and paintings. Fellow architect Leon Battista Alberti used "vanishing points" to create a sense of 3-D perspective and wrote about the use of perspective in art.

From maps to math

As Western explorers sailed to new lands, they needed accurate maps depicting the spherical world in two dimensions. In 1569, Flemish cartographer Gerardus Mercator devised a method now known as "cylindrical map projection." This can be envisaged as the surface of the globe transferred onto a surrounding cylinder. When the cylinder is cut from top to bottom and rolled out, it becomes a two-dimensional map.

In the 1630s, Desargues began investigating which properties were unchanged (invariant) when an image is projected onto a surface (perspective mapping). While its dimensions and angles may change, collinearity is

preserved; this means that if three points XYZ are on a straight line, with Y between X and Z, then their images xyz are also on a straight line with y between x and z. An image of any triangle is another triangle. The corresponding sides of each triangle can be extended to meet at three points on a line (axis of perspectivity), and a line from each vertex to its corresponding vertex and beyond will meet at a point (the center of perspectivity).

Desargues realized that all conic sections are equivalent in this way under projection. A single invariant property, such as collinearity, needs only to be proved for a single case, rather than tested on each conic. Pascal's "mystic hexagram" theorem, for instance, states that the intersections of lines connecting pairs of six points on a conic all lie on a straight line. It can be shown by connecting six points on a circle, a proof valid for other conics, too.

Desargues then considered what happens as the vertex of the projection cone moves further away. Parallel rays come from a point at

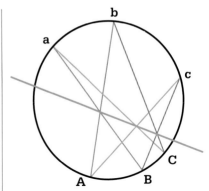

When six arbitrary points are drawn on a circle and connected as shown (Ab, aB; Ac, Ca; Cb, cB), a straight line can be drawn through the points where lines of the same color cross. Using projection, this is true for an ellipse, too.

infinity (such as the Sun). By adding these points at infinity to the Euclidean plane, each pair of lines meets at a point, including parallel lines, which meet at infinity.

The method was developed into a full geometry by Poncelet in 1822. Today, projective geometry is used by architects and engineers in CAD technology, and in computer animation for films and gaming. ▪

Girard Desargues

Born in 1591, Girard Desargues lived in Lyon all his life. He came from a family of wealthy lawyers who owned several properties, including a manor and a small chateau with fine vineyards. Desargues made several visits to Paris and, through Marin Mersenne, became friends with Descartes and Pascal.

Desargues worked initially as a tutor and later as an engineer and architect. He was an excellent geometer and shared his ideas with his mathematical friends. Some of his pamphlets were later expanded into published papers. He wrote on perspective and applied mathematics to practical projects, such as designing a spiral staircase and a new form of pump. Desargues died in 1661. His work was rediscovered and republished in 1864.

Key works

1636 *Perspective*
1639 *Rough Draft of Attaining the Outcome of Intersecting a Cone with a Plane*

SYMMETRY
IS WHAT WE SEE AT A
GLANCE

PASCAL'S TRIANGLE

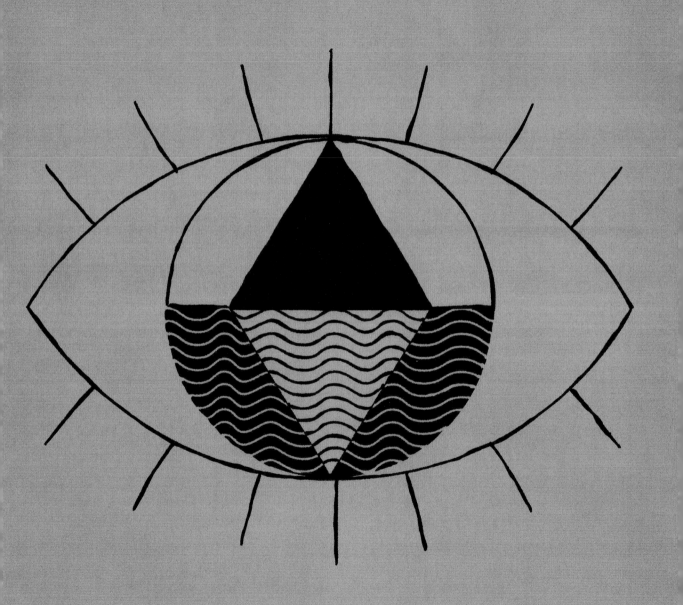

IN CONTEXT

KEY FIGURE
Blaise Pascal (1623–62)

FIELDS
Probability, number theory

BEFORE
975 Indian mathematician
Halayudha gives the first
surviving description of
numbers in Pascal's triangle.

c. 1050 In China, Jia Xian,
describes the triangle later
known as Yang Hui's triangle.

c. 1120 Omar Khayyam
creates an early version of
Pascal's triangle.

AFTER
1713 Jacob Bernoulli's
Ars Conjectandi (*The Art
of Conjecturing*) develops
Pascal's triangle.

1915 Wacław Sierpinski
describes the fractal pattern
of triangles later known as
Sierpinski triangles.

There are two types of mind…
the mathematical, and…
the intuitive. The former
arrives at its views slowly,
but they are… rigid; the
latter is endowed with
greater flexibility.
Blaise Pascal

Pascal's triangle is created
by adding together two adjacent
numbers (as shown by the
arrows) to give the sum in the
next row. Each row begins and
ends with the number 1.

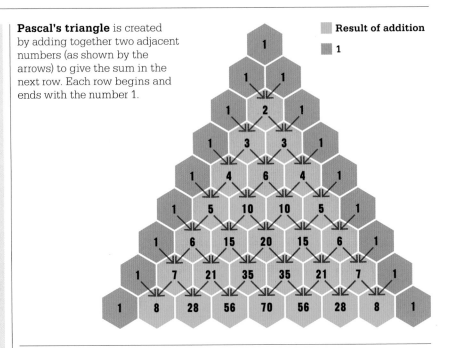

Result of addition

1

Mathematics is often
about the identification
of number patterns, and
one of the most remarkable number
patterns of all is Pascal's triangle.
Pascal's triangle is an equilateral
triangle built from a very simple
arrangement of numbers in ever-
widening rows. Each number
is the sum of the two adjacent
numbers in the row above. Pascal's
triangle can be any size, ranging
from just a few rows in depth to
any number.

While it might seem that such
a simple rule for arranging numbers
could only lead to simple patterns,
Pascal's triangle is fertile ground
for several branches of higher
mathematics, including algebra,
number theory, probability, and
combinatorics (the mathematics
of counting and arranging). Many
important sequences have been
found in the triangle, and
mathematicians believe that it
may reflect some truths about
relationships that we have yet to
understand between numbers.

The triangle is most commonly
named after French philosopher
and mathematician Blaise Pascal,
who explored it in detail in his
*Treatise on the Arithmetical
Triangle* in 1653. In Italy, however,
it is known as Tartaglia's triangle
after mathematician Niccolò
Tartaglia, who wrote about it in
the 1400s. In fact, the origins of the
triangle date back to ancient India
in 450 BCE (see box, p.160).

Probability theory
Pascal's contribution to the triangle
was notable because he set out a
clear framework for exploring its
properties. In particular, he used the
triangle to help lay the foundations
of probability theory in his
correspondence with fellow French
mathematician Pierre de Fermat.
Before Pascal, mathematicians
such as Luca Pacioli, Gerolamo
Cardano, and Tartaglia had written
about how to work out the chances
of dice rolling particular numbers
or hands of cards coming out a
certain way. Their understanding

was shaky at best, and it was Pascal's work with the triangle that pulled the strands together.

Dividing stakes

Pascal was asked to look into probability in 1652 by a notorious French gambler. Antoine Gombaud, the Chevalier de Méré, wanted to know how to divide stakes fairly if a game of chance was suddenly broken off. If a game would normally end only when one player had won a certain number of rounds, for instance, de Méré wanted to know if the division of the stakes should reflect how many rounds each player had won. Pascal combined the numbers step by step to represent the rounds played. The natural consequence was an ever-widening triangle. As Pascal showed, the numbers in the triangle count the number of ways various occurrences can combine to produce a given result.

The probability of an event is defined as the proportion of times it will happen. A dice has six faces, so the probability of it landing on

> **1** is at the **top** of Pascal's triangle and at the **start and end** of each row.

> Each row has **one more number** than the row above it.

> Every number is the **sum of the two above** it.

> **The resulting triangle of numbers could go on forever.**

any particular face when you roll it is $\frac{1}{6}$. In other words, it is a question of noting how many ways the event can occur, and dividing this by the total number of possibilities. While this is easy enough for a single dice, with multiple dice, or 52 playing cards, the calculations become complicated. However, Pascal found that the triangle could be used to find the number of possible combinations when

you choose a number of objects from a particular number of available options.

Binomial calculations

As Pascal realized, the answer lay in binomials—expressions with two terms, such as $x + y$. Each row of Pascal's triangle gives the binomial coefficients for a particular power. The zeroth row (the top of the triangle) is used »

Blaise Pascal

Born in Clermont-Ferrand, France, in 1623, Blaise Pascal was a mathematics prodigy. As a teenager, his father took him to Marin Mersenne's mathematical salon in Paris. Around the age of 21, Pascal developed a mechanical adding and subtraction machine, the first ever marketed. As well as his mathematical contributions, Pascal played an important role in many scientific developments of the 1600s, including explorations of fluids and the nature of a vacuum, which contributed to the understanding of the idea of air pressure: the

scientific unit of pressure is called the Pascal. In 1661, he launched what may have been the world's first public transportation service in Paris, with linked five-person coaches. He died from unexplained causes in 1662, aged just 39.

Key works

1653 *Traité du triangle arithmétique* (*Treatise on the Arithmetical Triangle*)
1654 *Potestatum Numericarum Summa* (*Sums of Powers of Numbers*)

The Bat Country, a jungle gym project by American artist Gwen Fisher, is a Sierpinski tetrahedron featuring softball bats and balls. This tetrahedron is a 3-D structure made of Sierpinski triangles.

children (the total number of objects), the probability of one girl and two boys is 3/8 (the sum of all the coefficients in the third row of the triangle is 8, and there are three ways of having one girl in a family of three children).

Pascal's triangle made it simple to find probabilities. As Pascal's triangle can continue forever, this works with any powers. The relationship between binomial coefficients and the numbers in Pascal's triangle reveals a fundamental truth about numbers and probability.

Visual patterns

Pascal's simple number pattern proved to be the launchpad, with Fermat's work, for the mathematics of probability, but its relevance does not stop there. For a start, it provides a quick way of multiplying out binomial expressions to high powers, which would otherwise take a very long time.

for the binomial to the power of 0: $(x + y)^0 = 1$. For the binomial to the power of 1, $(x + y)^1 = 1x + 1y$, so the coefficients (1 and 1) correspond to the first row of the triangle (the zeroth row is not counted as a row). The binomial $(x + y)^2 = 1x^2 + 2xy + 1y^2$ has the coefficients 1, 2, and 1, as on the second row of Pascal's triangle. As binomial expansion leads to ever longer expressions, the coefficients continue to match a corresponding line on the triangle. For example, in the binomial $(x + y)^3 = 1x^3 + 3x^2y + 3xy^2 + 1y^3$, the coefficients match the third row of the triangle. The probabilities are calculated by dividing the number of possibilities by the total of all the coefficients in the row that reflects the total number of objects: for example, in a family of three

The ancient triangle

Mathematicians knew about Pascal's triangle long before the 1600s. In Iran, it is known as Khayyam's triangle after Omar Khayyam, but he was just one of many Islamic mathematicians to have studied it between the 7th and 13th century—a golden age for learning. In China, too, c. 1050, Jia Xian created a similar triangle to show coefficients. His triangle was taken up and popularized by Yang Hui in the 1200s, which is why it is known in China as Yang Hui's triangle. It is illustrated in the 1303 book by Zhu Shijie entitled *Precious Mirror of the Four Elements*.

The most ancient references to Pascal's triangle, however, come from India. It appears in Indian texts from 450 BCE as a guide to poetic metre, by the name of "The Staircase of Mount Meru." The mathematicians of ancient India also realized that the shallow diagonal lines of numbers in the triangle showed what are now known as Fibonacci numbers (see right).

Myanmar's Hsinbyume pagoda represents the mythical Mount Meru, whose staircase inspired another name for Pascal's triangle.

Mathematicians are continually finding new surprises in it. Some of the patterns in Pascal's triangle are extremely simple. The outside edge is entirely made up of the number 1, and the next set of numbers, in the first diagonal, is a simple number line of 1, 2, 3, 4, 5, and so on.

One particularly appealing property of Pascal's triangle is the "hockey stick" pattern, which can be used for addition. If you take a diagonal down from any of the outer 1s, then stop anywhere, you can then find the total sum of all of the numbers in the diagonal by taking one step further in the opposite direction. For example, starting at the fourth 1 on the left edge and going down diagonally to the right, if you stop at the number 10, then the total of the numbers passed so far (1 + 4 + 10) can be found by going one diagonal step down to the left: 15.

Coloring in all of the numbers divisible by a particular number creates a fractal pattern, while coloring all of the even numbers

I cannot judge my work while I am doing it. I have to do as painters do, stand back and view it from a distance, but not too great a distance.
Blaise Pascal

creates a pattern of triangles identified by Polish mathematician Wacław Sierpinski in 1915. This pattern can be made without Pascal's triangle by breaking an equilateral triangle into ever smaller triangles by connecting the midpoints of each of the triangles' three sides. The division can continue indefinitely. Today, Sierpinski triangles are popular for use in knitting patterns and in

origami, where a Sierpinksi triangle is converted into three dimensions to create a Sierpinski tetrahedron.

Number theory

There are also many more complex patterns hidden within the triangle. One of the patterns found in Pascal's triangle is the Fibonacci sequence, which lies on a shallow diagonal (see below). Another link to number theory is the discovery that the sum of all the numbers in the rows above a given row is always one less than the sum of the numbers in the given row. When the sum of all the numbers above a given row is a prime, it is a Mersenne prime—a prime number that is one less than a power of 2, such as 3 ($2^2 - 1$), 7 ($2^3 - 1$), and 31 ($2^5 - 1$). The first list of these primes was made by Pascal's contemporary, Marin Mersenne. Today, the largest known Mersenne prime is $2^{82,589,933} - 1$. If Pascal's triangle were drawn at a sufficiently large scale, this number would be found there. ∎

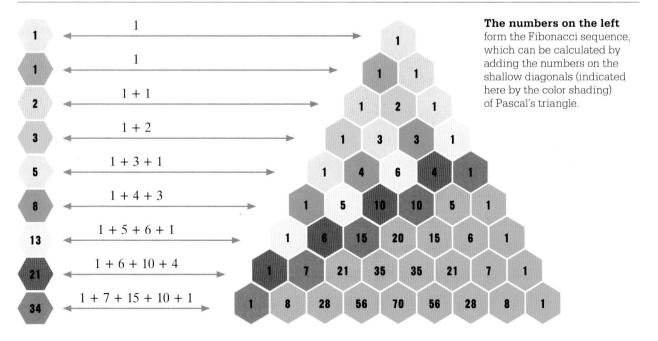

The numbers on the left form the Fibonacci sequence, which can be calculated by adding the numbers on the shallow diagonals (indicated here by the color shading) of Pascal's triangle.

Fibonacci	Sum
1	1
1	1
2	1 + 1
3	1 + 2
5	1 + 3 + 1
8	1 + 4 + 3
13	1 + 5 + 6 + 1
21	1 + 6 + 10 + 4
34	1 + 7 + 15 + 10 + 1

CHANCE IS BRIDLED AND GOVERNED BY LAW

PROBABILITY

IN CONTEXT

KEY FIGURES
Blaise Pascal (1623–62),
Pierre de Fermat (1601–65)

FIELD
Probability

BEFORE
1620 Galileo publishes *Sopra le Scoperte dei Dadi* (*On the Outcomes of Dice*), calculating the chances of certain totals when throwing dice.

AFTER
1657 Christiaan Huygens writes a treatise on probability theory and its applications to games of chance.

1718 Abraham de Moivre publishes *The Doctrine of Chances*.

1812 Pierre-Simon Laplace applies probability theory to scientific problems in *Théorie analytique des probabilités* (*Theory of Probabilities*).

B efore the 1500s, predicting the outcome of a future event with any degree of accuracy was thought to be impossible. However, in Renaissance Italy, scholar Gerolamo Cardano produced in-depth analyses of outcomes involving dice. In the 1600s, such problems attracted the attention of French mathematicians Blaise Pascal and Pierre de Fermat. More renowned for findings such as Pascal's triangle (see pp.156–61) and Fermat's last theorem (see pp.320–23), the two men took the mathematics of probability to a new level, laying the foundations for probability theory.

See also: The law of large numbers 184–85 ▪ Bayes' theorem 198–99 ▪ Buffon's needle experiment 202–03 ▪ The birth of modern statistics 268–71

> Probability theory is nothing but common sense reduced to calculation.
> **Pierre-Simon Laplace**

Forecasting the outcomes of games of chance proved a useful way of approaching probability, which, by definition, measures the likelihood of something occurring. For example, the chances of throwing a six with a die can be estimated by throwing the die a given number of times and dividing the amount of sixes thrown by the total number of throws. The result, called relative frequency, gives the probability of throwing a six, which can be expressed as a fraction, a decimal, or a percentage. This, however, is an observed finding, based on actual experiments. Theoretical probability of any single event is calculated by dividing the number of desired outcomes by the total number of possible outcomes. With one roll of a six-sided die, the probability of throwing a six is $\frac{1}{6}$; the probability of throwing any other number is $\frac{5}{6}$.

Estimating the odds

One popular game in 17th-century France involved two players taking turns to throw four dice in a bid to obtain at least one "ace," or six. The players contributed equal stakes and agreed, in advance, that the first one to win a certain number of rounds would take the whole stake. Writer and amateur mathematician Antoine Gombaud, who styled himself Chevalier »

Impossible	Unlikely	Evens	Likely	Certain
0		**0.5**		**1**

A blue candy will be pulled from a jar filled only with pink candies.

A blue candy will be pulled from a jar filled evenly with both blue and pink candies.

A blue candy will be pulled from a jar filled only with blue candies.

Probability is easily measured in the cases shown here. It is zero if the element in question (blue candies) is absent, and 0.5 (or $\frac{1}{2}$, or 50 percent) if half of all candies are blue. When events are certain, probability = 1 (or 100 percent).

Pierre de Fermat

Born in in 1601 in Beaumont-de-Lomagne in France, Pierre de Fermat moved to Orléans in 1623 to study law and soon began to pursue his interest in mathematics. Like other scholars of his day, he studied geometry problems from the ancient world and applied algebraic methods to try to solve them. In 1631, Fermat moved to Toulouse and worked as a lawyer.

In his spare time, Fermat continued his mathematical investigations, circulating his ideas in letters to friends, such as Blaise Pascal. In 1653, he was struck down by plague but survived to do some of his best work. As well as his ideas on probability, Fermat pioneered differential calculus, but is best remembered for his contribution to number theory and Fermat's last theorem. He died in Castres in 1665.

Key works

1629 *De tangentibus linearum curvarum* (*Tangents of Curves*)
1637 *Methodus ad disquirendam maximam et minimam* (*Methods of Investigating Maxima and Minima*)

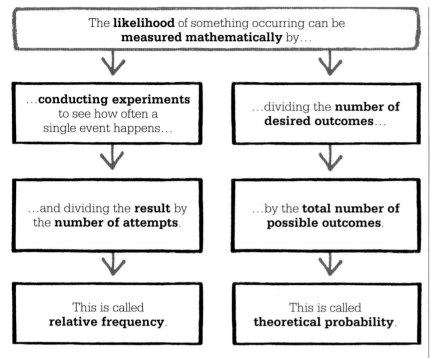

The **likelihood** of something occurring can be **measured mathematically** by…

↓

…**conducting experiments** to see how often a single event happens…

↓

…and dividing the **result** by the **number of attempts**.

↓

This is called **relative frequency**.

↓

…dividing the **number of desired outcomes**…

↓

…by the **total number of possible outcomes**.

↓

This is called **theoretical probability**.

de Méré, understood the $1/6$ odds of an ace with one throw of a die, and sought to calculate the odds of throwing a double ace using a pair of dice.

De Méré suggested that the chance of getting two aces from two throws of a dice was $1/36$, that is, $1/6$ as likely as getting an ace with one die in one roll. To make these odds the same, he argued that a pair of dice should be rolled six times for each roll of the single die. To have the same chance of rolling a double ace as you would from getting one ace when four dice are thrown, the pair should be thrown $6 \times 4 = 24$ times. De Méré consistently lost the bet and was compelled to deduce that a double

On a standard roulette wheel, there is a $1/37$ chance of the ball landing on any given number for a single spin of the wheel. This number gets closer to 1 the greater the number of spins.

ace from 24 throws of a pair of dice was less likely than one ace from four throws of a single die.

In 1654, de Méré consulted his friend Pascal about this problem, and about the further question of how a stake should be divided between the players when a game

was interrupted before completion. This was known as the "problem of points," and it had a long history. In 1494, Italian mathematician Luca Pacioli had suggested that the stakes should be divided in proportion to the number of rounds already won by each player.

In the mid-1500s, Niccolò Tartaglia, another prominent mathematician, had noted that such a division would be unfair if the game was interrupted, say, after only one round. His solution was to base the division of the pot on the ratio between the size of the lead and the length of the game, but this also gave unsatisfactory results for games with many rounds. Tartaglia remained unsure whether the problem was solvable in a way that would convince all players of its fairness.

The Pascal–Fermat letters
During the 1600s, it was common for mathematicians to meet at academies—scientific societies. In France, the leading academy was that of the Abbé Marin Mersenne, a Jesuit priest and mathematician who held weekly meetings at his

Choice means probability,
and probability means
mathematicians can
get to work.
Hannah Fry
British mathematician

Paris home. Pascal attended these meetings, but he and Fermat had never met. Nonetheless, having pondered de Méré's problems, Pascal chose to write to Fermat, communicating his thoughts on these and related issues and asking for Fermat's own views. This was the first of the letters between Pascal and Fermat in which the mathematical theory of probability was developed.

Player versus banker

The Pascal–Fermat letters were sent via Pierre de Carcavi, a mutual friend. Seven letters exchanged in 1654 reveal the two men's thoughts on the points problem, which they examined in different scenarios. They discuss a game between a player attempting to throw at least one ace in eight throws and a "banker" who takes the pot if the player is unsuccessful. If the game is interrupted before an ace has been thrown, Pascal seems to suggest that the stakes should be allocated according to the players' expectations of winning. At the start of the game, the probability of eight rolls of the die without success is $(5/6)^8 \approx 0.233$, and the probability of throwing at least one ace is $(1-0.233)$, or 0.7677. The game clearly favors the one who makes the throws, rather than the "banker."

Laying down the theory

In other letters, Pascal and Fermat discuss other cases of interrupted games, such as when the play alternates between two players until one is successful. Fermat notes that what matters is the number of throws remaining when the game stops. He points out that a player with a 7–5 lead in a game to 10 aces has the same chance of eventually winning as a player with a 17–15 lead in a game to 20.

Pascal gives an example with two opponents playing a sequence of games, each with an equal chance of winning, where the first to win three games wins the stake. Each player has staked 32 pistoles, so the stake is 64 pistoles. Over the course of three games, the first player wins twice and the other once. If they now play a fourth game and the first player wins, then he will take the 64 pistoles; if the other wins it, they will have each won two games and are equally likely to win the final game. If they stop at this point, each should take back his stake of 32 pistoles.

Pascal's step-by-step methods and Fermat's considered replies provide some of the earliest examples of using expectations when reasoning about probability. The correspondence between the two laid down basic principles of probability theory, and games of chance would continue to prove fertile ground for early theorists. Dutch physicist and mathematician Christiaan Huygens wrote a treatise translated as "On reasoning in games of chance," which was the first book on probability theory.

An early version of the law of large numbers (LLN)—a theorem examining the results of performing the same action (such as throwing a die) a number of times—was part of Swiss mathematician Jacob Bernoulli's *Ars Conjectandi* (*The Art of Conjecturing*, 1713). In the late 18th and early 19th century, Pierre-Simon Laplace applied probability theory to practical and scientific problems, setting out his methods in his *Théorie Analytique des Probabilités* (*Analytic Theory of Probabilities*) in 1812. ∎

Probability theory

While ancient and medieval law graded probability in the assessment of judicial evidence, there was no theory on which to base it. Similarly, in Renaissance times, when insurance was calculated for ships, premiums were based on an intuitive estimate of risk. Odds were a feature of gaming, but Gerolamo Cardano was the first to apply mathematics to the study of probability. Games of chance were the focus of such studies even after the deaths of both Pascal and Fermat, although their letters on the subject contributed much to subsequent theory.

In the late 1700s, Pierre-Simon Laplace extended the scope of probability theory to science, and introduced his mathematical tools for predicting the probability of many incidents, including natural phenomena. He also recognized its application in statistics. Probability theory is also used in many other fields, such as psychology, economics, engineering, and sports.

THE SUM OF THE DISTANCE EQUALS THE ALTITUDE
VIVIANI'S TRIANGLE THEOREM

IN CONTEXT

KEY FIGURE
Vincenzo Viviani
(1622–1703)

FIELD
Geometry

BEFORE
c. 300 BCE Euclid defines a triangle in his book *Elements* and proves many theorems concerning triangles.

c. 50 CE Heron of Alexandria defines a formula for finding the area of a triangle from its side lengths.

AFTER
1822 German geometer Karl Wilhelm Feuerbach publishes a proof for the nine-point circle, which passes through the midpoint of each side of a triangle.

1826 Swiss geometer Jakob Steiner describes the triangle center that has the minimum sum of distances from the triangle's three vertices.

I talian mathematician Vincenzo Viviani studied under Galileo in Florence. After Galileo's death in 1642, Viviani gathered together his master's work, editing the first collected edition in 1655–56.

Viviani's research included work on the speed of sound, which he measured to within 82 ft (25 m) per second of its true value. He is best known, however, for his triangle theorem, which states that the sum of the distances between any point in an equilateral triangle and that triangle's sides is equal to the altitude (height) of the triangle.

Proving the theorem
Starting with an equilateral triangle of base (side) *a*, and an altitude of *h* (see above right), a point is made inside the triangle. Perpendicular lines (*p*, *q*, and *r*) are drawn from that point to each of the three sides, meeting each side at 90°. The triangle is divided into three smaller triangles by drawing a line from the point to each corner of the main triangle.

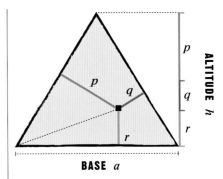

BASE *a*

The altitude in an equilateral triangle, such as the above, is always equal to the combined length of lines drawn from any point in the triangle perpendicular to its three sides.

The area of a triangle is $1/2 \times$ base $\times$ height, so if the lengths of the perpendiculars are *p*, *q*, and *r*, the areas of the triangles add up to $1/2 (p + q + r)a$. This is also the area of the large triangle, which is $1/2\, ha$, and so $h = p + q + r$. If you were to break a stick of length *h* into three, there would always be a point in the triangle from which the pieces form the perpendiculars *p*, *q*, and *r*. ∎

See also: Pythagoras 36–43 ▪ Euclid's *Elements* 52–57 ▪ Trigonometry 70–75
▪ Projective geometry 154–55 ▪ Non-Euclidean geometries 228–29

THE SWING OF A PENDULUM
HUYGENS'S TAUTOCHRONE CURVE

I n 1656, Dutch physicist and mathematician Christiaan Huygens created the pendulum clock, a clock with a swinging weight that was constant. He wanted to resolve the navigational problem of determining a ship's longitude. This was impossible without precise calculations of time, so it required an accurate clock to cope with the rolling motion of the waves, which caused wide variations in pendulum swing, leading to time discrepancies.

Seeking the right curve

The key lay in finding a curved path for the pendulum to follow (known as a tautochrone curve), whereby the time the pendulum takes to return to its lowest point is constant whatever its highest point. Huygens identified the cycloid, a curve that was steep at the top and shallow at the bottom. The curved path of any pendulum would have to be adjusted so it traveled in a cycloid. Huygens's idea was to constrain the pendulum by adding cycloid-shaped "cheeks." In theory, the time of each movement would now be the same from any starting point. However, friction introduced a larger error than the one Huygens was trying to resolve. It was only in the 1750s that the Italian Joseph-Louis Lagrange arrived at a solution, where the height of the curve needs to be in proportion to the square of the length of the arc traveled by the pendulum. ∎

I was… struck by the remarkable fact that in geometry all bodies gliding along the cycloid… descend from any point in precisely the same time.
Herman Melville
Moby Dick (1851)

WITH CALCULUS I CAN PREDICT THE FUTURE

CALCULUS

IN CONTEXT

KEY FIGURES
Isaac Newton (1642–1727),
Gottfried Leibniz (1646–1716)

FIELD
Calculus

BEFORE
287–212 BCE Archimedes
uses the method of exhaustion
to calculate areas and
volumes, introducing the
concept of infinitesimals.

c. 1630 Pierre de Fermat
uses a new technique for
finding tangents to curves,
locating their maximum and
minimum points.

AFTER
1740 Leonhard Euler applies
the ideas of calculus to
synthesize calculus, complex
algebra, and trigonometry.

1823 French mathematician
Augustin-Louis Cauchy
formalizes the fundamental
theorem of calculus.

The development of calculus,
the branch of mathematics
that deals with how things
change, was one of the most
significant advances in the history
of mathematics. Calculus can
show how the position of a moving
vehicle changes over time, how the
brightness of a light source dims as
it moves further away, or how the
position of a person's eyes alters
as they follow a moving object.
It can ascertain where changing
phenomena reach a maximum or
minimum value, and at what rate
they travel between the two.

Alongside rates of change,
another important aspect of
calculus is summation (the process
of adding things), which developed
from the need to calculate areas.
Eventually, the study of areas and
volumes was formalized into what
became known as integration,
while calculating rates of change
was termed differentiation.

By providing a better
understanding of the behavior
of phenomena, calculus can be
used to predict and influence their
future state. In much the same way
as algebra and arithmetic are tools
for working with numerical or

Nothing takes place in the
world whose meaning is
not that of some maximum
or minimum.
Leonhard Euler

generalized quantities, calculus
has its own rules, notations, and
applications, and its development
between the 17th and 19th
centuries led to rapid progress
in fields such as engineering
and physics.

Ancient origins
The ancient Babylonians and
Egyptians were particularly
interested in measurement. It was
important for them to be able to
calculate the dimensions of fields
for growing and irrigating crops
and to work out the volume of
buildings to store grain. They
developed early notions of area
and volume, although these tended
to be in the form of very specific
examples, such as in the Rhind
papyrus, where one problem
involves the area of a round field
with a diameter of 9 *khet* (a *khet*
being an ancient Egyptian unit
of length). The rules laid down in
the Rhind papyrus led ultimately
to what would become known
more than 3,000 years later as
integral calculus.

The concept of infinity is central
to calculus. In ancient Greece,
Zeno's paradoxes of motion, a set
of philosophical problems devised

Archimedes saw
the **circle** as having
an **infinite number
of sides**.

→

He approximated the
area of the **circle** by
placing it in **polygons
with infinitesimally
smaller sides**.

↓

**Ancient Greek
thought is at the
foundation of
modern calculus.**

←

Division into **infinite
parts** is essential
to integration (the study
of areas and volumes).

See also: The Rhind papyrus 32–33 ▪ Zeno's paradoxes of motion 46–47 ▪ Calculating pi 60–65 ▪ Decimals 132–37 ▪ The problem of maxima 142–43 ▪ The area under a cycloid 152–53 ▪ Euler's number 186–91 ▪ Euler's identity 197

by the philosopher Zeno of Elea in the 5th century BCE, posited that motion was impossible because there are an infinite number of halfway points in any given distance. In around 370 BCE, the Greek mathematician Eudoxus of Cnidus proposed a method of calculating the area of a shape by filling it with identical polygons of known area, and then making the

For by the ultimate velocity is meant that, with which the body is moved, neither before it arrives at its last place, when the motion ceases nor after but at the very instant when it arrives.
Isaac Newton

polygons infinitely smaller. It was thought that their combined area would eventually converge toward the true area of the shape.

This so-called "method of exhaustion" was taken up by Archimedes in around 225 BCE. He approximated the area of a circle by enclosing it within polygons with increasing numbers of sides. As the number of sides increases, the polygons (of known area) more closely resemble the circle. Taking this idea to the limit, Archimedes imagined a polygon with sides of infinitesimally smaller length. The recognition of infinitesimals was a pivotal moment in the development of calculus: previously insoluble puzzles, such as Zeno's paradoxes of motion, could now be solved.

Fresh ideas
Mathematicians in medieval China and India made further advances in dealing with infinite sums. In the Islamic world, too, the development of algebra meant that, rather than spelling out a calculation millions

As civilizations developed, accurate measurement became essential. This ancient Egyptian tomb painting shows surveyors using rope to calculate the dimensions of a wheat field.

of times for all possible variations, generalized symbols could be used to prove that a case is true for all numbers to infinity.

Mathematics had suffered a long period of stagnation in Europe but, as the Renaissance took hold in the 1300s, renewed interest in the subject led to fresh ideas about motion and the laws governing distance and speed. French mathematician and philosopher Nicole Oresme studied the velocity of an accelerating object against time, and he realized that the area under a graph depicting this relationship was equivalent to the distance traveled by the object. This notion would be formalized in the late 1600s by Isaac Newton and Isaac Barrow in England, Gottfried Leibniz in Germany, and Scottish mathematician James Gregory. »

Oresme's work was inspired by that of the "Oxford Calculators," a 14th-century group of scholars based at Merton College, Oxford, who developed the mean speed theorem, which Oresme later proved. It states that if one body is moving with a uniformly accelerated motion and a second body is moving with a uniform speed equal to the mean speed of the first body, and both bodies are moving for the same duration, they will cover the same distance. The Merton scholars were devoted to solving physical and philosophical problems using calculations and logic, and were interested in the quantitive analysis of phenomena such as heat, color, light, and velocity. They were inspired by the trigonometry of Arab astronomer al-Battani (858–929 CE) and the logic and physics of Aristotle.

New developments

The incremental steps toward the development of calculus gathered pace toward the end of the 16th century. In around 1600, French mathematician François Viète promoted the use of symbols in algebra (which had previously been described in words), while Flemish mathematician Simon Stevin initiated the concept of mathematical limits, whereby the sum of amounts could converge to a limiting value, much like the area of Archimedes' polygons converged to the area of a circle.

At much the same time, German mathematician and astronomer Johannes Kepler was researching the motion of the planets, including calculating the area enclosed by a planetary orbit, which he recognized as elliptical rather than circular. Using ancient Greek methods, he worked out the area by dividing the ellipse into strips of infinitesimal width.

A forerunner of the more formal integration to come, Kepler's method was further developed in 1635 by Italian mathematician Bonaventura Cavalieri in *Geometria indivisibilibus continuorum nova quadam ratione promota* (*Geometry, Advanced in a New Way by the Indivisibles of the Continua*). Cavalieri worked out a "method of indivisibles," which was a more rigorous method of determining the size of shapes. More developments followed in the 1600s with the work of English theologian and mathematician Isaac Barrow and Italian physicist Evangelista Torricelli, followed by that of Pierre de Fermat and René Descartes, whose analysis of curves advanced the new area of graphical algebra.

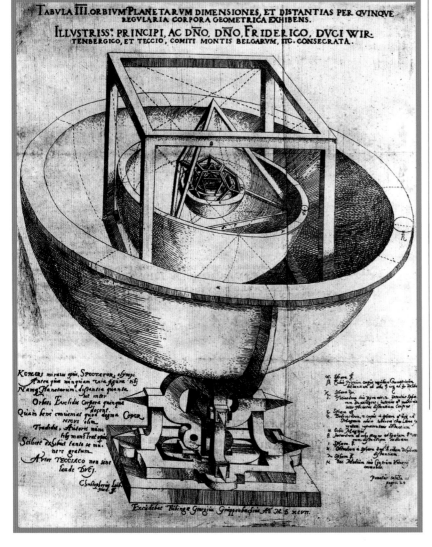

This illustration of Kepler's Platonic solid model of the Solar System appeared in a book published in 1596. Kepler used infinitesimally small strips to measure the distance covered in an orbit. This method was the forerunner of integration.

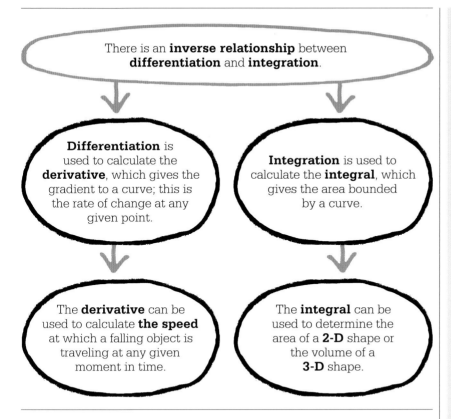

There is an **inverse relationship** between **differentiation** and **integration**.

Differentiation is used to calculate the **derivative**, which gives the gradient to a curve; this is the rate of change at any given point.

Integration is used to calculate the **integral**, which gives the area bounded by a curve.

The **derivative** can be used to calculate **the speed** at which a falling object is traveling at any given moment in time.

The **integral** can be used to determine the area of a **2-D** shape or the volume of a **3-D** shape.

The fundamental theorem of calculus

The study of calculus is underpinned by the fundamental theorem of calculus, specifying the relationship between differentiation and integration, both of which rely on the concept of infinitesimals. First articulated by James Gregory in his 1668 *Geometriae Pars Universalis* (*The Universal Part of Geometry*), it was then generalized by Isaac Barrow in 1670, and formalized in 1823 by Augustin-Louis Cauchy.

The theorem has two parts. The first states that integration and differentiation are opposites—for any continuous function (one that can be defined for all values), there exists an "anti-derivative" (or "integral"), whose derivative (a measure of the rate of change) is the function itself. The second part of the theorem states that if values are inserted into the anti-derivative $F(x)$, the result—the definite integral of the function $f(x)$—makes it possible to calculate areas under the curve of the function $f(x)$.

James Gregory (1638–75) was the first person to formulate the fundamental theorem of calculus.

Fermat also located maxima and minima, the greatest and least values of a curve.

Fluxion model

In 1665–66, English mathematician Isaac Newton developed his "method of fluxions," a method for calculating variables that changed over time, which was a milestone in the history of calculus. Like Kepler and Galileo, Newton was interested in studying moving bodies and was particularly eager to unify the laws governing the motion of celestial bodies with motion on Earth.

In Newton's fluxion model, he considered a point moving along a curve as being divided into two perpendicular components (x and y), and then considered the velocities of those components. This work laid the foundation for what became known as differential calculus (or differentiation), which together with the related field of integral calculus led to the fundamental theorem of calculus (see box, right). The idea of differential calculus is that the rate at which a variable changes at a point is equal to the gradient of a tangent at that point. This can be pictured by drawing a tangent »

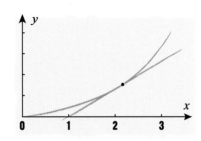

Differentiation can be used to find the rate of change at a given point in time. The blue line shows the rate of change overall and the orange tangent shows the rate of change at a given point.

(a straight line that touches a curve at only one point). The gradient or steepness of this line will be the rate of change of the curve at that point. Newton recognized that at the maxima and minima, the gradient of the curve was zero, because when something is at its highest or lowest point, it is momentarily not changing. Newton went on to develop his theory further by considering the converse problem—if the rate at which a variable changes is known, is it possible to calculate the shape of the variable itself? This "anti-differentiation" entailed working out areas under the curve.

Newton v. Leibniz

Around the time that Newton was developing his calculus, German mathematician Gottfried Leibniz was working on his own version, based on the consideration of infinitesimal changes in the two coordinates defining a point on a curve. Leibniz used very different notation from Newton's, and in 1684 published a paper on what would later become known as differential calculus. Two years later, he published another paper,

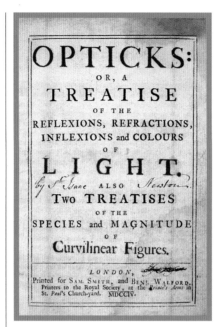

this time about integration, again using different notation from that of Newton. In an unpublished manuscript dated October 29, 1675, Leibniz was the first person to use the "integral" symbol ∫, which is used and recognized universally today.

There was much debate about who discovered modern calculus first: Newton or Leibniz. It led to protracted bitterness between the two rivals and across much of the mathematical community. Although Newton devised his theory of fluxions in 1665–66, he did not publish it until 1704, when it was added as an appendix to his work *Opticks*. Leibniz began to devise his version of calculus around 1673, and published it in 1684. Newton's subsequent *Principia* is said by some to have been influenced by Leibniz's work.

By 1712, Leibniz and Newton were openly accusing one another of plagiarism. The modern consensus is that Leibniz and Newton developed their ideas on the subject independently.

Isaac Newton's *Opticks*, a treatise about the reflections and refractions of light, published in 1704, contains the first details of his work in the area of calculus.

Significant contributions to calculus were also made in Switzerland by the brothers Jacob and Johann Bernoulli, who coined the term "integral" in 1690. Scottish mathematician Colin Maclaurin published his *Treatise on Fluxions* in 1742, promoting and furthering Newton's methods, and attempting to make them more rigorous. In this work, Maclaurin applies calculus to the study of infinite series of algebraic terms. Meanwhile Swiss mathematician Leonhard Euler, a close friend of Johann Bernoulli's sons, was influenced by their ideas on the subject. In particular, he applied the idea of infinitesimals to what is known as the exponential function, e^x. This ultimately led to "Euler's identity", $e^{i\pi} + 1 = 0$, an equation that connects five of the most fundamental mathematical quantities (e, i, π, 0, and 1) in a very simple way.

Assuming I know our instantaneous speed at every possible moment, can I then use that information to determine how far we've traveled? Calculus says I can.
Jennifer Ouellette
American science writer

When the values successively assigned to the same variable indefinitely approach a fixed value, so as to end by differing from it as little as desired, this fixed value is called the limit.
Augustin-Louis Cauchy

The notation of modern calculus	
$\dot{f}$	Invented by **Newton** for **differentiation**.
$\int$	Invented by **Leibniz** for **integration**.
dy/dx	Invented by **Leibniz** for **differentiation**.
f'	Invented by **Lagrange** for **differentiation**.

As the 18th century progressed, calculus proved increasingly useful as a tool for describing and understanding the physical world. In the 1750s, Euler, working in collaboration with French mathematician Joseph-Louis Lagrange, used calculus to provide an equation—the Euler–Lagrange equation—for understanding both fluid (gas and liquid) and solid mechanics. In the early 1800s, French physicist and mathematician Pierre-Simon Laplace developed electromagnetic theory with the help of calculus.

Formalizing the theories

The various developments in calculus were formalized in 1823 when Augustin-Louis Cauchy formally stated the fundamental theorem of calculus. In essence, this states that the process of differentiation (working out rates of change of a variable represented by a curve) is the inverse of the process of integration (calculating the area beneath a curve). Cauchy's formalization allowed calculus to be regarded as a unified whole, dealing with infinitesimals in a consistent way using universally agreed notation.

The field of calculus was further developed later in the 1800s. In 1854, German mathematician Bernhard Riemann formulated criteria for which functions would be integrable or not, based on defining finite upper and lower limits for the function.

Ubiquitous applications

Many advances in physics and engineering have relied on calculus. Albert Einstein used it in his theories of special and general relativity in the early 20th century, and it has been applied extensively in quantum mechanics (dealing with the motion of subatomic particles). Schrödinger's wave equation, a differential equation published in 1925 by Austrian physicist Erwin Schrödinger, models a particle as a wave whose state can only be determined by using probability. This was groundbreaking in a scientific world that had up until then been governed by certainty.

Calculus has many important applications today; it is used, for instance, in search engines, construction projects, medical advances, economic models, and weather forecasts. It is difficult to imagine a world without this all-pervasive branch of mathematics, as it would most certainly be one without computers. Many would argue that calculus is the most important mathematical discovery in the last 400 years. ∎

Gottfried Leibniz

Born in Leipzig, Germany, in 1646, Gottfried Leibniz was raised in an academic family. His father was a professor of moral philosophy, while his mother was the daughter of a professor of law. In 1667, after completing his university studies, Leibniz became an advisor on law and politics to the Elector of Mainz, a role that enabled him to travel and meet other European scholars. After his employer's death in 1673, he took up the role of librarian to the Duke of Brunswick in Hanover.

Leibniz was a celebrated philosopher as well as a mathematician. He never married and died in 1716 to little fanfare. His successes had been overshadowed by his calculus dispute with Newton and were only recognized several years after his death.

Key works

1666 *On the Art of Combination*
1684 *New Method for Maximums and Minimums*
1703 *Explanation of Binary Arithmetic*

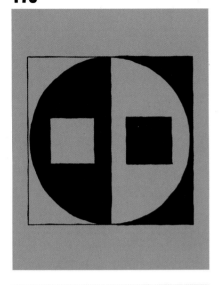

THE PERFECTION OF THE SCIENCE OF NUMBERS

BINARY NUMBERS

I n everyday life we are used to the base-10 counting system with its familiar ten digits, 0 to 9. When we count from 10 onward, we put a 1 in the "tens" column and a 0 in the "units" column, and so on, adding columns for hundreds, thousands, and beyond. The binary system is a base-2 counting system and employs just two symbols, 0 and 1. Rather than increasing in multiples of 10, each column represents a power of 2. So the binary number 1011 is not 1,011 but 11 (from right to left: one 1, one 2, no 4s, and one 8).

Decimal numbers	Binary number 16s	8s	4s	2s	1s	Binary visual 16s	8s	4s	2s	1s
1	0	0	0	0	1	☐	☐	☐	☐	■
2	0	0	0	1	0	☐	☐	☐	■	☐
3	0	0	0	1	1	☐	☐	☐	■	■
4	0	0	1	0	0	☐	☐	■	☐	☐
5	0	0	1	0	1	☐	☐	■	☐	■
6	0	0	1	1	0	☐	☐	■	■	☐
7	0	0	1	1	1	☐	☐	■	■	■
8	0	1	0	0	0	☐	■	☐	☐	☐
9	0	1	0	0	1	☐	■	☐	☐	■
10	0	1	0	1	0	☐	■	☐	■	☐

Binary numbers are written as 1s and 0s, using a base-2 system. This chart shows how to write the numbers 1 to 10, from the base-10 system, as both binary numbers and binary visuals—which is how a computer would process them—where 1 is "on" and 0 is "off."

Reckoning by twos, that is, by 0 and 1… is the most fundamental way of reckoning for science, and offers up new discoveries, which are… useful, even for the practice of numbers.
Gottfried Leibniz

Bacon's cipher

English philosopher and courtier Francis Bacon (1561–1626) was a great dabbler in cryptography, or the science of deciphering codes. He developed what he called a "biliteral" cipher, which used the letters a and b to generate the entire alphabet— a = aaaaa, b = aaaab, c = aaaba, d = aaabb, and so on. If you substitute 0 for a and 1 for b, this becomes a binary sequence. It is an easy code to break, but Bacon realized that a and b do not have to be letters—they can be any two different objects— "… as by bells, by trumpets, by lights and torches… and any instruments of like nature." It was an ingeniously adaptable cipher, which Bacon could use to "make anything signify anything." A secret message could be hidden in a group of objects or numbers, or even musical notation. Samuel Morse's dot–dash telegraph code, which revolutionized communication in the 1800s, and the on/off encoding in a modern computer both have parallels with Bacon's cipher.

Binary choices are black and white; in any column there is only ever 1 or 0. This simple "on or off" concept has proved vital in computing, for example, where every number can be represented by a series of switchlike on/off actions.

Binary power revealed
In 1617, Scottish mathematician John Napier announced a binary calculator based on a chessboard. Each square had a value, and that square's value was "on" or "off" depending on whether a counter was placed on the square. The calculator could multiply, divide, and even find square roots, but was considered a mere curiosity.

Around the same time, Thomas Harriot was experimenting with number systems, including the binary system. He was able to convert base-10 numbers to binary and back again, and could also calculate using binary numbers. However, Harriot's ideas remained unpublished until long after his death in 1621.

The potential of binary numbers was finally realized by German mathematician and philosopher Gottfried Leibniz. In 1679, he described a calculating machine that worked on binary principles, with open or closed gates to let marbles fall through. Computers work in a similar way, using switches and electricity rather than gates and marbles.

Leibniz outlined his ideas on the binary system in 1703 in *Explanation of Binary Arithmetic*, showing how 0s and 1s could represent numbers and so simplify even the most complex operations into a basic binary form. He had been influenced by correspondence with missionaries in China, who introduced him to the *I Ching*, an ancient Chinese book of divination. The book divided reality into the two opposing poles of yin and yang—one represented as a broken line, the other as an unbroken line. These lines were displayed as six-line hexagrams, combined into a total of 64 different patterns.

Leibniz saw links between this binary approach to divination and his work with binary numbers.

Above all, Leibniz was driven by his religious faith. He wanted to use logic to answer questions about God's existence and believed that the binary system captured his view of the Universe's creation, with 0 representing nothingness and 1 representing God. ∎

The teaching and commentaries on the *I Ching* of ancient Chinese philosopher Confucius (551–479 BCE) influenced the work of Leibniz and other 17th–18th-century scientists.

THE
ENLIGHT
1680–1800

ENMENT

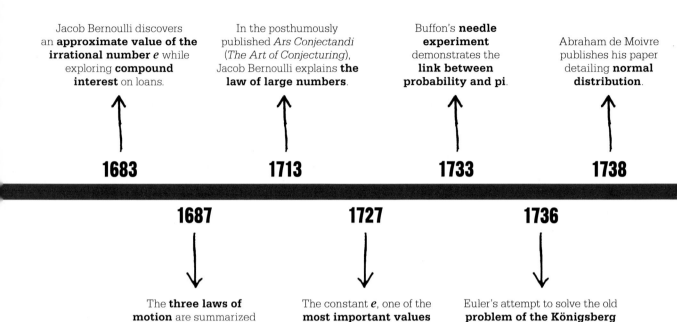

Jacob Bernoulli discovers an **approximate value of the irrational number *e*** while exploring **compound interest** on loans.

In the posthumously published *Ars Conjectandi* (*The Art of Conjecturing*), Jacob Bernoulli explains **the law of large numbers**.

Buffon's **needle experiment** demonstrates the **link between probability and pi**.

Abraham de Moivre publishes his paper detailing **normal distribution**.

1683 **1713** **1733** **1738**

1687 **1727** **1736**

The **three laws of motion** are summarized by Isaac Newton in his *Principia Mathematica*.

The constant *e*, one of the **most important values in math**, is given its notation by Leonhard Euler.

Euler's attempt to solve the old **problem of the Königsberg Bridges** leads to **graph theory** and important developments in **mathematical topology**.

By the late 1600s, Europe had become established as the cultural and scientific center of the world. The Scientific Revolution was well under way, inspiring a new, rational approach not only to the sciences, but to all aspects of culture and society. The Age of Enlightenment, as this period came to be known, was a time of significant sociopolitical change, and produced an enormous increase in the spread of knowledge and education during the 1700s. It was also a period of considerable progress in mathematics.

Swiss giants

Building on the work of Newton and Leibniz, whose ideas were finding practical application in physics and engineering, the brothers Jacob and Johann Bernoulli further developed the theory of calculus in their "calculus of variations" and several other mathematical concepts discovered in the 1600s. The elder brother, Jacob, is recognized for his work on number theory, but he also helped develop probability theory, introducing the law of large numbers.

Along with their mathematically gifted children, the Bernoullis were the leading mathematicians of the early 1700s, making their home town of Basel in Switzerland a center of mathematical study. It was here that Leonhard Euler, the next, and arguably greatest, Enlightenment mathematician, was born and educated. Euler was a contemporary and friend of Daniel and Nicholas Bernoulli, Johann's sons, and at an early age proved himself a worthy successor to Jacob and Johann. Aged only 20, he suggested a notation for the irrational number *e*, for which Jacob Bernoulli had calculated an approximate value.

Euler published numerous books and treatises, and worked in every field of mathematics, often recognizing the links between apparently separate concepts of algebra, geometry, and number theory, which were to become the basis for further fields of Mathematical study. For example, his approach to the seemingly simple problem of planning a route through the city of Königsberg, crossing each of its seven bridges only once, uncovered much deeper concepts of topology, inspiring new areas of research.

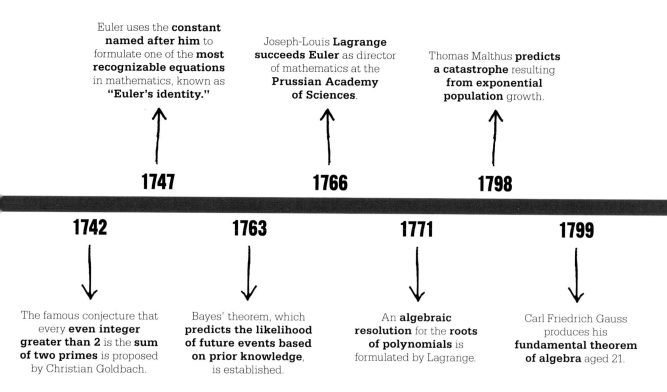

Euler uses the **constant named after him** to formulate one of the **most recognizable equations** in mathematics, known as **"Euler's identity."**

Joseph-Louis **Lagrange succeeds Euler** as director of mathematics at the **Prussian Academy of Sciences**.

Thomas Malthus **predicts a catastrophe** resulting **from exponential population** growth.

1747

1766

1798

1742

1763

1771

1799

The famous conjecture that every **even integer greater than 2** is the **sum of two primes** is proposed by Christian Goldbach.

Bayes' theorem, which **predicts the likelihood of future events based on prior knowledge**, is established.

An **algebraic resolution** for the **roots of polynomials** is formulated by Lagrange.

Carl Friedrich Gauss produces his **fundamental theorem of algebra** aged 21.

Euler's contributions to all fields of mathematics, but in particular calculus, graph theory, and number theory, were enormous, and he was also influential in standardizing mathematical notation. He is especially remembered for the elegant equation known as "Euler's identity," which highlights the connection between fundamental mathematical constants such as e and π.

Other mathematicians
The Bernoullis and Euler tended to eclipse the achievements of the many other mathematicians of the 1700s. Among them was Christian Goldbach, a German contemporary of Euler's. In the course of his career, Goldbach had befriended other influential mathematicians, including

Leibniz and the Bernoullis, and corresponded regularly with them about their theories. In a letter to Euler, he proposed the conjecture for which he is best known, that every even integer greater than 2 can be expressed as the sum of two primes, which remains unproven to this day.

Others contributed to the development of the growing field of probability theory. Georges-Louis Leclerc, Comte de Buffon, for example, applied the principles of calculus to probability, and demonstrated the link between pi and probability, while another Frenchman, Abraham de Moivre described the concept of normal distribution, and Englishman Thomas Bayes proposed a theorem of the probability of events based on knowledge of the past.

In the latter part of the 18th century, France became the European center of mathematical enquiry, with Joseph-Louis Lagrange in particular emerging as a significant figure. Lagrange had made his name working with Euler, but later made important contributions to polynomials and number theory.

New frontiers
As the century drew to a close, Europe was reeling from political revolutions that had toppled the monarchy in France and given birth to the United States of America. A young German, Carl Friedrich Gauss, published his fundamental theorem of algebra, marking the beginning of a spectacular career and a new period in the history of mathematics. ∎

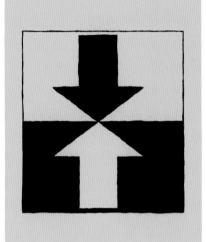

TO EVERY ACTION THERE IS AN EQUAL AND OPPOSITE REACTION
NEWTON'S LAWS OF MOTION

IN CONTEXT

KEY FIGURE
Isaac Newton (1642–1727)

FIELD
Applied mathematics

BEFORE
c.330 BCE Aristotle believes it takes force to maintain motion.

c.1630 Galileo Galilei conducts experiments on motion and finds that friction is a retarding force.

1674 Robert Hooke writes *An attempt to prove the motion of the Earth* and hypothesizes what will become Newton's first law.

AFTER
1905 Albert Einstein presents his theory of relativity, which challenges Newton's view of the force of gravity.

1977 Voyager 1 is launched. With no friction or drag in space, the craft keeps going due to Newton's first law, and exits the Solar System in 2012.

I n using mathematics to explain the movement of the planets and of objects on Earth, Isaac Newton fundamentally changed the way we see the Universe. He published his findings in 1687 in the three-volume *Philosophiae Naturalis Principia Mathematica* (*Mathematical Principles of Natural Philosophy*), often called the *Principia* for short.

How the planets move

By 1667, Newton had already developed early versions of his three laws of motion and knew about the force needed to enable a body to move in a circular path. He used his knowledge of forces and German astronomer Johannes Kepler's laws of planetary motion to deduce how elliptical orbits were related to the laws of gravitational attraction. In 1686, English astronomer Edmond Halley persuaded Newton to write up his new physics and its applications to planetary motion.

In his *Principia*, Newton used mathematics to show that the consequences of gravity were consistent with what had been

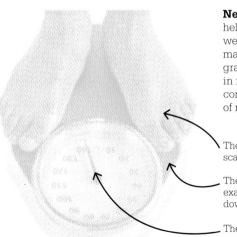

Newton's second and third law help explain how scales work. When we weigh ourselves, our weight (the mass of an object multiplied by gravity) is a force, now measured in newtons. Newtons can be converted into measurements of mass, such as pounds.

The body (mass) of the person on the scale is pushed down by gravity.

The scale pushes back up with exactly the same force as the downward pressure from gravity.

The weight is indicated on most scales in pounds. One pound is equal to 4.44 newtons.

See also: Syllogistic logic 50–51 ▪ The problem of maxima 142–43 ▪ Calculus 168–75 ▪ Emmy Noether and abstract algebra 280–81

Newton's three laws of motion

First law: Every body **continues in its state** of rest, or of uniform motion in a straight line, unless it is **compelled to change** that state by forces impressed upon it.

Second law: The **change** in motion is **proportional** to the **motive force** impressed, and is made in the direction of the straight line in which that force is impressed.

Third law: To every **action** there is an **equal** and **opposite reaction**.

observed experimentally. He analyzed the motion of bodies under the action of forces and posited gravitational attraction to explain the movement of the tides, projectiles, and pendulums, and the orbits of planets and comets.

Laws of motion

Newton began *Principia* by stating his three laws of motion. The first says that a force is needed to create motion, and that this force may be from the gravitational attraction between two bodies or an applied force (such as when a snooker cue strikes a ball). The second law explains what is happening when objects are in motion. Newton said that the rate of change of momentum (mass × velocity) of a body is equal to the force acting on it. If a graph is plotted showing velocity against time, then the gradient at any point is the rate of acceleration (any change in velocity).

Newton's third law says that if two objects are in contact, the reaction forces between them cancel out, each pushing on the other with an equal force, but in opposing directions. An object resting on a table pushes down on it, and the table pushes back with an equal force. If this were not true, the object would move. Until Einstein's theory of relativity, the whole of mechanical physics was based on Newton's three laws of motion. ▪

Isaac Newton

Isaac Newton was born on Christmas Day in 1642 in Lincolnshire, England, and was brought up in early childhood by his grandmother. Newton studied at Trinity College, Cambridge, where he showed a fascination for science and philosophy. During the Great Plague in 1665–1666, the university was forced to close, and it was during this period that he formulated his ideas on fluxions (rates of change at a given point in time).

Newton made significant discoveries in the fields of gravitation, motion, and optics, where he developed a rivalry with eminent English scientist Robert Hooke. One of several government positions he held was Master of the Royal Mint, where he oversaw the switch of the British currency from the silver to the gold standard. He was also President of the Royal Society. Newton died in 1727.

Key work

1687 *Philosophiae Naturalis Principia Mathematica* (*Mathematical Principles of Natural Philosophy*)

EMPIRICAL AND EXPECTED RESULTS ARE THE SAME
THE LAW OF LARGE NUMBERS

IN CONTEXT

KEY FIGURE
Jacob Bernoulli (1655–1705)

FIELD
Probability

BEFORE
c. 1564 Gerolamo Cardano
writes *Liber de ludo aleae* (*The
Book on Games of Chance*),
the first work on probability.

1654 Pierre de Fermat and
Blaise Pascal develop
probability theory.

AFTER
1733 Abraham de Moivre
proposes what becomes the
central limit theorem—as
a sample size increases, the
results will more closely match
normal distribution, or the
bell curve.

1763 Thomas Bayes develops
a way of predicting the chance
of an outcome by taking into
account the starting conditions
related to that outcome.

The law of large numbers is
one of the foundations of
probability theory and
statistics. It guarantees that, over
the long term, the outcomes of
future events can be predicted
with reasonable accuracy. This, for
example, gives financial companies
the confidence to set prices for
insurance and pension products,
knowing their chances of having to
pay out, and ensures that casinos
will always make a profit from their
gambling customers—eventually.

According to the law, as you make
more observations of an event
occurring, the measured probability
(or chance) of that outcome gets
ever closer to the theoretical
chance as calculated before any
observations began. In other words,
the average result from a large
number of trials will be a close
match to the expected value as
calculated using probability
theory—and increasing the number
of trials will result in that average
becoming an even closer match.

The expected **chance** of a **random event** can be
calculated using probability theory.

As the
number of trials
increases, the average
observed value gets **closer**
to the **expected** one.

In tests, the
observed results **do
not closely match**
the **expected** value
right away.

After a **large number** of trials, the
average observed value and the **expected
value** are **almost identical**.

See also: Probability 162–65 ▪ Normal distribution 192–93 ▪ Bayes' theorem 198–99 ▪ The Poisson distribution 220 ▪ The birth of modern statistics 268–71

We define the art of conjecture… as the art of evaluating… the probabilities of things, so that in our judgments and actions we can always base ourselves on what has been found to be the best.
Jacob Bernoulli

The law was named by French mathematician Siméon Poisson in 1835, but its origin is credited to Swiss mathematician Jacob Bernoulli. His breakthrough, which he called the "golden theorem," was published by his nephew in 1713 in the book *Ars Conjectandi* (*The Art of Conjecturing*).

Although not the first person to recognize the relationship between collecting data and predicting results, Bernoulli developed the first proof of this relationship by considering a game with two possible outcomes—a win or a loss. The theoretical chance of winning the game is W, and Bernoulli suspected that the fraction of games (f) that resulted in a win would converge on W as the number of games increased. He proved this by showing that

When a referee flips a coin, there is no advantage, according to the law of large numbers, in a team captain basing a heads or tails choice on what has been called in previous games.

the probability of f being greater or less than W by a specified amount approached 0 (meaning impossible) as the game was repeated.

The false probability

A coin toss is an example of the law of large numbers. Assuming that the chance of a heads or tails result is equal, the law dictates that after many tosses, half (or very near it) will have landed on heads, and half on tails. However, in the early stages, heads and tails are likely to be more unbalanced. For example, the first 10 tosses could be seven heads and three tails. It might then seem most likely that the next toss will produce a tail. That, however, is the "gambler's fallacy"—where a person assumes that the outcomes of each game (toss) are connected. A gambler might assume that toss number 11 is likely to be a tail because the number of heads and tails must balance out, but the probability of heads or tails is the same in every toss, and the outcome of one toss occurs independently of any other. This is the starting point of all probability theory. After 1,000 tosses, the imbalance apparent in those first 10 tosses becomes negligible. ▪

Jacob Bernoulli

Born in Basel, Switzerland, in 1655, Jacob Bernoulli studied theology, but developed an interest in mathematics. In 1687, he became a professor of mathematics at the University of Basel, a position he held for the rest of his life.

In addition to his work on probability, Bernoulli is remembered for discovering the mathematical constant e by calculating the growth of funds that received compound interest continuously in infinitesimal increments. He was also involved in the development of calculus, taking the side of Gottfried Leibniz against Isaac Newton in their rival claims to have invented a new mathematical field. Bernoulli worked on calculus with his younger brother Johann. However, Johann became jealous of his brother's achievements and their relationship broke down several years before Jacob died in 1705.

Key works

1713 *Ars Conjectandi* (*The Art of Conjecturing*)
1744 *Opera* (*Collected Works*)

ONE OF THOSE STRANGE NUMBERS THAT ARE CREATURES OF THEIR OWN

EULER'S NUMBER

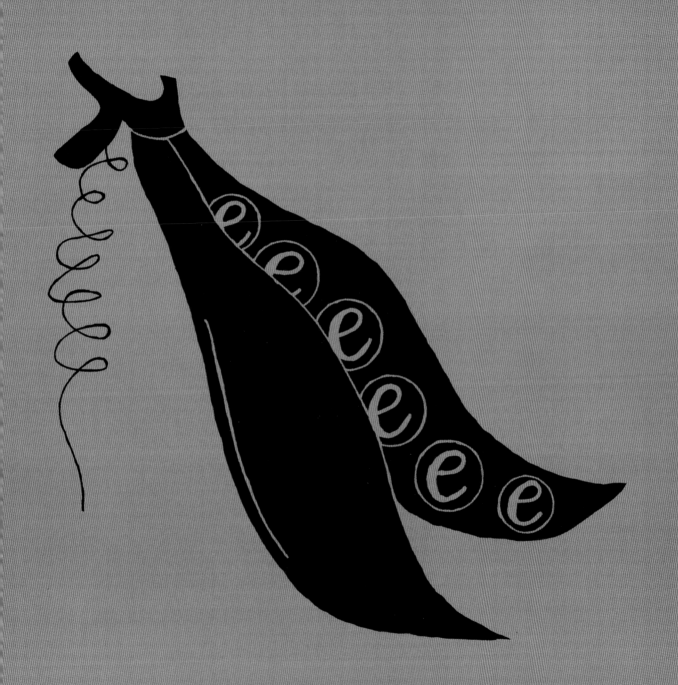

IN CONTEXT

KEY FIGURE
Leonhard Euler (1707–83)

FIELD
Number theory

BEFORE
1618 Logarithms calculated from the number now known as *e* are listed in an appendix to a book on logarithms by John Napier.

1683 Jacob Bernoulli uses *e* in his work on compound interest.

1733 Abraham de Moivre discovers "normal distribution": the way that values for most data cluster at a central point and taper off at the extremes. Its equation involves *e*.

AFTER
1815 Joseph Fourier's proof that *e* is irrational is published.

1873 French mathematician Charles Hermite proves that *e* is transcendental.

A mathematical **constant** is a significant, **well-defined** number. Its magnitude **never varies**.

The constant *e* (2.718…) has **special properties**.

It is **irrational**— it cannot be expressed as a **ratio of two integers** in a simple fraction.

It is **transcendental**— it is still **irrational** when **raised to any power**.

The mathematical constant that became known as *e*, or Euler's number—2.718… to an infinite number of decimal places—first appeared in the early 1600s, when logarithms were invented to help simplify complex calculations. Scottish mathematician John Napier compiled tables of logarithms to base 2.718…, which worked particularly well for calculations involving exponential growth. These were later dubbed "natural logarithms" because they can be used to mathematically describe many processes in nature, but with algebraic notation still in its infancy, Napier saw logarithms only as an aid to calculation involving the ratio of distances covered by moving points.

In the late 1600s, Swiss mathematician Jacob Bernoulli used 2.718… to calculate compound interest, but it was Leonhard Euler, a student of Bernoulli's brother Johann, who first called the number *e*. Euler calculated *e* to 18 decimal

---

Leonhard Euler

Born in 1707, in Basel, Switzerland, Euler grew up in nearby Riehen. Taught initially by his father, a Protestant minister who had some mathematical training and was also a friend of the Bernoulli family, Euler developed a passion for mathematics. Although he entered university to study for the ministry, he switched to mathematics with the support of Johann Bernoulli. Euler went on to work in Switzerland and Russia, and became the most prolific mathematician of all time, contributing greatly to calculus, geometry, and trigonometry, among other fields. This was despite steadily losing his sight from 1738 and becoming blind in 1771. Working to the very end, he died in 1783 in St. Petersburg.

Key works

1748 *Introductio in analysin infinitorium* (*Introduction to Analysis of the Infinite*)
1862 *Meditatio in experimenta explosione tormentorum nuper instituta* (*Meditation upon experiments made recently on the firing of Cannon*)

places, writing his first work on *e*, the *Meditatio* (*Meditation*), in 1727. However, it was not published until 1862. Euler explored *e* further in his 1748 *Introductio* (*Introduction*).

Compound interest

One of the earliest appearances of *e* was in calculating compound interest—where the interest on a savings account, for example, is paid into the account to increase the amount saved, rather than being paid out to the investor. If the interest is calculated on a yearly basis, an investment of $100 at an interest rate of 3% per year would produce $100 × 1.03 = $103 after one year. After two years, it would be 100 × 1.03 × 1.03 = $106.09, and after 10 years it would be $100 × 1.03^{10} = $134.39. The formula for this is $A = P(1 + r)^t$, where A is the final amount, P is the original investment (principal), r is the interest rate (as a decimal), and t is the number of years.

If interest is calculated more often than annually, the calculation changes. For example, if interest is calculated monthly, the monthly rate is $1/12$ of the yearly rate. 3 ÷ 12 = 0.25, so the investment after a year would be $100 × 1.0025^{12} = $103.04. If interest is calculated daily, the rate is 3 ÷ 365 = 0.008... and the amount after one year is $100 × $1.00008...^{365}$ = $103.05. The formula for this is $A = P(1 + {}^r/_n)^{nt}$, where n is the number of times the interest is calculated in each year. As the time intervals at which interest is calculated get smaller, the amount of interest yielded at the end of a year approaches $A = Pe^r$. Bernoulli came close to working this out in his calculations, when he identified e as the limit »

Compounding interest yields a bigger total sum. The examples below show how a $10 principal investment accrues interest if the yearly interest rate is 100 percent, versus compound interest paid at shorter intervals.

	1 year, 100% interest rate	6 months, 50% interest rate	3 months, 25% interest rate
January	**$10 principal deposit**	**$10 principal deposit**	**$10 principal deposit**
February			
March			↓
April			$10 × 0.25 = $2.50 $10 + $2.50 = **$12.50**
May			
June		↓	↓
July		$10 × 0.5 = $5 $10 + $5 = **$15**	$12.50 × 0.25 = $3.125 $12.50 + $3.125 = **$15.625**
August			
September			↓
October			$15.625 × 0.25 = $3.906 $15.625 + $3.906 = **$19.531**
November			
December	↓	↓	↓
January	$10 × 1 = $10 $10 + $10 = **$20**	$15 × 0.5 = $7.50 $15 + $7.50 = **$22.50**	$19.531 × 0.25 = $4.883 $19.531 + $4.883 = **$24.41**

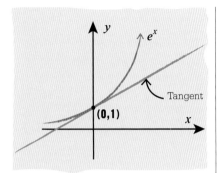

The exponential function can be used to calculate compound interest. The function produces the curve $y = e^x$, which cuts the y axis at (0,1), and gets exponentially steeper. This graph also shows the tangent to the curve.

of $(1 + 1/n)^n$ as n approaches infinity $(n \rightarrow \infty)$. The formula $(1 + 1/n)^n$ gives closer values for e as n increases. For example, $n = 1$ gives a value for e of 2, $n = 10$ gives a value for e of 2.5937… and $n = 100$ gives a value for e of 2.7048….

When Euler calculated a value for e correct to 18 decimal places, he probably used the sequence $e = 1 + 1 + 1/2 + 1/6 + 1/24 + 1/120 + 1/720$, going up to 20 terms. He arrived at these denominators by using the factorial for each integer.

The factorial of an integer is the product of the integer and all the integers below it: 2 (2 × 1), 3 (3 × 2 × 1), 4 (4 × 3 × 2 × 1), 5 (5 × 4 × 3 × 2 × 1) and so on, adding one more term in the product each time. This can be shown as $e = 1 + 1 + 1/2! + 1/3! + 1/4!$ in factorial notation.

Euler calculated e to 18 decimal places, but noted that the decimals continued indefinitely. This means that e is irrational. In 1873, French mathematician Charles Hermite proved that e is also non-algebraic—it is not a number with a terminating decimal that can be used in a regular polynomial equation. This makes it a "transcendental" number—a real number that cannot be computed by solving an equation.

The growth curve

Compound interest is an example of exponential growth. Such growth can be plotted on a graph and will appear as a curve. In the 1600s, English cleric Thomas Malthus posited that population also increases exponentially if there are no checks on its growth, such as war, famine, or food shortages. This means that the population continues to grow at the same

For the sake of brevity, we will always represent this number, 2.718281828… by the letter e.
Leonhard Euler

rate, leading to ever-larger totals. Constant population growth can be calculated with the formula $P = P_0 e^{rt}$ where P_0 is the original population number, r is the growth rate, and t is time.

Plotted on a graph, e shows other special properties. The graph of $y = e^x$ (the exponential function, see left) is a curve whose tangent (the straight line that touches but does not intersect the curve) at the coordinates (0,1) also has a gradient (steepness) of precisely 1. This is because the derivative (rate of change) of e^x is, in fact, e^x, and the derivative is used to find the

The Gateway Arch in St. Louis, Missouri is a flattened catenary arch, designed by Finnish-American architect Eero Saarinen in 1947.

The catenary

Sometimes defined as the shape a hanging chain takes if it is only supported at its ends, a catenary is a curve with the formula $y = 1/2 \times (e^x + e^{-x})$. Catenaries are often found in nature and in technology. For example, a square sail under pressure from the wind takes the form of a catenary. Arches in the shape of an inverted catenary are often used in architecture and construction due to their strength.

For a long time, the catenary's shape was believed to be the same as that of a parabola. Dutch

mathematician Christiaan Huygens—who coined the name catenary from the Latin *catena* ("chain") in 1690—showed that, unlike a parabola, a catenary curve could not be given by a polynomial equation. Three mathematicians—Huygens, Gottfried Leibniz, and Johann Bernoulli—calculated a formula for the catenary, coming to the same conclusion. Their results were published together in 1691. In 1744, Euler described a catenoid—shaped like a waisted cylinder and produced by rotating a catenary around an axis.

tangent. The tangent is used to calculate the rate of change at a specific point on a curve. Because the derivative is e^x, the slope (a measure of direction and steepness) of the tangent line will always be the same as the y value.

Derangements

The various ways in which a set of items can be ordered are called permutations. For example, the set 1, 2, 3 can be arranged as 1, 3, 2, or 2, 1, 3, or 2, 3, 1, or 3, 1, 2, or 3, 2, 1. There are six total ways, including the original, as the number of permutations in a set is equal to the factorial of the highest integer, in this case 3! (short for 3 × 2 × 1). Euler's number is also significant in a type of permutation called a derangement. In a derangement, none of the items can remain in their original position. For four items, the number of possible permutations is 24, but to find the derangements of 1, 2, 3, 4, all other arrangements beginning with 1 must first be eliminated. There are three derangements starting with 2: 2, 1, 4, 3; 2, 3, 4, 1; and 2, 4, 1, 3. There are also three derangements starting with 3 and three starting with 4, making nine in total. With

[Frederick the Great is] always at war; in summer with the Austrians, in winter with mathematicians.
Jean le Rond d'Alembert
French mathematician

five items, the total number of permutations is 120, and with six it is 720, making the task of finding all derangements a substantial one.

Euler's number makes it possible to calculate the number of derangements in any set. This number equals the number of permutations divided by e, rounded to the nearest whole number. For example, for the set of 1, 2, 3, where there are six permutations, $6 \div e = 2.207...$ or 2, to the nearest whole number. Euler analyzed derangements of 10 numbers for Frederick the Great of Prussia, who hoped to create a lottery to pay off his debts. For 10 numbers, Euler found that the probability of getting a derangement is $1/e$ to an accuracy of six decimal places.

Other uses

Euler's number is relevant in many other calculations—for example, in splitting up (partitioning) a number to discover which numbers in the partition have the largest product.

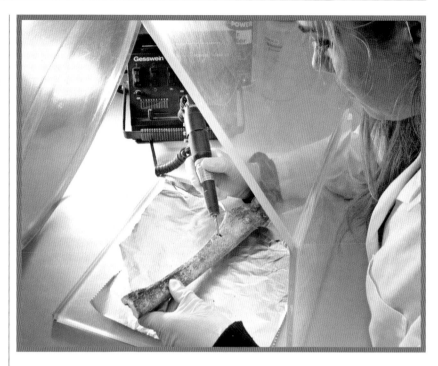

To carbon-date organic material, researchers test a sample—here from an ancient human bone—and use Euler's number to calculate its age from the rate of radioactive decay.

With the number 10, partitions include 3 and 7, with a product of 21; or 6 and 4 to produce 24; or 5 and 5 to give 25, which is the maximum product for a partition of 10 using two numbers. With three numbers, 3, 3, 4 has a product of 36, but moving into fractional numbers, $3^1/3 \times 3^1/3 \times 3^1/3 = {}^{1000}/27 = 37.037...$ the largest for three numbers. For a four-way partition, $2^1/2 \times 2^1/2 \times 2^1/2 \times 2^1/2 = 39.0625$, but in a five-way split, $2 \times 2 \times 2 \times 2 \times 2 = 32$. In short, $({}^{10}/2)^2 = 25$, $({}^{10}/3)^3 = 37.037...$, $({}^{10}/4)^4 = 39.0625$, and $({}^{10}/5)^5 = 32$. This smaller result for a five-way partition suggests that the optimal number of splits for 10 is between 3 and 4. Euler's number can help to find both the maximum product, as $e^{(10/e)} = 39.598...$, and number of partitions: $^{10}/e = 3.678...$. ∎

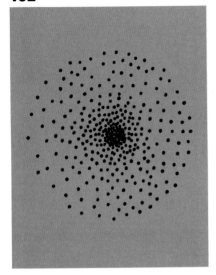

RANDOM VARIATION MAKES A PATTERN
NORMAL DISTRIBUTION

IN CONTEXT

KEY FIGURES
Abraham de Moivre
(1667–1754), **Carl Friedrich
Gauss** (1777–1855)

FIELDS
Statistics, probability

BEFORE
1710 British physician
John Arbuthnot publishes
a statistical proof of divine
providence in relation to the
number of men and women
in a population.

AFTER
1920 Karl Pearson, a British
statistician, expresses regret
about describing the Gaussian
curve as the "normal curve"
because it gives the impression
that all other probability
distributions were "abnormal."

1922 In the US, the New
York Stock Exchange
introduces the use of normal
distribution to model the
risks of investments.

I n the 18th century, French mathematician Abraham de Moivre made an important step forward in statistics; building on Jacob Bernoulli's discovery of binomial distribution, de Moivre showed that events cluster around the mean (**b** on graph below). This phenomenon is known as normal distribution.

Binomial distribution (used to describe outcomes based on one of two possibilities) was first shown by Bernoulli in *Ars Conjectandi* (*The Art of Conjecturing*), published in

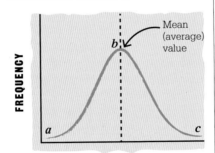

VALUE MEASURED

The bell curve is a visual illustration of normal distribution. The highest point of the curve (**b**) represents the mean, which the values cluster around. Values become less frequent the further they are from the mean, so are least frequent at points **a** and **c**.

1713. When a coin is flipped, there are two possible outcomes: "success" and "failure." This type of test, with two equally likely outcomes, is called a Bernoulli trial. Binomial probabilities arise when a fixed number, **n**, of such Bernoulli trials, each with the same success probability, **p**, are carried out and the total number of successes is counted. The resulting distribution is written as $b(n, p)$. Binomial distribution $b(n, p)$ can take values from 0 to **n**, centered on a mean of **np**.

Finding the mean

In 1721, Scottish baronet Alexander Cuming gave de Moivre a problem concerning the expected winnings in a game of chance. De Moivre concluded that it came down to finding the mean deviation (the average difference between the overall mean and each value in a set of figures) of binomial distribution. He wrote up his results in *Miscellanea Analytica*.

De Moivre had realized that binomial outcomes cluster around their mean—on a graph, they plot an uneven curve that gets closer to the shape of a bell (normal distribution) the more data is

See also: Probability 162–65 ▪ The law of large numbers 184–85 ▪ The fundamental theorem of algebra 204–09 ▪ Laplace's demon 218–19 ▪ The Poisson distribution 220 ▪ The birth of modern statistics 268–71

collected. In 1733, de Moivre was satisfied that he had found a simple way of approximating binomial probabilities using normal distribution, thus creating a bell curve for binomial distribution on a graph. He wrote up his findings as a short paper, then included it in the 1738 edition of his *Doctrine of Chances*.

Using normal distribution

From the mid-1700s, the bell curve cropped up as a model for all kinds of data. In 1809, Carl Friedrich Gauss pioneered normal distribution as a useful statistical tool in its own right. French mathematician Pierre-Simon Laplace used normal distribution to model curves for random errors, such as measurement errors, in one of the first applications of a normal curve.

In the 1800s, many statisticians studied variation in experimental results. British statistician Francis Galton used a device called the quincunx (or Galton board) to study random variation. The

Events cluster around the mean.

Normal distribution applies to continuous **data**, which can take any value within a given range. It produces a **bell curve**.

Binomial distribution applies to **discrete data**, meaning that it has distinct values.

De Moivre says that for a large enough sample size, the normal bell curve can be used to estimate binomial distribution.

board consisted of a triangular array of pegs through which beads dropped from top to bottom, where they collected in a series of vertical tubes. Galton measured how many beads were in each tube and described the resulting distribution as "normal." His work—along with that of Karl

Pearson—popularized the use of the term "normal" to describe what was also known as a "Gaussian" curve.

Today, normal distribution is widely used to model statistical data, with applications ranging from population studies to investment analysis. ▪

Abraham de Moivre

Born in 1667, Abraham de Moivre was raised as a Protestant in Catholic France, and lived there until 1685, when Louis XIV expelled the Huguenots. Briefly imprisoned for his religious beliefs, de Moivre emigrated to England upon his release. He became a private mathematics tutor in London. He had hoped for a university teaching position, but he still faced some discrimination as a Frenchman in England. Nevertheless, de Moivre impressed and befriended many eminent scientists of the time, including Isaac Newton, and was

elected as a fellow of the Royal Society in 1697. As well as his work on distribution, de Moivre was best known for his work on complex numbers. He died in London in 1754.

Key works

1711 *De Mensura Sortis (On the Measurement of Chance)*
1721–30 *Miscellanea Analytica (Miscellany of Analysis)*
1738 *The Doctrine of Chances* (1st edition)
1756 *The Doctrine of Chances* (3rd edition)

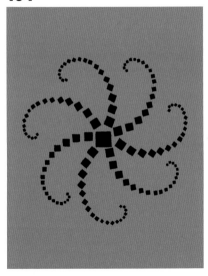

THE SEVEN BRIDGES OF KÖNIGSBERG

GRAPH THEORY

IN CONTEXT

KEY FIGURE
Leonhard Euler (1707–83)

FIELDS
Number theory, topology

BEFORE
1727 Euler develops the constant *e*, which is used in describing exponential growth and decay.

AFTER
1858 August Möbius extends Euler's graph theory formula to surfaces that are joined to form a single surface.

1895 Henri Poincaré publishes his paper *Analysis situs*, in which graph theory is generalized to create a new area of mathematics known as topology (the study of properties of geometrical figures that are not affected by continuous deformation).

Euler's graph theory focuses on the **connections between different points**.

A **graph** consists of a discrete set of **points** (called nodes or vertices) connected by **arcs** (curves or edges).

If a path reaches all nodes, traveling **each arc only once**, it is an **Eulerian** path.

An Eulerian path is impossible in the case of the Königsberg bridges.

Graph theory and topology began with Leonhard Euler's attempt to find a solution to a mathematical puzzle— whether it was possible to make a circuit of the seven bridges in Königsberg (now Kaliningrad, Russia) without crossing any bridge twice. The river flowed around an island and then forked. Realizing that the problem related to the geometry of position, Euler developed a new type of geometry to show that it was impossible to devise such a route. Distances between points were not relevant: the only thing that counted was the connections between points.

Euler modeled the Königsberg bridges problem by making each of the four land areas a point (node or vertex) and making the bridges arcs (curves or edges) that joined the various points. This gave him

See also: Coordinates 144–51 ▪ Euler's number 186–91 ▪ The complex plane 214–15 ▪ The Möbius strip 248–49
▪ Topology 256–59 ▪ The butterfly effect 294–99 ▪ The four-color theorem 312–13

Read Euler, read Euler.
He is our master
in everything.
Pierre-Simon Laplace

a "graph" that represented the relationships between the land and the bridges.

First graph theorem

Euler began from the premise that each bridge could be crossed only once and each time a land area was entered it also needed to be exited, which required two bridges in order to avoid crossing any bridge twice. Each land area therefore needed to connect to an even number of bridges, with the possible exception of the start and finish (if they were different locations). However, in the graph representing Königsberg (see right), A is the endpoint of five bridges and B, C, and D are each the endpoint of three. A successful route needs land areas (nodes or vertices) to have an even number of bridges (arcs) to enter and exit by. Only the start and end points can have an odd number. If more than two nodes have an odd number of arcs, then a route using each bridge only once is impossible. By showing this, Euler provided the first theorem in graph theory.

The word "graph" is most often used to describe a Cartesian system of coordinates with points plotted using x and y axes. More generally, a graph consists of a discrete set of nodes (or vertices) connected by arcs (or edges). The number of arcs meeting at a node is called its degree. For the Königsberg graph, A has degree 5 and B, C, and D each have degree 3. A path that travels each arc once and only once is called an Eulerian path (or a semi-Eulerian path if the start and end are at different nodes).

The Königsberg bridges problem can be expressed as the question: "Is there an Eulerian or a semi-Eulerian path for the graph of Königsberg?" Euler's answer is that such a graph must have at most two nodes of odd degree, but the Königsberg graph has four odd degree nodes.

Network theory

Arcs on a graph may be "weighted" (given degrees of significance) by assigning numerical values to them—for example, to represent the different lengths of roads on a map. A weighted graph is also called a network. Networks are used to model relationships between objects in many disciplines—including computer science, particle physics, economics, cryptography, sociology, biology, and climatology—usually with a view to optimizing a particular property, such as the shortest distance between two points.

One application of networks is to address the so-called "traveling salesperson problem." This involves finding the shortest route for a salesperson to travel from their home to a series of cities and back again. The puzzle was allegedly first set as a challenge on the back of a cereal box. In spite of advances in computing, no method exists that guarantees to always find the best solution, because the time this takes grows exponentially as the given number of cities increases. ▪

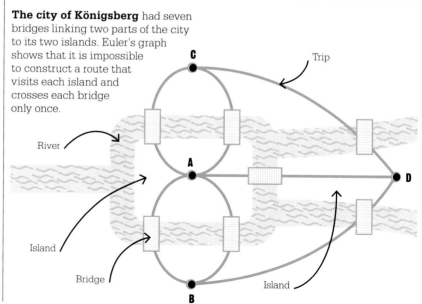

The city of Königsberg had seven bridges linking two parts of the city to its two islands. Euler's graph shows that it is impossible to construct a route that visits each island and crosses each bridge only once.

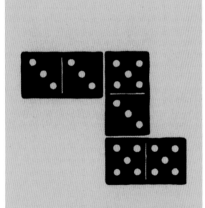

EVERY EVEN INTEGER IS THE SUM OF TWO PRIMES

THE GOLDBACH CONJECTURE

IN CONTEXT

KEY FIGURE
Christian Goldbach
(1690–1764)

FIELD
Number theory

BEFORE
c. 200 CE Diophantus of
Alexandria writes his
Arithmetica in which he lays
out key issues about numbers.

1202 Fibonacci identifies
what becomes known as
the Fibonacci sequence
of numbers.

1643 Pierre de Fermat
pioneers number theory.

AFTER
1742 Leonhard Euler refines
the Goldbach conjecture.

1937 Soviet mathematician
Ivan Vinogradov proves the
ternary Goldbach problem,
a version of the conjecture.

In 1742, Russian mathematician Christian Goldbach wrote to Leonhard Euler, the leading mathematician of the time. Goldbach believed he had observed something remarkable—that every even integer can be split into two prime numbers, such as 6 (3 + 3) or 8 (3 + 5). Euler was convinced that Goldbach was right, but he could not prove it. Goldbach also proposed that every odd integer above 5 is the sum of three primes, and concluded that every integer from 2 upward can be created by adding together primes; these additional proposals are dubbed "weak" versions of the original "strong" conjecture, as they would follow naturally if the strong conjecture were true.

Manual and electronic methods have, as yet, failed to find any even number that does not conform to the original strong conjecture. In 2013, a computer tested every even number up to 4×10^{18} without finding one. The bigger the number, the more pairs of primes can create it, so it seems highly likely that the conjecture is valid and no exception will be found. Mathematicians, however, require a definitive proof.

Over centuries, different "weak" versions of the conjecture have been proved, but no one to date has proved the strong conjecture, which seems destined to defeat even the brightest minds. ∎

UCLA's Terence Tao, winner of the Fields Medal in 2006 and the Breakthrough Prize in mathematics in 2015, published a rigorous proof of a weak Goldbach conjecture in 2012.

See also: Mersenne primes 124 ▪ The law of large numbers 184–85 ▪ The Riemann hypothesis 250–51 ▪ The prime number theorem 260–61

THE MOST BEAUTIFUL EQUATION

EULER'S IDENTITY

IN CONTEXT

KEY FIGURE
Leonhard Euler (1707–83)

FIELD
Number theory

BEFORE
1714 Roger Cotes, the English mathematician who proofread Newton's *Principia*, creates an early formula similar to Euler's, but using imaginary numbers and a complex logarithm (a type of logarithm used when the base is a complex number).

AFTER
1749 Abraham de Moivre uses Euler's formula to prove his theorem, which links complex numbers and trigonometry.

1934 Soviet mathematician Alexander Gelfond shows that e^π is transcendental, that is, irrational and still irrational when raised to any power.

Formulated by Leonhard Euler in 1747, the equation known as Euler's identity, $e^{i\pi} + 1 = 0$, encompasses the five most important numbers in mathematics: 0 (zero), which is neutral for addition and subtraction; 1, which is neutral for multiplication and division; e (2.718..., the number at the heart of exponential growth and decay); i ($\sqrt{-1}$, the fundamental imaginary number); and π (3.142..., the ratio of a circle's circumference to its diameter, which occurs in many equations in mathematics and physics). Two of these numbers, e and i, were introduced by Euler himself. His genius lay in combining all five milestone numbers with three simple operations: raising a number to a power (for example, 5^4, or $5 \times 5 \times 5 \times 5$), multiplication, and addition.

Complex powers

Mathematicians such as Euler asked themselves if it would be meaningful to raise a number to a complex power—a complex number being a number that combines a

It is simple... yet incredibly profound; it comprises the five most important mathematical constants.
David Percy
British mathematician

real number with an imaginary one, such as $a + bi$, where a and b are any real numbers. When Euler raised the constant e to the power of the imaginary number i multiplied by π, he discovered that it equals -1. Adding 1 to both sides of the equation produces Euler's identity, $e^{i\pi} + 1 = 0$. The equation's simplicity has led mathematicians to describe it as "elegant," a description reserved for proofs that are profound yet also unusually succinct. ■

See also: Calculating pi 60–65 ▪ Trigonometry 70–75 ▪ Imaginary and complex numbers 128–31 ▪ Logarithms 138–41 ▪ Euler's number 186–91

NO THEORY IS PERFECT

BAYES' THEOREM

IN CONTEXT

KEY FIGURE
Thomas Bayes (1702–61)

FIELD
Probability

BEFORE
1713 Jacob Bernoulli's
Ars Conjectandi (*The Art of
Conjecturing*), published after
his death, sets out his new
mathematical theory
of probability.

1718 Abraham de Moivre
defines the statistical
independence of events in his
book *The Doctrine of Chances*.

AFTER
1774 In his *Memoir on the
Probability of the Causes of
Events*, Pierre-Simon Laplace
introduces the principle of
inverse probability.

1992 The International Society
for Bayesian Analysis (ISBA)
is founded to promote the
application and development
of Bayes' theorem.

Bayes' theorem is used to **calculate probabilities** of events
based on prior knowledge.

Conditions related to the event can help us **assess its
probability more accurately**.

**The theorem can be used to predict more
accurately the likelihood of future events.**

In 1763, Richard Price, a Welsh minister and mathematician, published a paper called "An Essay Towards Solving a Problem in the Doctrine of Chances." Its author, the Reverend Thomas Bayes, had died two years earlier, leaving the paper to Price in his will. It was a breakthrough in the modeling of probability and is still used today in areas as diverse as locating lost aircraft and testing for disease.

Jacob Bernoulli's book *Ars Conjectandi* (1713) showed that as the number of identically distributed, randomly generated variables increases, so their observed average gets closer to their theoretical average. For example, if you toss a coin for long enough, the number of times it comes up heads will get closer and closer to half the total of tosses —a probability of 0.5.

In 1718, Abraham de Moivre grappled with the mathematics underpinning probability. He demonstrated that, provided the sample size was large enough, the distribution of a continuous random variable—people's heights, for example—averaged out into a bell-

See also: Probability 162–65 ▪ The law of large numbers 184–85 ▪ Normal distribution 192–93 ▪ Laplace's demon 218–19 ▪ The Poisson distribution 220 ▪ The birth of modern statistics 268–71 ▪ The Turing machine 284–89 ▪ Cryptography 314–17

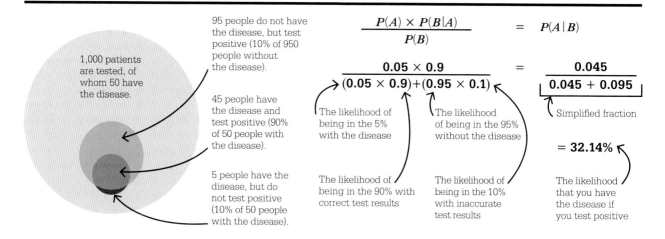

1,000 patients are tested, of whom 50 have the disease.

95 people do not have the disease, but test positive (10% of 950 people without the disease).

45 people have the disease and test positive (90% of 50 people with the disease).

5 people have the disease, but do not test positive (10% of 50 people with the disease).

$$\frac{P(A) \times P(B|A)}{P(B)} = P(A|B)$$

$$\frac{0.05 \times 0.9}{(0.05 \times 0.9)+(0.95 \times 0.1)} = \frac{0.045}{0.045 + 0.095}$$

The likelihood of being in the 5% with the disease

The likelihood of being in the 95% without the disease

Simplified fraction

The likelihood of being in the 90% with correct test results

The likelihood of being in the 10% with inaccurate test results

$$= 32.14\%$$

The likelihood that you have the disease if you test positive

If a disease affects 5 percent of the population (event A) and is diagnosed using a test with 90 percent accuracy (event B), you might assume that the probability (P) of having the disease if you test positive—$P(A|B)$—is 90 percent. However, Bayes' theorem factors in the false results produced by the test's 10 percent inaccuracy—$P(B)$.

shaped curve, later named the "normal distribution" by German mathematician Carl Gauss.

Working out probabilities

Most real-world events, however, are more complicated than the toss of a coin. For probability to be useful, mathematicians needed to determine how an event's outcome could be used to draw conclusions about the probabilities that led to it. This reasoning based on the causes of observed events—rather than using direct probabilities, such as the 50 percent chance of a heads coin toss—became known as inverse probability. Problems that deal with the probabilities of causes are called inverse probability problems and might involve, for example, observing a bent coin landing on heads 13 times out of 20 and then trying to determine whether the probability of that coin landing on heads lies somewhere between 0.4 and 0.6.

To show how to calculate inverse probabilities, Bayes considered two interdependent events—"event A" and "event B". Each has a probability of occurring—$P(A)$ and $P(B)$—with P for each being a number between 0 and 1. If event A occurs, it alters the probability of event B happening, and vice versa. To denote this, Bayes introduced "conditional probabilities." These are given as $P(A|B)$, the probability of A given B, and $P(B|A)$, the probability of B given A. Bayes managed to solve the problem of how all four probabilities related to one another with the equation: $P(A|B) = P(A) \times P(B|A)/P(B)$. ▪

Thomas Bayes

The son of a Nonconformist minister, Thomas Bayes was born in 1702 and grew up in London. He studied logic and theology at the University of Edinburgh and followed his father into the ministry, spending much of his life leading a Presbyterian chapel in Tunbridge Wells, Kent.

Although little is known of Bayes' life as a mathematician, in 1736 he anonymously published *An Introduction to the Doctrine of Fluxions, and a Defence of the Mathematicians Against the Objections of the Author of the Analyst*, in which he defended Isaac Newton's calculus foundations against the criticisms of the philosopher Bishop George Berkeley. Bayes was made a fellow of the Royal Society in 1742 and died in 1761.

Key works

1736 *An Introduction to the Doctrine of Fluxions, and a Defence of the Mathematicians Against the Objections of the Author of the Analyst*

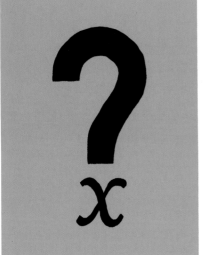

SIMPLY A QUESTION OF ALGEBRA
THE ALGEBRAIC RESOLUTION OF EQUATIONS

IN CONTEXT

KEY FIGURE
Joseph-Louis Lagrange
(1736–1813)

FIELD
Algebra

BEFORE
628 Brahmagupta publishes a formula for solving many quadratic equations.

1545 Gerolamo Cardano creates formulae for resolving cubic and quartic equations.

1749 Leonhard Euler proves that polynomial equations of degree n have exactly n complex roots (where n = 2, 3, 4, 5, or 6).

AFTER
1799 Carl Gauss publishes the first proof of the fundamental theorem of algebra.

1824 In Norway, Niels Henrik Abel completes Paolo Ruffini's 1799 proof that there is no general formula for the quintic equation.

Polynomial equations involving numbers and a single unknown quantity (x, and powers of x such as x^2 and x^3) are a powerful tool for solving real-world problems. An example of a polynomial equation is $x^2 + x + 41 = 0$. While such equations can be solved approximately by repeated numerical calculations, solving them exactly (algebraically) was not achieved until the 1700s. The quest led to many mathematical innovations, including new types of numbers—such as negative and complex numbers—as well as modern algebraic notation and group theory.

Searching for solutions

The Babylonians and ancient Greeks used geometrical methods to solve problems that are now usually expressed by quadratic equations. In medieval times, more abstract algorithmic approaches

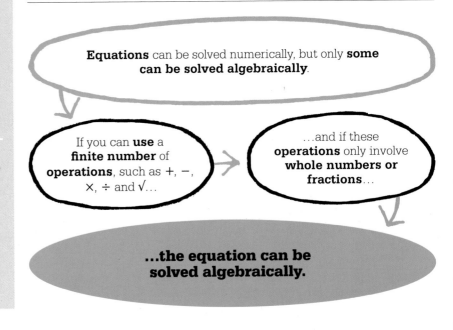

Equations can be solved numerically, but only **some can be solved algebraically**.

If you can **use a finite number of operations**, such as +, −, ×, ÷ and √…

…and if these **operations** only involve **whole numbers or fractions**…

…the equation can be solved algebraically.

See also: Quadratic equations 28–31 ▪ Algebra 92–99 ▪ The binomial theorem 100–01 ▪ Cubic equations 102–05
▪ Huygens's tautochrone curve 167 ▪ The fundamental theorem of algebra 204–09 ▪ Group theory 230–33

were established, and by the 1500s, mathematicians knew certain relations between the coefficients of a polynomial equation and its roots, and had devised formulas to solve cubic (highest power 3) and quartic equations (highest power 4). In the 1600s, a general theory of polynomial equations, now called the fundamental theorem of algebra, took shape. This stated that an equation of degree n (where the highest power of x is x^n) has exactly n roots or solutions, which may be real or complex numbers.

Roots and permutations

In his *Reflections on the algebraic resolution of equations* (1771), French-Italian mathematician Joseph-Louis Lagrange introduced a general approach for solving polynomial equations. His work was theoretical—he investigated the structure of polynomial equations to find the circumstances under which a formula could be found for solving them. Lagrange combined the technique of using

These are the coefficients of the equation.

The fundamental theorem of algebra says that a cubic equation has three solutions—which are three numbers which, when each is used to replace x, make the equation equal to zero.

$$\mathbf{m}x^3 + \mathbf{n}x^2 + \mathbf{p}x + \mathbf{q} = 0$$

The highest power in this equation is x^3, so it is a cubic equation.

x is the variable of the equation.

An algebraic equation is made up of variables and coefficients. The highest power of the equation determines how many solutions it has: in this case, there are three solutions.

a related, lower-degree polynomial equation whose coefficients were related to the coefficients of the original equation with a striking innovation—he considered the possible permutations (reorderings) of the roots. Lagrange's insight into the symmetries that arose from these permutations showed why the cubic and quartic equations could be solved by formulas, and showed (due to the different permutations of symmetries and roots) why a formula for the quintic equation needed a different approach.

Within 20 years of Lagrange's work, Italian mathematician Paolo Ruffini began to prove that there was no general formula for the quintic equation. Lagrange's investigation into permutations (and symmetries) formed the basis of the even more abstract and general group theory advanced by French mathematician Évariste Galois, who used it to prove why it is impossible to resolve equations of degree 5 or higher algebraically—that is, why there is no general formula for solving such equations. ▪

Joseph-Louis Lagrange

Born Giuseppe Lodovico Lagrangia in Turin in 1736, Lagrange embraced his family's French heritage and went by the French version of his name. As a young mathematician, self-taught, he worked on the tautochrone problem and developed a new formal method to find the function that solved such problems. At the age of 19, he wrote to Leonhard Euler, who recognized his talent. Lagrange applied his method, which Euler named the calculus of variations, to study a wide range of physical phenomena, including the vibration of strings. In 1766, at

Euler's recommendation, he was made Director of Mathematics at the Berlin Academy, and in 1787 he moved to the Académie des Sciences in Paris. Despite being an academic and a foreigner, Lagrange survived the French Revolution and Reign of Terror, and died in Paris in 1813.

Key works

1771 *Reflections on the algebraic resolution of equations*
1788 *Analytical Mechanics*
1797 *Theory of analytic functions*

LET US GATHER FACTS
BUFFON'S NEEDLE EXPERIMENT

IN CONTEXT

KEY FIGURE
**Georges Louis Leclerc,
Comte de Buffon** (1707–88)

FIELD
Probability

BEFORE
1666 *Liber de ludo aleae* (*On
Games of Chance*) by Italian
mathematician Gerolamo
Cardano is published.

1718 Abraham de Moivre
publishes *The Doctrine of
Chances*, the first textbook
on probability.

AFTER
1942–46 The Manhattan
Project, a US-led body for
developing nuclear weapons,
makes extensive use of Monte
Carlo methods (computational
processes that model risk by
generating random variables).

Late 1900s Quantum Monte
Carlo methods are used to
explore particle interactions
in microscopic systems.

In 1733, the mathematician and naturalist George Leclerc, the Comte de Buffon, raised—and answered—a fascinating question. If a needle is dropped onto a series of parallel lines, all the same width apart, what is the likelihood that the needle will cross one of the lines? Now known as Buffon's needle experiment, it was one of the earliest probability calculations.

An elegant illustration

Buffon originally used the needle experiment to estimate π (pi)—the ratio of a circle's circumference to its diameter. He did this by dropping a needle of length l many times onto a series of parallel lines distance d apart, where d is greater than the needle's length ($d > l$). Buffon then counted the number of times the needle crossed the line as a proportion of total attempts (p) and came up with the formula that π is approximately equal to twice the length of the needle l, divided by the distance (d) multiplied by the proportion of needles crossing the line: $\pi \approx 2l \div dp$. The probability of the needle crossing a line can be calculated by multiplying each side of the formula by p, then dividing each side by π to get $p \approx 2l \div \pi d$.

The relationship with π can be used in a number of probability problems. One example involves a quarter circle, inscribed in a square, which curves from the top left corner of the square to the bottom right (see below). The bottom horizontal edge of the square is the x axis and the left vertical edge is the y axis, with a value of 0 in the lower left corner and 1 in the corners at each end of the curve. When two numbers between 0 and 1 are chosen at random as the x and y coordinates, whether the point will lie inside the quarter circle (success) or outside it (failure) can be deduced by examining $\sqrt{a^2 + b^2}$, where a is the x coordinate

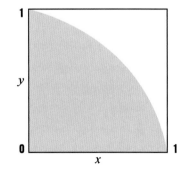

Using pi, the probability of a randomly chosen point falling within the quarter circle in this square can be calculated as roughly 78 percent.

See also: Calculating pi 60–65 ▪ Probability 162–65 ▪ The law of large numbers 184–85 ▪ Bayes' theorem 198–99 ▪ The birth of modern statistics 268–71

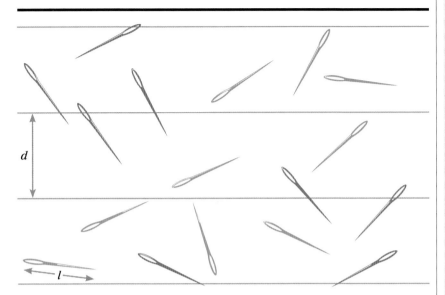

d = distance between the lines
l = length of the needle

Buffon's needle experiment demonstrated how probability can be connected to pi. Buffon classed needles as "successful" (pink) if they crossed a line when dropped, or "unsuccessful" (blue) if they didn't, then calculated the probability of "success."

and b is the y coordinate. The result is > 1 for points outside the curve and < 1 for points within it. The point is chosen at random, so could be anywhere in the square. Points on the line of the quarter circle can be counted as a success. The chance of "success" is πr^2 (the area of a circle) ÷ 4. If the radius is 1, $r^2 = 1$, so the area is just π; for a quarter circle, divide π by 4 to get approximately 0.78. The whole area is the area of the square, which is $1 \times 1 = 1$, so the probability of landing in the shaded area is approximately 0.78 ÷ 1 = 0.78.

The Monte Carlo method

This problem is an example of a wider class of experiments that employ a statistical approach called the Monte Carlo method, a code name coined by Polish-American scientist Stanislaw Ulam and his colleagues for the random sampling used during secret work on nuclear weapons in World War II. Monte Carlo methods went on to be useful in modern applications, especially once computers made it far less time-consuming to repeat a probability experiment over and over again. ▪

In wind energy yield analysis, the predicted energy output of a wind farm during its lifetime is calculated, giving different levels of uncertainty, by using Monte Carlo probability methods.

Georges-Louis Leclerc, Comte de Buffon

Born in Montbard, France, in 1707, Georges-Louis Leclerc was urged by his parents to pursue a career in law, but was more interested in botany, medicine, and mathematics, which he studied at the University of Angers, France. At the age of 20, he explored the binomial theorem.

Independently wealthy, Buffon was able to write and study tirelessly, corresponding with many of the scientific elite of his day. His interests were wide-ranging, and his output was immense—on topics ranging from ship-building to natural history and astronomy. The comte also translated a number of scientific works.

Appointed keeper of the Jardin du Roi, the royal botanical gardens in Paris, in 1739, Buffon enriched its collections and doubled its size. He held the post until his death in Paris in 1788.

Key works

1749–1786 *Histoire naturelle* (*Natural History*)
1778 *Les époques de la nature* (*The Epochs of Nature*)

ALGEBRA OFTEN GIVES MORE THAN IS ASKED OF HER

THE FUNDAMENTAL THEOREM OF ALGEBRA

IN CONTEXT

KEY FIGURE
Carl Gauss (1777–1855)

FIELD
Algebra

BEFORE
1629 Albert Girard states that a polynomial of degree **n** will have **n** roots.

1746 The first attempt at a proof of the fundamental theorem of algebra (FTA) is made by Jean d'Alembert.

AFTER
1806 Robert Argand publishes the first rigorous proof of the FTA that allows polynomials with complex coefficients.

1920 Alexander Ostrowski proves the remaining assumptions in Gauss's proof of the FTA.

1940 Hellmuth Kneser gives the first constructive variant of the Argand FTA proof that allows for the roots to be found.

This method of solving problems by honest confession of one's ignorance is called Algebra.
Mary Everest Boole
British mathematician

A n equation asserts that one quantity is equal to another, and provides a means of determining an unknown number. Since Babylonian times, scholars have sought solutions to equations, periodically encountering seemingly insoluble examples. In the 5th century BCE, Hippasus's attempts to resolve $x^2 = 2$ and his realization that $\sqrt{2}$ was irrational (neither a whole number nor a fraction) are said to have led to his death for betraying Pythagorean beliefs. Some 800 years later, Diophantus had no knowledge of negative numbers, so could not accept an equation where x is negative, such as $4 = 4x + 20$, where x is -4.

Polynomials and roots

In the 1700s, one of the most studied areas of mathematics involved polynomial equations. These are often used to solve problems in mechanics, physics, astronomy, and engineering, and involve powers of an unknown value, such as x^2. The "root" of a polynomial equation is a specific numerical value that will replace the unknown value to make the polynomial equal 0. In 1629, French mathematician Albert Girard

Gerolamo Cardano encountered negative roots while working on cubic equations in the 1500s. His acceptance of these as valid solutions was an important step in algebra.

A **polynomial equation** is an expression made up of **variables** (such as x and y) and **coefficients** (such as 4) along with operations (such as $+$ and $-$) to form an equation (such as $x^2 + 4x - 12 = y$).

A **root is a number that replaces a variable (such as $x = -6$) to make the equation equal zero.**

All polynomial equations have **roots** that are either **real** or **complex**.

This is known as the **fundamental theorem of algebra** (FTA).

See also: Quadratic equations 28–31 ▪ Negative numbers 76–79 ▪ Algebra 92–99 ▪ Cubic equations 102–05 ▪ Imaginary and complex numbers 128–31 ▪ The algebraic resolution of equations 200–01 ▪ The complex plane 214–15

Finding the root(s) of an equation

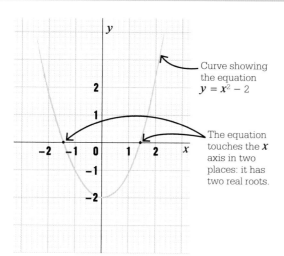

Curve showing the equation $y = x^2 - 2$

The equation touches the x axis in two places: it has two real roots.

An equation of power 2, such as $y = x^2 - 2$, always has two real or complex roots.

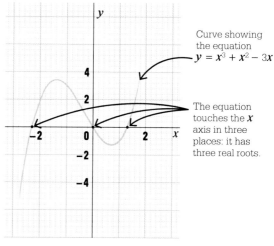

Curve showing the equation $y = x^3 + x^2 - 3x$

The equation touches the x axis in three places: it has three real roots.

An equation of power 3, such as $y = x^3 + x^2 - 3x$, always has three real or complex roots.

showed that a polynomial of degree n will have n roots. The quadratic equation $x^2 + 4x - 12 = 0$, for example, has two roots, $x = 2$ and $x = -6$, both of which produce the answer 0. It has two roots because of the x^2 term – 2 is the equation's highest power. If any quadratic equation is drawn on a graph (as shown above), these roots can be easily found: they are where the line touches the x axis. Although his theorem was useful, Girard's work was hindered by the fact that he had no concept of complex numbers. These would be key to producing a fundamental theorem of algebra (FTA) for solving all possible polynomials.

Complex numbers

The collection of all positive and negative, rational and irrational numbers together make up the real numbers. Some polynomials,

however, do not have real-number roots. This was a problem faced by Italian mathematician Gerolamo Cardano and his peers in the 1500s; while working on cubic equations, they found that some of their solutions involved square roots of negative numbers. This seemed impossible, because a negative number multiplied by itself produces a positive result.

The problem was solved in 1572 when another Italian, Rafael Bombelli, set out the rules for an extended number system that included numbers such as $\sqrt{-1}$ alongside the real numbers. In 1751, Leonhard Euler investigated the imaginary roots of polynomials, and called $\sqrt{-1}$ the "imaginary unit," or i. All imaginary numbers are multiples of i. Combining the real and the imaginary, such as $a + bi$ (where a and b are any real numbers, and $i = \sqrt{-1}$), creates

what is called a complex number. Once mathematicians had accepted the necessity of negative and complex numbers for solving certain equations, the question remained as to whether finding roots of higher-degree polynomials would require the introduction of yet new types of number. Euler and other mathematicians, »

Imaginary numbers are a fine and wonderful refuge of the divine spirit.
Gottfried Leibniz

Carl Gauss

Born in Brunswick, Germany, in 1777, Carl Gauss showed his mathematical talents early: aged only three, he corrected an error in his father's payroll calculations, and by the age of five he was taking care of his father's accounts. In 1795, he entered Göttingen University and in 1798, he constructed a regular heptadecagon (a polygon with 17 sides) using only a ruler and compasses—the biggest advance in polygon construction since Euclid's geometry some 2,000 years earlier. Gauss's *Arithmetical Investigations*, written at the age of 21 and published in 1801, was key to defining number theory. Gauss also made advances in astronomy (such as the rediscovery of the astroid Ceres), cartography, the study of electromagnetism, and the design of optical instruments. However, he kept many of his ideas to himself; a great number were discovered in his unpublished papers after his death in 1855.

Key work

1801 *Disquisitiones Arithmeticae* (*Arithmetical Investigations*)

notably Carl Gauss in Germany, would seek to address this question, eventually concluding that the roots of any polynomial are either real or complex numbers—no further types of number are needed.

Early research

The FTA can be stated in a number of ways, but its most common formulation is that every polynomial with complex coefficients will have at least one complex root. It can also be stated that all polynomials of degree n containing complex coefficients have n complex roots.

The first significant attempt at proving the FTA was made in 1746 by French mathematician Jean le Rond d'Alembert in his "Recherches sur le calcul intégral" ("Research on integral calculus"). D'Alembert's proof argued that if a polynomial $P(x)$ with real coefficients has a complex root, $x = a + ib$, then it also has a complex root, $x = a - ib$. To prove this theorem, he used a complicated idea now known as "d'Alembert's lemma." In mathematics, a lemma is an intermediary proposition used to solve a bigger theorem. However, d'Alembert did not prove his lemma satisfactorily; his proof was correct, but contained too many holes to

There are only two kinds of certain knowledge: awareness of our own existence and the truths of mathematics.
Jean d'Alembert

Jean d'Alembert was the first to attempt to prove the FTA. In France, it is called the d'Alembert–Gauss theorem, acknowledging the influence of d'Alembert on Gauss.

satisfy his fellow mathematicians. Later attempts to prove the FTA included those of Leonhard Euler and Joseph-Louis Lagrange. While useful to later mathematicians, these were also unsatisfactory. In 1795, Pierre-Simon Laplace tried an FTA proof using the polynomial's "discriminant," a parameter determined from its coefficients which indicates the nature of its roots, such as real or imaginary. His proof contained an unproved assumption that d'Alembert had avoided—that a polynomial will always have roots.

Gauss's proof

In 1799, at the age of 21, Carl Friedrich Gauss published his doctoral thesis. It began with a summary and criticism of d'Alembert's proof, among others. Gauss pointed out that each of these earlier proofs had assumed part of what they were trying to prove. One such assumption was that polynomials of odd degree (such as cubics and quintics) always have a real root. This is true, but Gauss argued that the point needed to be proved. His first proof was based on assumptions about algebraic curves. Although these were plausible, they were not rigorously proved in Gauss's work. It was not until 1920, when Ukrainian mathematician Alexander Ostrowski published his proof, that Gauss's assumptions could all be justified. Arguably, Gauss's first, geometric proof suffered for being premature—in 1799, the concepts of continuity and of the complex plane, which would have helped him explain his ideas, had not yet been developed.

Argand's additions

Gauss published an improved proof of the FTA in 1816 and a further refinement at his golden jubilee lecture (celebrating 50 years since his doctorate) in 1849. Unlike his first geometric approach, his second and third proofs were more algebraic and technical in nature. Gauss published four proofs of the FTA, but did not fully resolve the problem. He failed to address the obvious next step: although he had

established that every real number equation would have a complex number solution, he had not considered equations built from complex numbers such as $x^2 = i$.

In 1806, Swiss mathematician Jean-Robert Argand found a particularly elegant solution. Any complex number, z, can be written in the form $a + bi$, where a is the real part of z and bi is the imaginary part. Argand's work allowed complex numbers to be represented geometrically. If the real numbers are drawn along the x axis and the imaginary numbers are drawn along the y axis, then the whole plane between them becomes the realm of the complex numbers. Argand proved that the solution for every equation built from complex numbers could be found among the complex numbers on his diagram and that there was therefore no need to extend the number system. Argand's was the first truly rigorous proof of the FTA.

Legacy of the theorem

The proofs by Gauss and Argand established the validity of complex numbers as roots of polynomials. The FTA stated that anyone faced

> I have had my results for a long time, but I do not yet know how to arrive at them.
> **Carl Gauss**

with solving an equation built from real numbers could be sure that they would find their solution within the realm of complex numbers. These groundbreaking ideas formed the foundations of complex analysis.

Mathematicians since Argand have continued to work on proving the FTA using new methods. In 1891, for example, German Karl Weierstrass created a method—now known as the Durand–Kerner method, due to its rediscovery by mathematicians in the 1960s—for simultaneously finding all of the roots of a polynomial. ∎

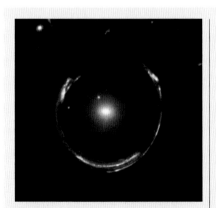

An Einstein ring, first discovered in 1998, is the deformation of light from a source into a ring through gravitational lensing.

Applications of the FTA

Research on the fundamental theorem of algebra has led to breakthroughs in other fields. In the 1990s, British mathematicians Terrence Sheil-Small and Alan Wilmshurst extended the FTA to harmonic polynomials. These may have an infinite number of roots, but in some cases, there are a finite number. In 2006, American mathematicians Dmitry Khavinson and Genevra Neumann proved that there was an upper limit to the number of roots of a certain class of harmonic polynomials.

After publishing their results, they were told that their proof settled a conjecture by South Korean astrophysicist Sun Hong Rhie. Her conjecture concerned images of distant astronomical light sources. Massive objects in the Universe bend passing light rays from distant sources in a phenomenon called gravitational lensing, creating multiple images seen through a telescope. Rhie posited that there would be a maximum number of images produced; this turned out to be exactly the upper bound found by Khavinson and Neumann.

THE 19
CENTU
1800–1900

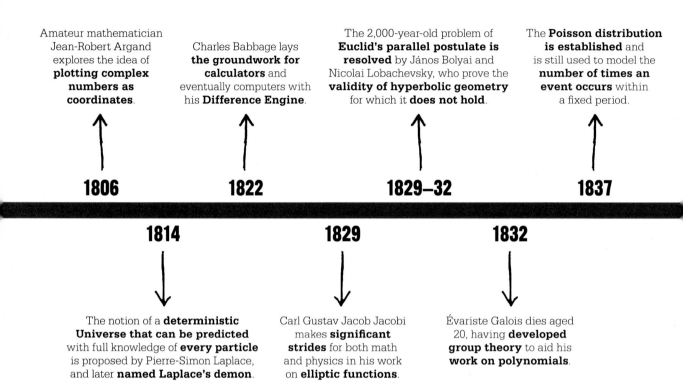

Amateur mathematician Jean-Robert Argand explores the idea of **plotting complex numbers as coordinates**.

Charles Babbage lays **the groundwork for calculators** and eventually computers with his **Difference Engine**.

The 2,000-year-old problem of **Euclid's parallel postulate is resolved** by János Bolyai and Nicolai Lobachevsky, who prove the **validity of hyperbolic geometry** for which it **does not hold**.

The **Poisson distribution is established** and is still used to model the **number of times an event occurs** within a fixed period.

1806   **1822**   **1829–32**   **1837**

1814   **1829**   **1832**

The notion of a **deterministic Universe that can be predicted** with full knowledge of **every particle** is proposed by Pierre-Simon Laplace, and later **named Laplace's demon**.

Carl Gustav Jacob Jacobi makes **significant strides** for both math and physics in his work on **elliptic functions**.

Évariste Galois dies aged 20, having **developed group theory** to aid his **work on polynomials**.

Progress in mathematics accelerated through the 1800s, with science and mathematics now becoming respected academic studies. As the Industrial Revolution spread and 1848's "Year of Revolution" saw nationalism surge across old empires, there was a renewed drive to understand the workings of the Universe in scientific terms, rather than through religion or philosophy. Pierre-Simon Laplace, for example, applied the theories of calculus to celestial mechanics. He proposed a form of scientific determinism, saying that with the relevant knowledge of moving particles, the behavior of everything in the Universe could be predicted.

Another characteristic of 19th-century mathematics was its increasing tendency toward the theoretical. This trend was fostered by the influential work of Carl Friedrich Gauss, regarded by many in the field as the greatest of all mathematicians. He dominated the study of mathematics for much of the first half of the century, making contributions to the fields of algebra, geometry, and number theory, and giving his name to such concepts as Gaussian distribution, Gaussian function, Gaussian curvature, and Gaussian error curve.

New fields

Gauss was also a pioneer of non-Euclidean geometries, which epitomized the revolutionary spirit of 19th-century mathematics. The subject was taken up by Nicolai Lobachevsky and János Bolyai, who independently developed theories of hyperbolic geometry and curved spaces, resolving the problem of Euclid's parallel postulate. This opened up a completely new approach to geometry, paving the way for the nascent field of topology (the study of space and surfaces) which was also influenced by the possibility of more than three dimensions offered by William Hamilton's discovery of quaternions.

Perhaps the best known of the pioneers of topology is August Möbius, inventor of the Möbius strip, which had the unusual property of being a two-dimensional surface with only one side. Non-Euclidean geometries were further developed by Bernhard Riemann, who identified and defined different types of geometry in multiple dimensions.

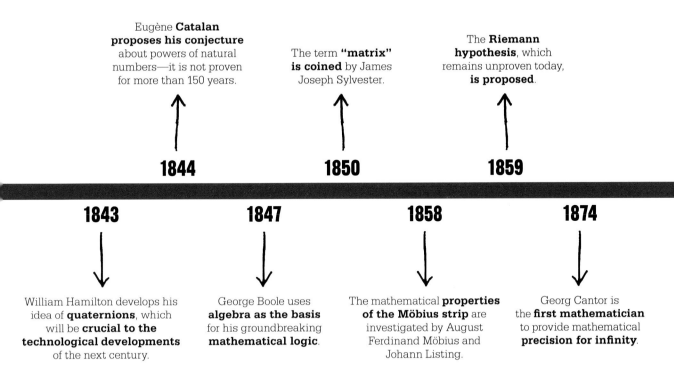

Eugène **Catalan proposes his conjecture** about powers of natural numbers—it is not proven for more than 150 years.

The term **"matrix" is coined** by James Joseph Sylvester.

The **Riemann hypothesis**, which remains unproven today, **is proposed**.

1844　　**1850**　　**1859**

1843　　**1847**　　**1858**　　**1874**

William Hamilton develops his idea of **quaternions**, which will be **crucial to the technological developments** of the next century.

George Boole uses **algebra as the basis** for his groundbreaking **mathematical logic**.

The mathematical **properties of the Möbius strip** are investigated by August Ferdinand Möbius and Johann Listing.

Georg Cantor is the **first mathematician** to provide mathematical **precision for infinity**.

Riemann did not confine his studies to geometry, however. As well as his work on calculus, he made important contributions to number theory, following in the footsteps of Gauss. The Riemann hypothesis, derived from the Riemann zeta function concerning complex numbers, is as yet unsolved. Other notable discoveries in number theory at this time include the creation of set theory and the description of an "infinity of infinities" of Georg Cantor, Eugène Catalan's conjecture about the powers of natural numbers, and the application of elliptic functions to number theory proposed by Carl Gustav Jacob Jacobi.

Jacobi was, like Riemann, multi-talented, often linking different fields of mathematics in new ways. His primary interest was in algebra, another area of mathematics that was becoming increasingly abstract during the 1800s. The groundwork for the growing field of abstract algebra was laid by Évariste Galois, who, although he died young, also developed group theory while determining a general algebraic method for solving polynomial equations.

New technologies

Not all mathematics in this period was purely theoretical—and even some of the abstract concepts soon found more practical applications. Siméon Poisson, for example, used his knowledge of pure mathematics to develop ideas such as the Poisson distribution, a key concept in the field of probability theory. Charles Babbage, on the other hand, responded to practical demand for a means of accurate and quick calculation with his mechanical calculating device, the "Difference Engine." In so doing, he laid the groundwork for the invention of computers. Babbage's work in turn inspired Ada Lovelace to devise the forerunner of modern computer algorithms.

Meanwhile, there were other developments in mathematics that were to have far-reaching implications for later technological progress. Using algebra as his starting point, George Boole devised a form of logic based on a binary system, and using the operators AND, OR, and NOT. These became the foundation of modern mathematical logic, but just as importantly paved the way for the language of computers almost a century later. ∎

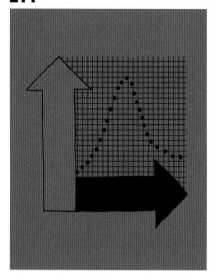

COMPLEX NUMBERS ARE COORDINATES ON A PLANE
THE COMPLEX PLANE

IN CONTEXT

KEY FIGURE
Jean-Robert Argand
(1768–1822)

FIELD
Number theory

BEFORE
1545 Italian scholar Gerolamo uses negative square roots to solve cubic equations in his book *Ars Magna*.

1637 French philosopher and mathematician René Descartes develops a way to plot algebraic expressions as coordinates on a grid.

AFTER
1843 Irish mathematician William Hamilton extends the complex plane by adding two more imaginary units to create quaternions—expressions that are plotted in a 4-D space.

1859 By merging two complex planes, Bernhard Riemann develops a 4-D surface to help him analyze complex functions.

Some equations **cannot be solved** without using **complex numbers**.

↓

Complex numbers have **two components**: a **real number** and an **imaginary number**.

↓

Real numbers (−1, 0, 1, etc.) are traditionally expressed on a horizontal **number line**.

→

Imaginary numbers can be plotted on a line perpendicular to the number line—the two lines make an *x* and *y* axis.

↓

This creates a plane of complex numbers, with real numbers plotted along the *x* axis, and imaginary numbers along the *y* axis.

After centuries of suspicion, mathematicians finally embraced the concept of negative numbers in the 1700s. They did so by using imaginary numbers in algebra. In 1806, the key contribution of Swiss-born mathematician Jean-Robert Argand was to plot complex numbers (made up of a real and imaginary component) as coordinates on a plane created by two axes—*x* for real numbers and *y* for imaginary numbers. This complex plane

See also: Quadratic equations 28–31 ▪ Cubic equations 102–05 ▪ Imaginary and complex numbers 128–31 ▪ Coordinates 144–51 ▪ The fundamental theorem of algebra 204–09

There can be very little… science and technology that is not dependent on complex numbers.
Keith Devlin
British mathematician

provided the first geometrical interpretation of the distinctive properties of complex numbers.

Algebraic roots

Imaginary numbers had emerged in the 1500s when Italian mathematicians such as Gerolamo Cardano and Niccolò Fontana Tartaglia found that solving cubic equations required a square root of a negative number. The square of a real number cannot be negative— any real number multiplied by itself results in a positive—so they decided to treat $\sqrt{-1}$ as a new unit that operated separately from the real numbers. Leonhard Euler first used i to denote the imaginary unit ($\sqrt{-1}$) in his attempts to prove the fundamental theorem of algebra (FTA). This theorem states that all polynomial equations of degree n have n roots. This means that if x^2 is the highest power in an algebraic expression made up of a single variable (such as x) and real coefficients (numbers multiplying the variable), the expression has a degree of two, and also two roots; roots are values of x that make the

polynomial equal to zero. Many seemingly simple polynomials, however, such as $x^2 + 1$, do not equal zero if x is a real number. Plotting $x^2 + 1$ on a graph with an x and y axis creates a neat curve that never passes through the origin, or (0,0) point. To make the FTA work for $x^2 + 1$, Gauss and others used real numbers combined with imaginary numbers to create complex numbers. All numbers are in essence complex. For example, the real number 1 is the complex number $1 + 0i$, while the number i is $0 + i$. The equation $x^2 + 1$ can equal zero when x is i or $-i$.

Argand's discovery

As Argand began to plot complex numbers, he discovered that the imaginary number i does not get bigger if raised to higher powers. Instead, it follows a four-step pattern that repeats infinitely: $i^1 = i$; $i^2 = -1$; $i^3 = -i$, $i^4 = 1$; $i^5 = i$, and so on. This can be visualized on the complex plane. Multiplying real numbers by imaginary numbers produces 90° rotations through the complex

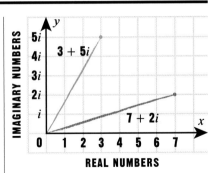

An Argand diagram uses the x and y axes to represent real numbers and imaginary numbers, combining them to plot complex numbers. This diagram shows two numbers: $3 + 5i$ and $7 + 2i$.

plane. So $1 \times i = i$, which does not appear on the real number x axis at all, but on the imaginary y axis. Continuing to multiply by i results in more 90° rotations, which is why every four multiplications arrive back at the start point.

Plots of complex numbers— or Argand diagrams—make complicated polynomials easier to solve. The complex plane is now a powerful tool that works far beyond the interests of number theory. ▪

Jean-Robert Argand

Little is known of Jean-Robert Argand's early life. He was born in Geneva in 1768, but appears to have had no formal education in mathematics. In 1806, he moved to Paris to manage a bookshop, and self-published the work containing the geometrical interpretation of complex numbers for which he is known. (Norwegian cartographer Casper Wessel is now known to have used similar constructions in 1799.) Argand's essay was republished in a mathematics

journal in 1813, and in the next year, he used the complex plane to produce the first rigorous proof of the fundamental theorem of algebra. Argand published eight more articles before his death in Paris in 1822.

Key work

1806 *Essai sur une manière de représenter les quantités imaginaires dans les constructions géométriques* (*Essay on a method of representing imaginary quantities geometrically*)

NATURE IS THE MOST FERTILE SOURCE OF MATHEMATICAL DISCOVERIES
FOURIER ANALYSIS

IN CONTEXT

KEY FIGURE
Joseph Fourier (1768–1830)

FIELD
Applied mathematics

BEFORE
1701 In France, Joseph Sauveur suggests that vibrating strings oscillate with many waves of different lengths at the same time.

1753 Swiss mathematician Daniel Bernoulli shows that a vibrating string consists of an infinite number of harmonic oscillations.

AFTER
1965 In the US, James Cooley and John Tukey develop the Fast Fourier Transform (FFT), an algorithm that is able to speed up Fourier analysis.

2000s Fourier analysis is used to create a number of speech recognition programs for computers and smartphones.

The sound created by vibrating strings has been a topic of research for more than 2,500 years. In about 550 BCE, Pythagoras discovered that if you take two taut strings of the same material and the same tension, but one is twice the length of the other, the short string will vibrate with twice the frequency of the longer string and the resulting notes will be an octave apart.

Two centuries later, Aristotle suggested that sound traveled through the air in waves, although he incorrectly thought that higher-pitched sounds traveled faster than lower-pitched ones. In the 1600s, Galileo recognized that sounds are produced by vibrations: the higher the frequency of the vibrations, the higher the pitch of the sound we perceive.

Heat and harmony
By the end of the 1600s, physicists including Joseph Sauveur were making great strides in studying the relationships between the waves in stretched strings and the pitch and frequency of sounds that they produced. In the course of their research, mathematicians showed that any string will

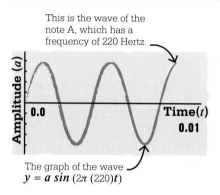

This is the wave of the note A, which has a frequency of 220 Hertz.

The graph of the wave $y = a \sin (2\pi (220)t)$

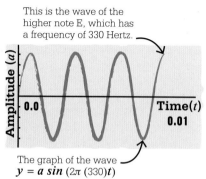

This is the wave of the higher note E, which has a frequency of 330 Hertz.

The graph of the wave $y = a \sin (2\pi (330)t)$

Sounds are made of a complex series of tones. Fourier analysis can separate out pure tones, which can be represented as sine waves on a graph, from each other. Tones have frequency, which determines pitch, and amplitude, which determines volume.

See also: Pythagoras 36–43 ▪ Trigonometry 70–75 ▪ Bessel functions 221 ▪ Elliptic functions 226–27 ▪ Topology 256–59 ▪ The Langlands Program 302–03

Joseph Fourier

Jean-Baptiste Joseph Fourier was born in Auxerre, France, in 1768. A tailor's son, he went to military school, where his keen interest in mathematics led him to become a successful teacher of the subject.

Fourier's career was disrupted by two arrests—one for criticizing the French Revolution, the other for supporting it—but in 1798, he accompanied Napoleon's forces into Egypt as a diplomat. Napoleon later made him a baron, and then a count. After Napoleon's fall in 1815, Fourier moved to Paris to become director of the Statistical Bureau of the Seine, where he pursued his studies in mathematical physics, including work on the Fourier series (a series of sine waves that characterize sounds). In 1822, Fourier was made the secretary of the French Academy of Sciences, a post he held until his death in 1830. Fourier is one of 72 scientists whose names are inscribed on the Eiffel Tower.

Key works

1822 *Théorie analytique de la chaleur* (*The Analytical Theory of Heat*)

support a potentially infinite series of vibrations, starting from the fundamental (the string's lowest natural frequency) and including its harmonics (integer multiples of the fundamental). The pure tone of a single pitch is produced by a smooth repetitive oscillation called a sine wave (see graph). The sound quality of a musical instrument results principally from the number and relative intensities of the harmonics present in the sound, or its harmonic content. The result is a variety of waves interfering with each other.

Joseph Fourier was attempting to solve the problem of how heat diffused through a solid object. He developed an approach that would allow him to calculate the temperature at any location within

an object, at any time after a source of heat had been applied to one of its edges.

Fourier's studies of heat distribution showed that no matter how complex a waveform, it could be broken down into its constituent sine waves, a process that is now called Fourier analysis. Since heat in the form of radiation is a wave, Fourier's discoveries about heat distribution had applications to the study of sound. A sound wave can be understood in terms of the amplitudes of its constituent sine waves, a set of numbers that is sometimes referred to as the harmonic spectrum.

Today, Fourier analysis plays a key role in many applications including digital file compression, analyzing MRI scans, speech recognition software, musical pitch correction software, and determining the composition of planetary atmospheres. ▪

Fourier analysis of the way materials vibrate allows engineers to construct buildings that resonate at different frequencies from a typical earthquake and thus avoid the kind of damage that occurred in Mexico City in 2017.

THE IMP THAT KNOWS THE POSITIONS OF EVERY PARTICLE IN THE UNIVERSE
LAPLACE'S DEMON

In 1814, Pierre-Simon Laplace, a French mathematician who combined mathematics and science with philosophy and politics, presented a thought experiment now known as Laplace's demon. Laplace never used the word "demon" himself; it was introduced in later retellings, evoking a supernatural being made godlike by mathematics.

Laplace imagined an intellect that could analyze movements of all atoms in the Universe in order to accurately predict their future paths. His experiment was an exploration of determinism, a philosophical concept that says that all future events are determined by causes in the past.

Mechanical analysis

Laplace was inspired by classical mechanics—a field of mathematics describing the behavior of moving bodies, based on Isaac Newton's laws of motion. In a Newtonian universe, atoms (and even light particles) follow the laws of motion, and bounce around in a jumble of trajectories. Laplace's "intellect" would be capable of capturing and analyzing all of their movements; it would create a single formula that uses present movements to ascertain past and predict future ones.

Laplace's theory had a startling philosophical consequence. It can only work if the Universe follows a predictable mechanical path, so that everything from the spin of galaxies to the tiny atoms in nerve cells controlling thoughts could be mapped out into the future. This

The orrery, a "clockwork universe" showing the movement of the celestial bodies in the Solar System, became a popular device after the publication of Newton's universal theory of gravity.

See also: Probability 162–65 ■ Calculus 168–75 ■ Newton's laws of motion 182–83 ■ The butterfly effect 294–99

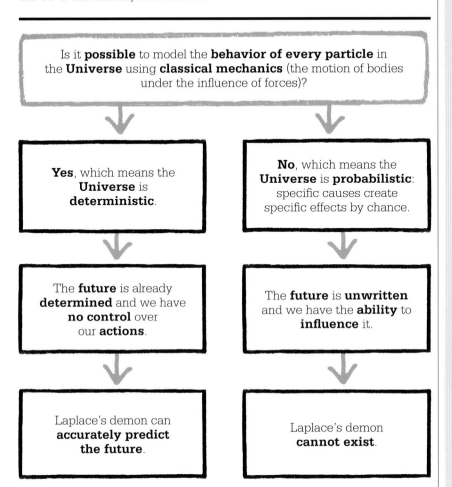

Is it **possible** to model the **behavior of every particle** in the **Universe** using **classical mechanics** (the motion of bodies under the influence of forces)?

Yes, which means the **Universe** is **deterministic**.

No, which means the **Universe** is **probabilistic**: specific causes create specific effects by chance.

The **future** is already **determined** and we have **no control** over our **actions**.

The **future** is **unwritten** and we have the **ability** to **influence** it.

Laplace's demon can **accurately predict the future**.

Laplace's demon **cannot exist**.

Pierre-Simon Laplace

Born into an aristocratic family in 1749, Laplace lived through the French Revolution and the Reign of Terror, in which many of his friends were killed. In 1799, he became Minister of the Interior under Napoleon Bonaparte, but was dismissed after only six weeks for being too analytical and ineffectual. Laplace later sided with the Bourbons (the French royal family) and was rewarded with the return of his original title of marquis when the monarchy was restored.

Laplace's demon was a side note to a career that also encompassed physics and astronomy, where Laplace was the first to postulate the concept of a black hole. His many contributions to mathematics were in classical mechanics, probability theory, and algebraic transformations. Laplace died in Paris in 1827.

Key works

1798–1828 *Celestial Mechanics*
1812 *Analytic Theory of Probability*
1814 *A Philosophical Essay on Probabilities*

would mean that every aspect of a person's life up until their death has already been predetermined; they have no free will and no agency over their thoughts and deeds.

Chance and statistics

Although mathematics helped to create such a crushing vision of reality, it also helped to dismiss it. By the 1850s, the study of heat and energy—thermodynamics—was ushering in a new model, the atomic world. To do this, it needed to describe the motion of atoms and molecules inside matter. Classical mechanics was not up to the task.

Instead, physicists used a technique invented by Swiss mathematician Daniel Bernoulli in 1738, which used probability theory to model the movement of independent units within a space. Refined by Austrian physicist Ludwig Boltzmann, this technique became known as statistical mechanics. It described the atomic world in terms of random chance—something at odds with the mechanical determinism of Laplace's demon. By the 1920s, the idea of a probabilistic Universe was solidified with the development of quantum physics, which has uncertainty at its heart. ■

WHAT ARE THE CHANCES?
THE POISSON DISTRIBUTION

IN CONTEXT

KEY FIGURE
Siméon Poisson (1781–1840)

FIELD
Probability

BEFORE
1662 English merchant John Graunt publishes *Natural and Political Observations upon the Bills of Mortality*, marking the birth of statistics.

1711 Abraham de Moivre's *De Mensura Sortis* (*On the Measurement of Chance*), describes what is later known as the Poisson distribution.

AFTER
1898 Russian statistician Ladislaus Bortkiewicz uses the Poisson distribution to study the number of Prussian soldiers killed by horse kicks.

1946 British statistician R. D. Clarke publishes a study, based on the Poisson distribution, of patterns of V-1 and V-2 flying bomb impacts on London.

In statistics, the Poisson distribution is used to model the number of times a randomly occurring event happens in a given interval of time or space. Introduced in 1837 by French mathematician Siméon Poisson, and based on the work of Abraham de Moivre, it can help to forecast a wide range of possibilities.

Take, for example, a chef who needs to forecast the number of baked potatoes that will be ordered in her café. She needs to decide how many potatoes to pre-cook each day. She knows the daily average order, and decides to prepare n potatoes where there is at least 90 percent certainty that n will match demand.

To use the Poisson distribution to calculate n, conditions must be met: orders must occur randomly, singly, and uniformly—on average, the same number of potatoes are ordered each day. If these conditions apply, the chef can find the value of n—how many potatoes to pre-bake. The average number of events per unit of space or time (lambda, or λ) is key. If $\lambda = 4$ (the average number of potatoes ordered in one day), and the number of potato orders on any one day is B, the probability that B is less than or equal to 6 is 89 percent, while the probability that B is less than or equal to 7 is 95 percent. The chef must be at least 90 percent sure that demand will be met, so n will be 7 here. ∎

Siméon Poisson is credited with finding the Poisson distribution, but this may be an example of Stigler's Law—no scientific discovery is credited to the true discoverer.

See also: Probability 162–65 ▪ Euler's number 186–91 ▪ Normal distribution 192–93 ▪ The birth of modern statistics 268–71

AN INDISPENSABLE TOOL IN APPLIED MATHEMATICS
BESSEL FUNCTIONS

IN CONTEXT

KEY FIGURE
Friedrich Wilhelm Bessel
(1784–1846)

FIELD
Applied geometry

BEFORE
1609 Johannes Kepler shows that the orbits of the planets are ellipses.

1732 Daniel Bernoulli uses what later become known as Bessel functions to study the vibrations of a swinging chain.

1764 Leonhard Euler analyzes a vibrating membrane using what are later understood to be Bessel functions.

AFTER
1922 British mathematician George Watson writes his hugely influential *A treatise on the theory of Bessel functions.*

I n the early 1800s, German mathematician and astronomer Friedrich Wilhelm Bessel gave solutions to a particular differential equation, the so-called Bessel equation. He systematically investigated these functions (solutions) in 1824. Now known as Bessel functions, they are useful to scientists and engineers. Central to the analysis of waves, such as electromagnetic waves moving along wires, they are also used to describe the diffraction of light, the flow of electricity or heat in a solid cylinder, and the motions of fluids.

Movement of the planets

The origins of Bessel functions lie in the pioneering work of German mathematician and astronomer Johannes Kepler in the early 1600s on the motions of the planets. His meticulous analysis of observations led him to realize that the orbits of the planets around the Sun are elliptical, not circular, and he described the three key laws of planetary motion. Mathematicians later used Bessel functions to make

Bessel's functions are very beautiful functions, in spite of their having practical applications.
E. W. Hobson
British mathematician

breakthroughs in various fields. Daniel Bernoulli found equations for the oscillations of a pendulum, and Leonhard Euler developed corresponding equations for the vibration of a stretched membrane. Euler and others also used Bessel functions to find solutions to the "three-body problem," concerned with the motion of a body, such as a planet or moon, being acted upon by the gravitational fields of two other bodies. ■

See also: The problem of maxima 142–43 ▪ Calculus 168–75 ▪ The law of large numbers 184–85 ▪ Euler's number 186–91 ▪ Fourier analysis 216–17

IT WILL GUIDE THE FUTURE COURSE OF SCIENCE

THE MECHANICAL COMPUTER

IN CONTEXT

KEY FIGURES
Charles Babbage (1791–1871),
Ada Lovelace (1815–52)

FIELD
Computer science

BEFORE
1617 Scottish mathematician
John Napier invents a manual
calculating device.

1642–44 In France,
Blaise Pascal creates a
calculating machine.

1801 French weaver Joseph-
Marie Jacquard demonstrates
the first programmable
machine – a loom controlled
by a punchcard.

AFTER
1944 British codebreaker
Max Newman builds Colossus,
the first digital electronic
programmable computer.

British mathematician and
inventor Charles Babbage
anticipated the computer
age by more than a century
with two ideas for mechanical
calculators and "thinking"
machines. The first he called the
Difference Engine, a calculating
machine that would work
automatically, using a combination
of brass cogs and rods. Babbage
only managed to part-build the
machine, but even this was able
to process complex calculations
accurately in moments.

The second, more ambitious,
idea was the Analytical Engine. It
was never built, but was envisaged
as a machine that could respond
to new problems and solve them
without human intervention.

The project received crucial input from Ada Lovelace, a brilliant young mathematician. Lovelace anticipated many of the key mathematical aspects of computer programming and foresaw how the machine could be used to analyze any kinds of symbol.

Automatic calculation

In the 17th and 18th centuries, mathematicians such as Gottfried Leibniz and Blaise Pascal had created mechanical calculating aids, but these were limited in power and also prone to error as human input was needed at every step. Babbage's idea was to create a calculating machine that worked automatically, eliminating human error. He called his machine the Difference Engine because it allowed complex multiplications and divisions to be reduced to additions and subtractions—"differences"—that could be handled by scores of interlocking cogs. It would even print out the results.

No previous calculator had ever worked with numbers larger than four digits. Yet the Difference

Charles Babbage was spurred to start his work on a mechanical calculator by the errors he found in astronomical tables produced by poorly paid and unreliable workers.

Engine was designed to handle numbers of up to 50 digits by means of more than 25,000 moving parts.

To set up the machine for a calculation, each number was represented by a column of cogwheels, and each cogwheel was marked with digits from 0 to 9. A number was set by turning the cogwheels in the column to show the correct digit on each. The machine would then work through the entire calculation automatically.

Babbage built several small working models with just seven number columns but remarkable

At each increase of knowledge, as well as on the contrivance of every new tool, human labor becomes abridged.
Charles Babbage

calculating power. In 1823, he managed to persuade the British government to part-fund the project, with the promise that it would make producing official tables much quicker, cheaper, »

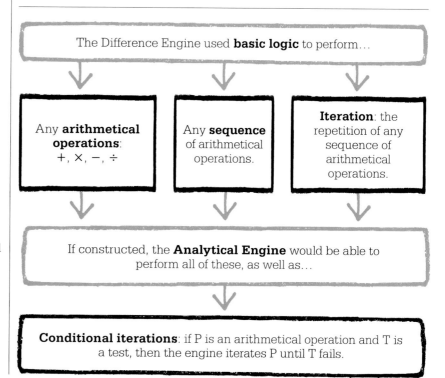

The Difference Engine used **basic logic** to perform…

| Any **arithmetical operations**: +, ×, −, ÷ | Any **sequence** of arithmetical operations. | **Iteration**: the repetition of any sequence of arithmetical operations. |

If constructed, the **Analytical Engine** would be able to perform all of these, as well as…

Conditional iterations: if P is an arithmetical operation and T is a test, then the engine iterates P until T fails.

and more accurate. However, the full machine was hugely expensive to develop, and tested the technological capability of the day to its limits. After two decades' work, the government canceled the project in 1842.

Meanwhile, in drawings and calculations, Babbage had also been working on his idea for an

This replica of the demonstration model Babbage made in 1832 of Difference Engine No. 1 has three columns, each with its numbered cogwheels. Two are for calculation, one for the result.

Analytical Engine. His papers suggest that the machine, if built, could have been close to what we now call a computer. His design anticipated virtually all of the key components of the modern computer, including the central processing unit (CPU), memory storage, and integrated programs.

One problem facing Babbage was what to do with numbers carried over into the next column when adding up columns of digits. At first, he used a separate mechanism for each carryover, but that proved too complicated. Then he split his machine into two parts, the "Mill" and the "Store," which made it possible to separate the addition and carryover

processes. The Mill was where the arithmetical operations were performed; the Store was where numbers were held before processing and then received back from the Mill after processing. The Mill was Babbage's version of a computer's CPU, while the Store acted as its memory.

The idea of telling a machine what it should do—programming— came from a French weaver, Joseph-Marie Jacquard. He developed a loom that used cards punched with holes to tell it how to weave complex patterns in silk. In 1836, Babbage realized he too could use punched cards—to control his own machine but also to record results and calculation sequences.

A supporting genius
One of the greatest advocates for Babbage's work was his fellow mathematician Ada Lovelace, who wrote of the Analytical Engine that it would "weave algebraic patterns

Types of punched card to program the Analytical Engine

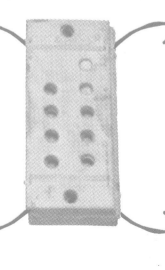

Number cards
specified the value of numbers entered into the Store, or received numbers back from the Store for external storage.

Variable cards
specified which data—held in "axes," or storage units— should be fed into the Mill, and where returned data should be stored.

Operational cards
determined the arithmetical operations to be performed by the Mill.

Combinatorial cards
controlled how variable cards and operational cards turned backward or forward after specific operations were completed.

> The object of the Analytical Engine is twofold. First, the complete manipulation of number. Second, the complete manipulation of algebraical symbols.
> **Charles Babbage**

just as the Jacquard loom weaves flowers and leaves." As a teenager in 1832, Lovelace had seen one of the Difference Engine models working and had been instantly entranced. In 1843, she arranged the publication of her translation of a pamphlet about the Analytical Engine written by Italian engineer Luigi Menabrea, to which she added extensive explanatory notes.

Many of these notes covered systems that would become part of modern computing. In "Note G," Lovelace described possibly the first computer algorithm, "to show an implicit function can be worked out by the engine without human head and hands first." She also theorized that the engine could solve problems by repeating a series of instructions—a process known today as "looping." Lovelace envisaged a program card, or set of cards, that returned repeatedly to its original position to work on the next data card or set. In this way, Lovelace argued, the machine could solve a system of linear equations or generate extensive tables of prime numbers. Perhaps the greatest insight in her notes was Lovelace's vision of machines

as mechanical brains with wide applications. "The engine can arrange and combine its numerical quantities exactly as if they were letters or any other general symbols," she wrote, realizing that any kind of symbol, not just numbers, could be manipulated and processed by machines. This is the difference between calculation and computation—and the basis of the modern computer. Lovelace also foresaw how such machines would be limited by the quality of the input. Arguably, the first programmable computer—rather than calculator—was created by Konrad Zuse in 1938.

Delayed legacy

Lovelace's plans to develop Babbage's work were curtailed by her early death, by which time Babbage himself was tired, ill, and disillusioned by the lack of support for his Difference Engine. The high-precision mechanics required to build the machine were beyond what any engineer could achieve at the time. Largely forgotten until they were republished in 1953, Lovelace's notes confirm that she and Babbage foresaw many of the features of the computer now found in every home and office. ∎

> The more I study [the Analytical Engine], the more insatiable I feel my genius for it to be.
> **Ada Lovelace**

Ada Lovelace

Born Augusta Byron in London in 1815, Ada, Countess of Lovelace, was the only legitimate child of the poet Lord Byron. Byron left England a few months after her birth, and Lovelace never saw her father again. Her mother, Lady Byron, was mathematically gifted—Byron called her his "Princess of Parallelograms"— and insisted Lovelace study mathematics, too.

Lovelace became renowned for her talents in mathematics and languages. She met Charles Babbage when she was 17 and was fascinated by his work. Two years later, she married William King, Earl of Lovelace, with whom she had three children, but she continued to study mathematics and follow the progress of Babbage, who called her "the Enchantress of Number."

Lovelace wrote exhaustive notes on Babbage's Analytical Engine. She set out many ideas about what was to become computing, earning herself a reputation as the first computer programmer. Lovelace died in 1852 of uterine cancer; in line with her wishes, she was buried next to her father.

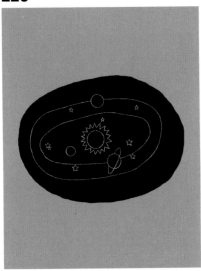

A NEW KIND OF FUNCTION

ELLIPTIC FUNCTIONS

IN CONTEXT

KEY FIGURE
Carl Gustav Jacob Jacobi
(1804–51)

FIELDS
Number theory, geometry

BEFORE
1655 John Wallis applies
calculus to the length of an
elliptic curve; the elliptic
integral he derives is defined
by an infinite series of terms.

1799 Carl Gauss determines
the key characteristics of
elliptic functions, but his work
is not published until 1841.

1827–28 Niels Abel
independently derives and
publishes the same findings
as Gauss.

AFTER
1862 German mathematician
Karl Weierstrass develops
a general theory of elliptic
functions, showing that they
can be applied to problems in
both algebra and geometry.

Physics—to
calculate the charge
of a particle from its
curved path through
a magnetic field.

Astronomy—
the orbits of planets
are elliptical.

Mechanics—to
make calculations about
the motion of
pendulums.

Some **uses**
for **elliptic
functions**
include…

Trigonometry—
functions in spherical
trigonometry based on
the circle are special cases
of elliptic functions.

Cryptography—
to obscure the keys
involved in encrypting
private information
in public.

The "squashed circle" of
an ellipse is one of the
most recognizable curves
in math. Ellipses have a long
history in mathematics. They were
studied by the ancient Greeks as
one of the conic sections. Slicing
through a cone horizontally creates
a circle; slicing at a steeper angle
creates an ellipse (and then open
curves called a parabola and
a hyperbola). An ellipse is a
closed curve that is defined as
the set of all points in a plane, the
sum of whose distances from two
fixed points—each one called a
focus—is always the same number.
(A circle is a special ellipse with

See also: Huygens's tautochrone curve 167 ▪ Calculus 168–75 ▪ Newton's laws of motion 182–83 ▪ Cryptography 314–17 ▪ Proving Fermat's last theorem 320–23

> I learnt with as much astonishment as satisfaction that two young geometers… succeeded in their own individual work in considerably improving the theory of elliptic functions.
> **Adrien-Marie Legendre**

just one central focus, not two.) In 1609, German astronomer and mathematician Johannes Kepler demonstrated that the orbits of the planets were elliptical, with the Sun being located at one of the foci.

New tools

Just as the mathematics of a circle could be used to model and predict natural phenomena that varied and repeated in a rhythmic (or periodic) way, such as the up-and-down motion of a simple sound wave, the mathematics of the ellipse can be used to do the same for phenomena that follow more complex periodic patterns, such as electromagnetic fields or the orbital motion of planets.

The genesis of such tools, the elliptic functions, began in England with 17th-century mathematicians John Wallis and Isaac Newton. Working independently, they developed a method for calculating the arc length, or length of a section, of any ellipse. With later contributions, their technique was developed into the elliptic functions and became a way of analyzing many kinds of complex curves and oscillating systems beyond the simple ellipse.

Practical applications

In 1828, Norwegian Neils Abel and German Carl Jacobi, again working independently, showed wider applications for elliptic functions in both mathematics and physics. For example, these functions appear in the 1995 proof of Fermat's last theorem, and the latest public-key cryptography systems. Since Abel died at 26, just months after making his major discoveries, many of these applications were developed by Jacobi. Jacobi's elliptic functions are complex, but a more simple form, the *p*-function, was introduced in 1862 by German mathematician Karl Weierstrass. *P*-functions are used in classical and quantum mechanics. ▪

Elliptic functions are used to define the trajectories of spacecraft such as the Dawn probe, which explored the dwarf planet Ceres and the asteroid Vesta in the asteroid belt.

Carl Gustav Jacob Jacobi

Born in Potsdam, Prussia, in 1804, Carl Gustav Jacob Jacobi was initially tutored by an uncle. Having learned all that school could teach him by the age of 12, he had to wait until he was 16 to be allowed to attend Berlin University, and spent the intervening years teaching himself mathematics. He continued to do so when he found the university courses too basic. He graduated within a year, and in 1832 he became a professor at the University of Königsberg. Falling ill in 1843, Jacobi returned to Berlin, where he was supported by a pension from the king of Prussia. In 1848, he ran unsuccessfully for parliament as a liberal candidate and the offended king temporarily withdrew his support. In 1851, aged just 46, Jacobi contracted smallpox and died.

Key work

1829 *Fundamenta nova theoria functionum ellipticarum* (*The foundations of a new theory of elliptic functions*)

I HAVE CREATED ANOTHER WORLD OUT OF NOTHING

NON-EUCLIDEAN GEOMETRIES

IN CONTEXT

KEY FIGURE
János Bolyai (1802–60)

FIELD
Geometry

BEFORE
1733 In Italy, mathematician Giovanni Saccheri fails to prove Euclid's parallel postulate from his other four postulates.

1827 Carl Friedrich Gauss publishes his *Disquisitiones generales circa superficies curvas* (*General Investigations of Curved Surfaces*), defining the "intrinsic curvature" of a space, which can be deduced from within the space.

AFTER
1854 Bernhard Riemann describes the kind of surface that has hyperbolic geometry.

1915 Einstein describes gravity as curvature in spacetime in his general theory of relativity.

The parallel postulate (PP) is the fifth of five postulates from which Euclid deduced his theorems of geometry in his *Elements*. The PP was controversial among the ancient Greeks, since it did not seem as self-evident as Euclid's other postulates, nor was there an obvious way of verifying it. However, without the PP, many elementary theorems in geometry could not be proved. Over the next 2,000 years, mathematicians would stake their reputations on attempts to resolve the issue. In the 5th century CE, the philosopher Proclus

Euclidean and non-Euclidean geometries

In Euclidean geometry (see right) the surface is assumed to be flat. In non-Euclidean forms of geometry (see below), this is not the case. In hyperbolic geometry, the surface curves inward like a saddle, while an elliptic surface curves outward like a sphere.

The parallel postulate (PP) can be expressed by Scottish mathematician John Playfair's Axiom: given a plane containing a line A and a point P not on A, there exists exactly one line B through P that does not intersect A. These lines A and B are parallel.

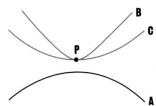

In hyperbolic geometry, there are infinitely many lines (e.g. B and C) through point P that do not intersect line A. Surfaces in hyperbolic geometry exhibit "negative curvature"—for example, the bell of a trumpet.

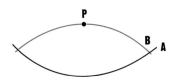

In elliptic geometry, such as on the surface of a sphere, the PP does not hold and every line (e.g. B) through point P intersects line A. For example, Earth's meridians are parallel lines that intersect at the poles.

> Leave the science of parallels alone. I was ready to… remove the flaw from geometry [but] turned back when I saw that no man can reach the bottom of this night.
> **Wolfgang Bolyai**
> Father of János Bolyai

argued that the PP was a theorem that could be derived from the other postulates and should therefore be struck out.

During the Golden Age of Islam (8th–14th century), mathematicians attempted to prove the PP. Persian polymath Nasir al-Din al-Tusi showed that the PP is equivalent to stating that the sum of angles in any triangle is 180°, but the PP nonetheless remained controversial. In the 1600s, new translations of *Elements* reached Europe, and Giovanni Saccheri showed that if the PP was untrue, then the sum of angles in a triangle was always either less than or greater than 180°.

By the early 1800s, Hungarian János Bolyai and Russian Nicolai Lobachevsky independently proved the validity of a "hyperbolic" non-Euclidean geometry (see opposite) in which the PP did not hold but the other four of Euclid's postulates did. Bolyai claimed to have "created another world out of nothing," but the idea was not well received in its time. Gauss acknowledged its

validity, but claimed to have discovered it first. Gauss's idea of the "intrinsic curvature" of a surface or space was an important tool in establishing this new world, but he left little evidence of having developed non-Euclidean geometry himself. He did, however, consider that the Universe might be non-Euclidean. Subsequent advances by Bernhard Riemann, Eugenio Beltrami, Felix Klein, David Hilbert, and others mean that today, non-Euclidean geometries are no longer seen as exotic, and physicists have given serious consideration to whether our Universe is indeed flat (Euclidean) or curved.

Artistic explorations

Hyperbolic geometry also features in art. Models devised by Henri Poincaré inspired many graphic works by M. C. Escher, while some mathematicians, notably Daina Taimina, have used art and craft techniques to make these "new worlds" intuitively graspable. ▪

Crochet models of hyperbolic surfaces created by Daina Taimina are more tactile than paper models. She claims that the crocheting process helps develop geometrical intuition.

Daina Taimina

Born in Latvia in 1954, Daina Taimina began her career in the fields of computer science and the history of mathematics. After teaching for 20 years at the University of Latvia, she moved to Cornell University in the United States in 1996, where a chance encounter opened up a new area of interest. Taimina attended a geometry workshop led by David Henderson in which he demonstrated how to make a paper model of a hyperbolic surface. Henderson himself had learned the technique from pioneering American topologist William Thurston.

Taimina went on to make her own models of hyperbolic surfaces using crochet to assist in her teaching. The models were a success, breaking the stereotype of mathematics as a field unrelated to arts and crafts. Taimina has since embarked on a second career as a mathematician–artist.

Key work

2004 *Experiencing Geometry* with David W. Henderson

ALGEBRAIC STRUCTURES HAVE SYMMETRIES

GROUP THEORY

IN CONTEXT

KEY FIGURE
Évariste Galois (1811–32)

FIELDS
Algebra, number theory

BEFORE
1799 Italian mathematician
Paolo Ruffini considers the sets
of permutations of roots as an
abstract structure.

1815 Augustin-Louis Cauchy,
a French mathematician,
develops his theory of
permutation groups.

AFTER
1846 Galois' work is published
posthumously by fellow
Frenchman Joseph Liouville.

1854 British mathematician
Arthur Cayley extends the
work of Galois to a full theory
of abstract groups.

1872 German mathematician
Felix Klein defines geometry
in terms of group theory.

Group theory is a branch
of algebra that pervades
modern mathematics.
Its genesis was largely due to
French mathematician Évariste
Galois, who developed it in order
to understand why only some
polynomial equations could be
solved algebraically. In so doing, he
not only gave a definitive answer to
a historical quest that had begun in
ancient Babylon, but also laid the
foundations of abstract algebra.

Galois' approach to this problem
was to relate it to a question in
another area of mathematics. This
can be a powerful strategy when the
other area is well understood. In this
case, however, Galois first had to

See also: The algebraic resolution of equations 200–01 ▪ Emmy Noether and abstract algebra 280–81 ▪ Finite simple groups 318–19

A **group** is a **set of elements**, such as numbers or shapes…

⬇

…along with an **operation** (such as addition or rotation) **that acts upon them**.

⬇

To be labeled a group, a set must satisfy four axioms.

It must have an **identity**: an element that leaves any other element **unchanged** when it acts on it.

It must have an **inverse**: every element has a **corresponding** element that combines to give the identity.

It must be **associative**: the **order** in which the operations are performed on the elements does not matter.

It must be **closed**: performing the operation will **not introduce** elements outside of the set.

develop the theory of the "simpler" area (the theory of groups) in order to tackle the more difficult problem (solubility of equations). The link he made between the two areas is now called Galois theory.

Arithmetic of symmetries

A group is an abstract object—it consists of a set of elements and an operation that combines them, subject to some axioms. When these elements include shapes, groups can be thought of as encoding symmetry. Simple symmetries—such as those of a regular polygon—are intuitively graspable. For example, an equilateral triangle with the vertices A, B, and C (see next page) can be rotated in three ways (through 120°, 240°, or 360°) »

Évariste Galois

Born in 1811, Évariste Galois lived a brief but fiery and brilliant life. He was already familiar as a teenager with the works of Lagrange, Gauss, and Cauchy, but failed (twice) to enter the prestigious École Polytechnique—possibly due to his mathematical and political impetuousness, though no doubt affected by the suicide of his father.

In 1829, Galois enrolled at the École Préparatoire, only to be expelled in 1830 for his politics. A staunch republican, he was arrested in 1831 and imprisoned for eight months. Shortly after his release in 1832, he became involved in a duel—it is unclear whether this was over a love affair or politics. Badly wounded, he died the next day, leaving behind just a handful of mathematical papers which contain the foundations of group theory, finite field theory, and what is now called Galois theory.

Key works

1830 *Sur la théorie des nombres* (*On Number Theory*)
1831 *Premier Mémoire* (*First Memoir*)

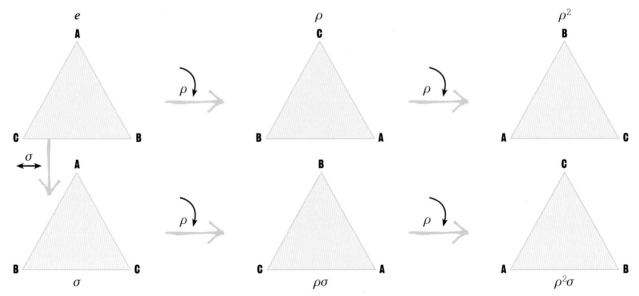

The equilateral triangle has six symmetries. They are rotation (ρ) through 120°, 240°, and 360° and reflection (σ) through a vertical line through A, B, or C. The diagram above shows the results of applying one symmetry after another to e, the identity element (rotation through 0°), and how they are written—$\rho^2\sigma$ (the last equilateral triangle in the diagram) means "rotate through 120 degrees twice and reflect."

around its center, and be reflected in three different lines. Each of these six transformations fits the triangle onto itself—it looks exactly the same, except that the vertices are permuted (rearranged). A clockwise rotation of 120° sends vertex A to where B was, B to C, and C to A, while a reflection in the vertical line through A swaps vertices B and C. The three rotations and the reflections give all possible symmetries of the triangle ABC.

One way to see the symmetries of the triangle is to consider all of the possible permutations of the vertices. A rotation or reflection can send the vertex A to one of three points (including itself). From each of these possibilities, the vertex B has two available destinations. The destination of the third vertex is now determined because the triangle is rigid, so there are 3 × 2 = 6 possibilities. The symmetry groups of polygons can be thought of as permutations of a set of elements. The symmetry group of the equilateral triangle is a member of a small group called D_3.

Axioms of group theory

Group theory has four main axioms. The first is the identity axiom; it states that a unique element exists that does not change any element in the group when combined with it. With the ABC triangle, the identity is the rotation of 0°. The second axiom is the inverse axiom. It says that every element has a unique inverse element; combining the two yields the identity element.

The third axiom concerns associativity, which means that the result of operations on elements does not depend on the order in which they are applied. For example, if you combine any set of three elements with a multiplication operator, you can perform the operations in any order. So if the elements 1, 2, and 3 are members of a group, then (1 × 2) × 3 = 2 × 3 = 6, and 1 × (2 × 3) = 1 × 6 = 6, all giving the same result.

The fourth axiom is closure, meaning that a group should have no elements outside the group as a

The possible rotations of a Rubik's Cube form a mathematical group with 43,252,003,274,489,856,000 elements, but solving the cube from any position requires no more than 26 turns of 90°.

The ATLAS detector at the CERN accelerator is designed to study subatomic particles, including those predicted by group theory.

Group theory in physics

The Universe, as we understand it through physics, is full of symmetries, and group theory is proving a powerful tool for both understanding and prediction. Physicists use the Lie groups, named after the 19th-century Norwegian mathematician Sophus Lie. Lie groups are continuous, not discrete—for example, they model the infinite number of rotational symmetries, such as those associated with a circle, rather than the finite number of transformations of a polygon.

In 1915, German algebraist Emmy Noether demonstrated how Lie groups related to conservation laws (such as the conservation of energy). By the 1960s, physicists began to use group theory to classify subatomic particles. But the mathematical groups they used included a combination of symmetries that no known particles had. Scientists tried looking for a particle with that combination of symmetries, and found the Omega minus particle. More recently, the Higgs boson has filled another such gap.

result of performing the operations. One example of a group obeying all four axioms is the set of integers {…, −3, −2, −1, 0, 1, 2, 3, …} with the operation of addition. The unique identity element is 0, and the inverse of any integer n is $−n$ as $n + −n = 0 = −n + n$. The addition of integers is associative, and the set is also closed, because adding any of the integers together gives another integer.

Groups can also have a further attribute known as commutativity. If a group is commutative, it is known as an Abelian group. This means that its elements can be swapped around without changing the result. Integers added in any order will give the same result ($6 + 7 = 13$ and $7 + 6 = 13$), so the set of integers with the operation of addition is an Abelian group.

Galois groups and fields

Groups are just one kind of abstract algebraic structure among many. Closely related structures include rings and fields, which are also defined in terms of a set with operations and axioms. A field contains two operations; complex numbers (with the operations of addition and multiplication) are a field. The field of complex numbers is the territory in which solutions to polynomial equations are found.

Galois theory relates the solvability of a polynomial equation (whose roots are elements of a field) to a group—specifically, to the permutation group that encodes possible rearrangements of its roots. Galois showed that this group, now called a Galois group, must have one kind of structure if the equation is algebraically solvable, and a different kind of structure if it is not. Galois groups of quartic

equations and simpler polynomials are solvable, but those of higher degree polynomials are not. Modern algebra is an abstract study of groups, rings, fields, and other algebraic structures.

Group theory continues to develop in its own right and has many applications. Group theory is used to study symmetries in chemistry and physics, for example, and can be used in public key cryptography, which secures much of today's digital communication. ∎

Wherever groups disclosed themselves, or could be introduced, simplicity crystallized out of comparative chaos.
Eric Temple Bell
Scottish mathematician

We need a super-mathematics in which the operations are as unknown as the quantities they operate on… such a super-mathematics is the Theory of Groups.
Arthur Eddington
British astrophysicist

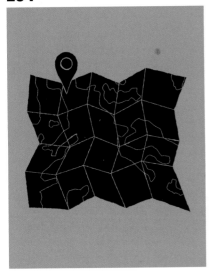

JUST LIKE A POCKET MAP

QUATERNIONS

An extension of complex numbers, quaternions are used to model, control, and describe motion in three dimensions, which is essential in, for example, creating the graphics of a video game, planning a space probe's trajectory, and calculating the direction in which a smartphone is pointing. Quaternions were the brainchild of William Rowan Hamilton, an Irish mathematician who was interested in how to model movement mathematically in three-dimensional space. In 1843, in a flash of inspiration, he realized

Complex numbers (sums of real and imaginary numbers) have **two dimensions** and describe motion in two dimensions.

To describe motion in **three dimensions**, we need an **extended version** of complex numbers.

A **three-dimensional number** is **not sufficient** to describe motion in three dimensions.

A full description of motion in three-dimensional space requires a four-dimensional number, or quaternion.

Since quaternions can model and control the motion of objects in three dimensions, they are particularly useful in virtual reality games.

that the "third dimension problem" could not be solved with a three-dimensional number, but needed a four-dimensional one (a quaternion).

Movements and rotations
Complex numbers are two-dimensional: they are made up of a real and an imaginary part, for example, $1 + 2i$. As a result, the two parts of any complex number can act as coordinates, and the number can be plotted on a surface or plane. The two-dimensional complex plane extends the one-dimensional number line by combining real numbers with imaginary units. The plotting of complex numbers then enables the calculation of motion and rotation in two dimensions. Any linear motion from point A to B can be expressed as the addition of two complex numbers. Adding more numbers creates a sequence of movements across the plane. To describe rotation, complex numbers are multiplied together. Every multiplication by i, the imaginary unit, results in a 90° rotation, and a rotation of any other angle is due to some factor or fraction of i.

Once complex numbers were understood, the next challenge for mathematicians was to create a number that worked the same way in a three-dimensional space. The logical answer was to add a third number line, j, which ran at 90 degrees to both the real and imaginary number lines, but no one could figure out how such a number added, multiplied, and so on.

Four dimensions
Hamilton's solution was to add a fourth nonreal unit, k. This created a quaternion, with a basic structure of $a + bi + cj + dk$, where a, b, c, and d are real numbers. The two additional quaternion units, j and k, share similar properties to i and are imaginary. A quaternion can define a vector, or a line in three-dimensional space, and can describe an angle and direction of rotation around that vector. Like the complex plane, simple quaternion mathematics, combined with basic trigonometry, offers a way to describe all kinds of movements within three-dimensional space. ▪

An undercurrent of thought was going on in my mind which gave at last a result… An electric circuit seemed to close; and a spark flashed forth, the herald of many long years.
William Rowan Hamilton

William Rowan Hamilton

Born in Dublin in 1805, Hamilton became interested in mathematics from the age of eight after meeting Zerah Colburn, a touring American mathematical child prodigy. At the age of 22, while still studying at Trinity College, Dublin, he was appointed both professor of astronomy at the university and Royal Astronomer of Ireland.

Hamilton's expertise in Newtonian mechanics enabled him to calculate the paths of heavenly bodies. He later updated Newtonian mechanics into a system that enabled further advances to be made in electromagnetism and quantum mechanics. In 1856, he tried to capitalize on his skills by launching the icosian game, in which players search for a path connecting the points of a dodecahedron without returning to the same point twice. Hamilton sold the rights to the game for £25. He died in 1865, following a severe attack of gout.

Key works

1853 *Lectures on Quaternions*
1866 *Elements of Quaternions*

POWERS OF NATURAL NUMBERS ARE ALMOST NEVER CONSECUTIVE
CATALAN'S CONJECTURE

IN CONTEXT

KEY FIGURE
Eugène Catalan (1814–94)

FIELD
Number theory

BEFORE
c. 1320 French philosopher and mathematician Levi ben Gershon (Gersonides) shows that the only powers of 2 and 3 that differ by 1 are $8 = 2^3$ and $9 = 3^2$.

1738 Leonhard Euler proves that 8 and 9 are the only consecutive square or cube numbers.

AFTER
1976 Dutch number theorist Robert Tijdeman proves that, if more consecutive powers exist, there are only a finite number of them.

2002 Preda Mihăilescu proves Catalan's conjecture, 158 years after it was formulated in 1844.

Many problems in number theory are easy to pose, but extremely difficult to prove. Fermat's last theorem, for example, remained a conjecture (unproven claim) for 357 years. Like Fermat's conjecture, Catalan's conjecture is a deceptively simple claim about powers of positive integers that was proved long after its initial statement.

In 1844, Eugène Catalan claimed that there is only one solution to the equation $x^m - y^n = 1$, where x, y, m, and n are natural numbers (positive integers) and m and n are greater than 1. The solution is $x = 3$, $m = 2$, $y = 2$, and $n = 3$, since $3^2 - 2^3 = 1$. In other words, squares, cubes, and higher powers of natural numbers are almost never consecutive. Five hundred years before, Gersonides had proved a special case of the claim. He used only powers of 2 and 3, solving the equations $3^n - 2^m = 1$ and $2^m - 3^n = 1$. In 1738, Leonhard Euler similarly proved a case in which the only powers allowed were squares and cubes. Euler did this by solving the equation $x^2 - y^3 = 1$. This was closer to Catalan's conjecture, but

Using **natural numbers** (positive integers), the smallest **difference between two powers** is 1.

Catalan expressed this as the **formula** $x^m - y^n = 1$, where m and n must be **greater than 1**.

There is only one solution to this equation using natural numbers: $3^2 - 2^3 = 1$.

See also: Pythagoras 36–43 ▪ Diophantine equations 80–81 ▪ The Goldbach conjecture 196 ▪ Taxicab numbers 276–77 ▪ Proving Fermat's last theorem 320–23

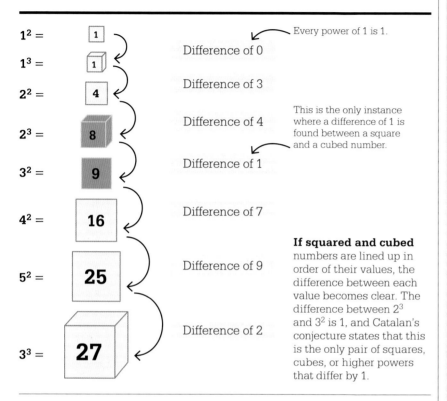

$1^2 =$ — 1 — Every power of 1 is 1.

Difference of 0

$1^3 =$ — 1

Difference of 3

$2^2 =$ — 4

Difference of 4

$2^3 =$ — 8 — This is the only instance where a difference of 1 is found between a square and a cubed number.

Difference of 1

$3^2 =$ — 9

Difference of 7

$4^2 =$ — 16

Difference of 9

$5^2 =$ — 25

Difference of 2

$3^3 =$ — 27

If squared and cubed numbers are lined up in order of their values, the difference between each value becomes clear. The difference between 2^3 and 3^2 is 1, and Catalan's conjecture states that this is the only pair of squares, cubes, or higher powers that differ by 1.

did not allow for the possibility that larger powers or exponents could result in consecutive numbers.

Becoming a theorem

Catalan himself said that he could not prove his conjecture completely. Other mathematicians tackled the problem, but it was only in 2002 that Romanian mathematician Preda Mihăilescu solved the outstanding issues and turned conjecture into theorem.

It might seem that Catalan's conjecture must be false, since simple calculations quickly yield examples of powers that are almost consecutive. For example, $3^3 - 5^2 = 2$, and $2^7 - 5^3 = 3$. On the other hand, even these near-solutions are rare. One approach to proving the conjecture appeared to involve making many calculations: in 1976,

Robert Tijdeman found an upper bound (maximum size) for x, y, m, and n. This proved that there is only a finite number of powers that can be consecutive. The truth of Catalan's conjecture could now be tested by checking each of these powers. Unfortunately, Tijdeman's upper bound is astronomically large, making such computation practically unfeasible even for modern computers.

Mihăilescu's proof of Catalan's conjecture does not involve any such computation. Mihăilescu built on 20th-century advances (by Ke Zhao, J. W. S. Cassels, and others) that had proved m and n must be odd primes for any further solutions of $x^m - y^n = 1$. His proof is not as formidable as Andrew Wiles's proof of Fermat's last theorem, but it is still highly technical. ▪

Eugène Catalan

Born in Bruges, Belgium, in 1814, Eugène Catalan studied under French mathematician Joseph Liouville at the École Polytechnique in Paris. Catalan was a republican from an early age and a participant in the 1848 revolution. His political beliefs led to his expulsion from a number of academic posts.

Catalan was particularly interested in geometry and combinatorics (counting and arranging), and his name is associated with the Catalan numbers. This sequence (1, 2, 5, 14, 42…) counts, among other things, the ways that polygons can be divided into triangles.

Although he considered himself French, Catalan won recognition in Belgium, where he lived from his appointment as professor of analysis at the University of Liège in 1865 until his death in 1894.

Key works

1860 *Traité élémentaire des séries* (*Elementary Treatise on Series*)
1890 *Intégrales eulériennes ou elliptiques* (*Eulerian or Elliptic Integrals*)

THE MATRIX IS EVERYWHERE

MATRICES

IN CONTEXT

KEY FIGURE
James Joseph Sylvester
(1814–97)

FIELDS
Algebra, number theory

BEFORE
200 BCE The ancient Chinese text *The Nine Chapters on the Mathematical Art* presents a method for solving equations using matrices.

1545 Gerolamo Cardano publishes techniques using determinants.

1801 Carl Friedrich Gauss uses a matrix of six simultaneous equations to compute the orbit of the asteroid Pallas.

AFTER
1858 Arthur Cayley formally defines matrix algebra, and proves results for 2 × 2 and 3 × 3 matrices.

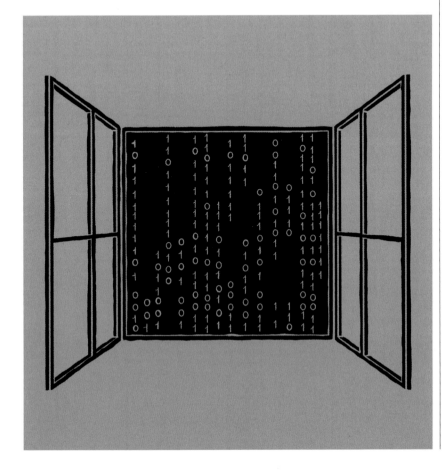

Matrices are rectangular arrays (grids) of elements (numbers or algebraic expressions), arranged in rows and columns enclosed by square brackets. The rows and columns can be extended indefinitely, which enables matrices to store vast amounts of data in an elegant and compact manner. Although a matrix contains many elements, it is treated like one unit. Matrices have applications in mathematics, physics, and computer science, such as in computer graphics and describing the flow of a fluid.

The earliest known evidence for such arrays comes from the ancient Mayan civilization of

The dimensions of a matrix are important, as operations such as addition and subtraction require the matrices involved to have the same dimensions. The 2 × 2 matrices below are square matrices, meaning that they have the same number of rows as they have columns. The graphic below shows how matrices are added together by adding the elements in corresponding positions.

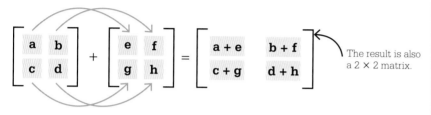

$$\begin{bmatrix} a & b \\ c & d \end{bmatrix} + \begin{bmatrix} e & f \\ g & h \end{bmatrix} = \begin{bmatrix} a+e & b+f \\ c+g & d+h \end{bmatrix}$$

The result is also a 2 × 2 matrix.

James Joseph Sylvester

Born in 1814, James Joseph Sylvester began his studies at University College London, but left when he was accused by another student of wielding a knife. He then went to Cambridge and came second in the university examinations, but was not allowed to graduate because, as a Jew, he would not swear allegiance to the Church of England.

Sylvester taught briefly in the US, but faced similar difficulties there. Returning to London, he studied law and was admitted to the bar in 1850. He also began to work on matrices with fellow British mathematician Arthur Cayley. In 1876, Sylvester returned to the US as a math professor at Johns Hopkins University, Maryland, where he founded the *American Journal of Mathematics*. Sylvester died in London in 1897.

Key works

1850 *On a New Class of Theorems*
1852 *On the principle of the calculus of forms*
1876 *Treatise on elliptic functions*

Central America, c. 300 BCE. Some historians believe the Maya people manipulated numbers in rows and columns to solve equations, and cite gridlike decorations on their monuments and priestly robes as evidence. Others, however, doubt these patterns represent actual matrices.

The first verified instance of the use of matrices comes from ancient China. In the second century BCE, the textbook *The Nine Chapters on the Mathematical Art* described how to set out a counting board and use a matrixlike method to solve linear simultaneous equations with several unknown values. This method was similar to the elimination system introduced by German mathematician Carl Gauss in the 1800s, which is still used today for solving simultaneous equations.

Matrix arithmetic

In 1850, British mathematician James Joseph Sylvester first used the term "matrix" to describe an array of numbers. Shortly after Sylvester introduced the term, his friend and colleague Arthur Cayley formalized the rules for manipulating matrices. Cayley showed that the rules of matrix algebra are different from those in standard algebra. Two matrices of the same size (with the same number of elements in their respective rows and columns) are added by simply adding corresponding elements. Matrices with different dimensions cannot be added. Matrix multiplication is, however, quite different from multiplication of numbers. Not all »

The arrays found in Mayan relics suggest to some historians that the Maya used matrices to solve linear equations. However, others believe they were merely replicating patterns in nature, such as on a turtle's shell.

Multiplying two matrices together is achieved by multiplying the horizontals in the first matrix by the vertical numbers in the second (the centered dot indicates multiplication) and adding the results. In matrix algebra, switching around the order in which the two matrices are multiplied produces a different result as shown here with the multiplication of two square matrices (A and B).

$$
\begin{matrix} A \end{matrix} \quad \begin{matrix} B \end{matrix}
$$

$$
\begin{bmatrix} 4 & 8 \\ 1 & 3 \end{bmatrix} \times \begin{bmatrix} 2 & 9 \\ 7 & 0 \end{bmatrix} = \begin{bmatrix} 4 \cdot 2 + 8 \cdot 7 & 4 \cdot 9 + 8 \cdot 0 \\ 1 \cdot 2 + 3 \cdot 7 & 1 \cdot 9 + 3 \cdot 0 \end{bmatrix} = \begin{bmatrix} 64 & 36 \\ 23 & 9 \end{bmatrix}
$$

$$
\begin{matrix} B \end{matrix} \quad \begin{matrix} A \end{matrix}
$$

$$
\begin{bmatrix} 2 & 9 \\ 7 & 0 \end{bmatrix} \times \begin{bmatrix} 4 & 8 \\ 1 & 3 \end{bmatrix} = \begin{bmatrix} 2 \cdot 4 + 9 \cdot 1 & 2 \cdot 8 + 9 \cdot 3 \\ 7 \cdot 4 + 0 \cdot 1 & 7 \cdot 8 + 0 \cdot 3 \end{bmatrix} = \begin{bmatrix} 17 & 43 \\ 28 & 56 \end{bmatrix}
$$

matrices can be multiplied together; in matrix multiplication, AB (see above) can only be calculated if the row count of B is the same as the column count of A. Matrix multiplication is noncommutative, meaning that even where both A and B are square matrices, AB is not equal to BA.

Square matrices

Because of their symmetry, square matrices have particular properties. For example, a square matrix can be repeatedly multiplied by itself. A square matrix of size $n \times n$ with the value 1 along the diagonal starting top left, and the value 0 everywhere else, is called the identity matrix (I_n).

Every square matrix has an associated value called its determinant, which encodes many of the matrix's properties and can be computed by arithmetic operations on the matrix's elements. Square matrices whose elements are complex numbers, and whose determinants are not zero, form an algebraic structure called a group. Theorems that are true for groups are therefore also true for such matrices, and advances in group theory can be applied to matrices. Groups can also be represented as matrices, enabling difficult problems in group theory to be expressed in terms of matrix algebra, which is more easily solved. Representation theory, as this field is known, is applied in number theory and analysis, and in physics.

Determinants

The determinant of a matrix was named by Gauss, due to the fact that it determines whether the system of equations represented by the matrix has a solution. As long as the determinant is not zero, the system will have a unique solution. If the determinant is zero, the system may have either no solution or many.

In the 1600s, Japanese mathematician Seki Takakaze had shown how to calculate the determinants of matrices up to size 5×5. Over the following century,

A linear transformation in 2 dimensions maps lines through the origin to other lines through the origin, and parallel lines to parallel lines. Linear transformations include rotations, reflections, enlargements, stretches, and shears (lines that slide parallel to a fixed line, in proportion to their distance from the fixed line). The image of any point (x, y) is found by multiplying the matrix by the column vector representing the point (x, y). In the examples below, the original shape is the pink square, with vertices (0, 0), (2, 0), (2, 2) and (0, 2), and the image is the green quadrilateral.

Horizontal shear with shear factor 1

$$
\begin{bmatrix} 1 & 1 \\ 0 & 1 \end{bmatrix} \times \begin{bmatrix} x \\ y \end{bmatrix}
$$

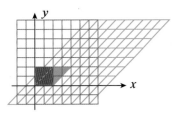

Reflection in the vertical axis

$$
\begin{bmatrix} -1 & 0 \\ 0 & 1 \end{bmatrix} \times \begin{bmatrix} x \\ y \end{bmatrix}
$$

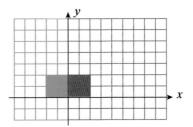

Enlargement by factor 1.5

$$
\begin{bmatrix} 1.5 & 0 \\ 0 & 1.5 \end{bmatrix} \times \begin{bmatrix} x \\ y \end{bmatrix}
$$

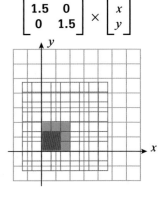

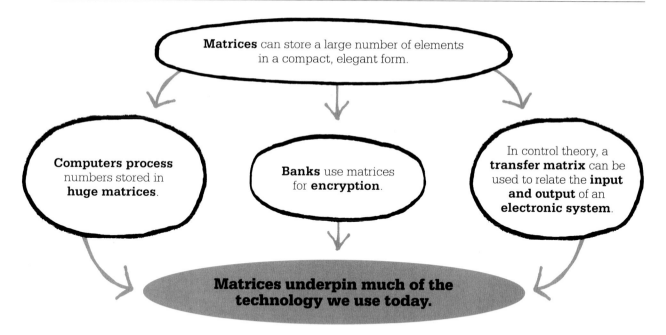

Matrices can store a large number of elements in a compact, elegant form.

Computers process numbers stored in **huge matrices**.

Banks use matrices for **encryption**.

In control theory, a **transfer matrix** can be used to relate the **input and output** of an **electronic system**.

Matrices underpin much of the technology we use today.

mathematicians uncovered the rules for finding determinants of larger and larger arrays. In 1750, Swiss mathematician Gabriel Cramer stated a general rule (now called Cramer's rule) for the determinant of a matrix with *m* rows and *n* columns, but he failed to give the proof of this rule.

In 1812, French mathematicians Augustin-Louis Cauchy and Jacques Binet proved that when two square matrices of the same size are multiplied, the determinant of this product is, in fact, the same as the product of their individual determinants: $detAB = (detA) \times (detB)$. This rule simplified the process of finding the determinant of a very large matrix by breaking it down into the determinants of two smaller matrices.

Transformation matrices

Matrices can be used to represent linear geometric transformations (see opposite) such as reflections, rotations, translations, and scalings. Transformations in two dimensions are encoded by 2 × 2 matrices, while 3-D transformations involve 3 × 3 matrices. The determinant of a transformation matrix contains information about the area or volume of the transformed figure. Today, computer aided design (CAD) software makes extensive use of matrices for this purpose.

Modern applications

Matrices can store vast amounts of data compactly, making them essential across math, physics, and computing. Graph theory uses matrices to encode how a set of vertices (points) is connected by edges (lines). One formulation of quantum physics, called matrix mechanics, makes extensive use of matrix algebra, and particle physicists and cosmologists use transformation matrices and group theory to study the symmetries of the Universe.

Matrices are used to represent electrical circuits for solving problems about voltage and current. They are also important in computer science and cryptography. Stochastic matrices, whose elements represent probabilities, are used by search engine algorithms for ranking web pages. Programmers use matrices as keys when encrypting messages; letters are assigned individual numerical values, which are then multiplied by the numbers in the matrix. The larger the matrix used, the more secure the encryption is. ∎

I have not thought it necessary to undertake the labor of a formal proof of the theorem in the general case of a matrix of any degree.
Arthur Cayley

AN INVESTIGATION INTO
THE LAWS OF
THOUGHT

BOOLEAN ALGEBRA

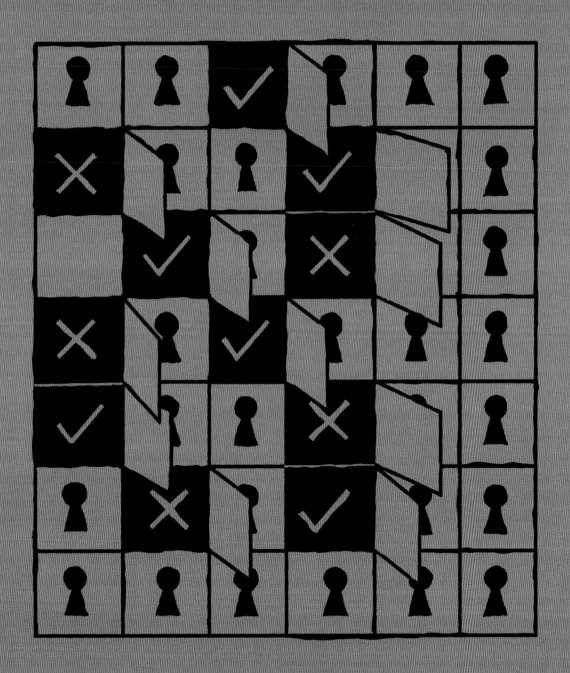

IN CONTEXT

KEY FIGURE
George Boole (1815–64)

FIELD
Logic

BEFORE
350 BCE Aristotle's philosophy discusses syllogisms.

1697 Gottfried Leibniz tries, unsuccessfully, to use algebra to formalize logic.

AFTER
1881 John Venn introduces Venn diagrams to explain Boolean logic.

1893 Charles Sanders Peirce uses truth tables to show outcomes of Boolean algebra.

1937 Claude Shannon uses Boolean logic as the basis for computer design in his *A Symbolic Analysis of Relay and Switching Circuits*.

Mathematics had never more than a secondary interest for him, and even logic he cared for chiefly as a means of clearing the ground.
Mary Everest Boole
British mathematician and wife of George Boole

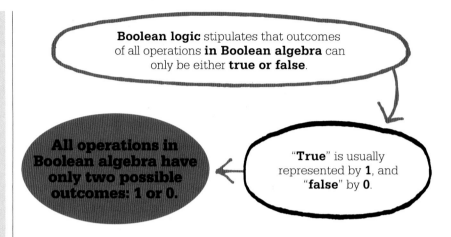

Boolean logic stipulates that outcomes of all operations **in Boolean algebra** can only be either **true or false**.

All operations in Boolean algebra have only two possible outcomes: 1 or 0.

"**True**" is usually represented by **1**, and "**false**" by **0**.

L ogic is the bedrock of mathematics. It provides us with the rules of reasoning and gives us a basis for deciding on the validity of an argument or proposition. A mathematical argument uses the rules of logic to ensure that if a basic proposition is true, then any and all statements constructed from that proposition will also be true.

The earliest attempt to set out the principles of logic was carried out by the Greek philosopher Aristotle around 350 BCE. His analysis of the various forms of arguments marked the beginning of logic as a subject for study in its own right. In particular, Aristotle looked at a type of argument known as a syllogism, consisting of three propositions. The first two propositions, called the premises, logically entail the third proposition, the conclusion. Aristotle's ideas about logic were unrivaled and unchallenged in Western thought for more than 2,000 years.

Aristotle approached logic as a branch of philosophy, but in the 1800s, scholars began to study logic as a mathematical discipline. This involved moving from arguments expressed in words to a symbolic logic where arguments could be expressed using abstract symbols. One of the pioneers of this shift to mathematical logic was British mathematician George Boole, who sought to apply methods from the emerging field of symbolic algebra to logic.

Algebraic logic

Boole's investigations into logic began in an unconventional way. In 1847, a friend, British logician Augustus De Morgan, became involved in a dispute with a philosopher about who deserved the credit for a particular idea. Boole was not directly involved, but the event spurred him to set down his ideas concerning how logic could be formalized with mathematics, in his 1847 essay *Mathematical Analysis of Logic*.

Boole wanted to discover a way to frame logical arguments so that they could be manipulated and solved mathematically. In order to achieve this, he developed a type of linguistic algebra, in which the operations of ordinary algebra, such as addition and multiplication, were replaced by the connectors that were used in logic. As in algebra, Boole's use of symbols and connectives allowed for the simplification of logical expressions.

The three key operations of Boole's algebra were AND, OR, and NOT; Boole believed these were the only operations necessary to perform comparisons of sets, as well as basic mathematical functions. For example, in logic, two statements may be connected by AND, as in "this animal is covered in hair" AND "this animal feeds its young with milk," or by OR, as in "this animal can swim" OR "this animal has feathers." The statement "*A* AND *B*" is true when *A* and *B* are both individually true, whereas the statement "*A* OR *B*" is true if one or both of *A* and *B* is true. In Boolean terms, such statements can be given as, for example: $(A$ OR $B) = (B$ OR $A)$; NOT $($NOT $A) = A$; or even NOT $(A$ OR $B) = ($NOT $A)$ AND $($NOT $B)$.

Boole's binaries
In 1854, Boole published his most important work, *An investigation into the laws of thought*. Boole had studied the algebraic properties

of numbers and realized that the set {0, 1}, together with operations such as addition and multiplication, could be used to form a consistent algebraic language. Boole proposed that logical propositions could have only two values—true or false—and could not be anything in between.

In Boole's logical algebra, truth and falsity were reduced to binary values: 1 for true and 0 for false. Starting out with an initial statement that was either true or false, Boole could then construct further statements and use the AND, OR, and NOT operations in order to determine whether or not these further statements were true.

One plus one is one
Despite the resemblance, Boole's true and false binary of 1 and 0 is not the same as binary numbers. Boolean numbers are entirely different from the mathematics of real numbers. The "laws" of Boole's algebra allow statements that would not be permitted by other forms of algebra. In Boole's

Boolean algebra makes it possible to prove logical statements by performing algebraic calculations.
Ian Stewart
British mathematician

algebra, there are only two possible values for any quantity, either 1 or 0. There is also no such thing as subtraction in Boole's algebra. For example, if statement *A*, "my dog is hairy," is true, it has a value of 1, and if statement *B*, "my dog is brown" is true, it also has a value of 1. *A* and *B* can be combined to make the statement "my dog is hairy OR my dog is brown," which is also true, and also has a value of 1. In Boolean »

George Boole

Born in Lincoln in 1815, George Boole was the son of a shoemaker who passed his love of science and mathematics on to him. When his father's business collapsed, the 16-year-old George took up a post as an assistant schoolmaster to support his family. He began to study mathematics seriously, starting by reading a book on calculus. He later published work in the *Cambridge Mathematical Journal*, but still could not afford to study for a degree.

In 1849, as a result of his correspondence with Augustus De Morgan, Boole was appointed professor of mathematics at the new Queen's College in Cork, Ireland, where he remained until his premature death at the age of 49.

Key works

1847 *Mathematical Analysis of Logic*
1854 *An investigation into the laws of thought*
1859 *Treatise on differential equations*
1860 *Treatise on the calculus of finite differences*

> The furthest thing from my mind has been those efforts which try to establish an artificial similarity [between logic and algebra].
> **Gottlob Frege**

algebra, OR behaves like + (aside from 1 + 1 = 1) and AND behaves like × (see table on p.247).

Visualizing results

One way of visualizing Boole's algebra is in the form of diagrams invented by British logician John Venn. In his work *Symbolic Logic* (1881), Venn developed Boole's theories employing what became known as Venn diagrams. These depict relations of inclusion (AND) and exclusion (NOT) between sets. They consist of intersecting circles, each one representing a distinct set. A two-circle Venn diagram represents propositions such as: "All *A* are *B*," while a three-circle diagram represents propositions involving three sets (such as *X*, *Y*, and *Z* below).

The results of a statement in Boolean algebra can also be assessed using a truth table (see opposite), in which all possible input combinations are tried and written out. These truth tables were first used by American logician Charles Saunders Peirce in 1893, nearly 30 years after Boole's death. For example, the statement *A* AND *B* can only be considered true if both *A* and *B* are true. If one or both of *A* and *B* are false, then *A* AND *B* is false. Therefore, out of the four possible combinations of *A* and *B*, only one results in a true answer. On the other hand, for *A* OR *B*, there are three possible combinations in which that statement is true, as it will only be false if both *A* and *B* are false. More complex statements can also be assessed by drawing truth tables. For example, *A* AND (*B* OR NOT *C*) is true when *A* and *B* are both true and *C* is false, and is false when *A* is false and both *B* and *C* are true. Out of eight possible combinations of true and false, there are three in which the statement is true and five in which it is false.

Limitations

One drawback in Boole's system of algebra was that it contained no method of quantification: there was no simple way of expressing a statement such as "for all *X*," for example. The first symbolic logic with quantification was produced in 1879 by German logician Gottlob Frege, who objected to Boole's attempts to turn logic into algebra. Frege's work was followed by Charles Sanders Peirce and another German logician, Ernst Schröder, who introduced

These Venn diagrams represent three of the most basic functions in Boolean algebra: the functions for AND, OR, and NOT. The three-circle diagram represents a combination of two functions: (*X* AND *Y*) OR *Z*.

The region showing the output of the function

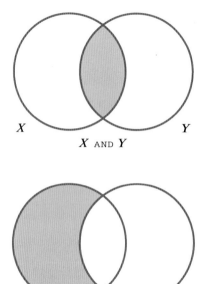

X AND *Y*

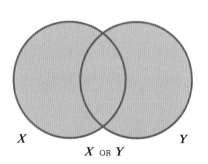

X OR *Y*

X NOT *Y*

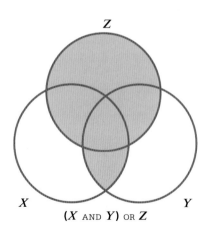

(*X* AND *Y*) OR *Z*

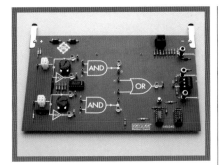

This logic module is used for teaching how logic gates function in electronic circuits. The gates can be connected to lights or buzzers which go on and off depending on the output.

quantification into Boole's algebra and produced substantial works using Boole's system.

Boole's legacy

It was not until some 70 years after Boole's death that the potential of his ideas was fully grasped. American engineer Claude Shannon used Boole's *Mathematical Analysis of Logic* to establish the basis of modern digital computer circuits. While working on the electrical circuitry for one of the world's first computers, Shannon realized that Boole's two-value binary system could be the basis of logic gates (physical devices that move based on Boolean functions) in the circuitry. Aged just 21, Shannon published the ideas that would form the basis of future computer design in *A Symbolic Analysis of Relay and Switching Circuits*, published in 1937.

The building blocks of codes now used to program computer software are based on the logic formulated by Boole. Boolean logic is also at the heart of how internet search engines work. In the early days of the internet, the AND, OR, and NOT commands were commonly used to filter results to find the specific thing being searched for, but advances in technology allow people today to search using more natural language. The Boolean commands have simply become silent: a search for "George Boole," for example, has an implied AND between the two words, so that only web pages containing both names will appear in the results. ■

Gate	Symbol	Truth table		

NOT
A NOT gate's output is the opposite of its input.

INPUT	OUTPUT
1	0
0	1

AND
An AND gate's output is 1 only if both its inputs are 1.

INPUT		OUTPUT
A	B	A AND B
0	0	0
0	1	0
1	0	0
1	1	1

OR
An OR gate's output is 0 only if both its inputs are 0.

INPUT		OUTPUT
A	B	A AND B
0	0	0
0	1	1
1	0	1
1	1	1

NAND
A NAND gate is an AND gate followed by a NOT gate.

INPUT		OUTPUT
A	B	A AND B
0	0	1
0	1	1
1	0	1
1	1	0

NOR
A NOR gate is an OR gate followed by a NOT gate.

INPUT		OUTPUT
A	B	A AND B
0	0	1
0	1	0
1	0	0
1	1	0

Logic gates, which are physical electronic devices implementing Boolean functions, form an important part of computer circuitry. This table shows the various symbols for each type of logic gate. Truth tables show the possible outcomes of various inputs into the gate.

A SHAPE WITH JUST ONE SIDE
THE MÖBIUS STRIP

IN CONTEXT

KEY FIGURE
August Möbius (1790–1868)

FIELD
Applied geometry

BEFORE
3rd century CE A Roman mosaic of Aion, Greek god of eternal time, features a zodiac shaped like a Möbius strip.

1847 Johann Listing publishes *Vorstudien zur Topologie* (*Introductory Studies in Topology*).

AFTER
1882 Felix Klein describes the Kleinsche Flasche (Klein bottle), a shape composed of two Möbius strips.

1957 In the US, the B. F. Goodrich Company produces a patent for a conveyor belt based on the Möbius strip.

2015 Möbius strips are used in laser beam research, with potential application in nanotechnology.

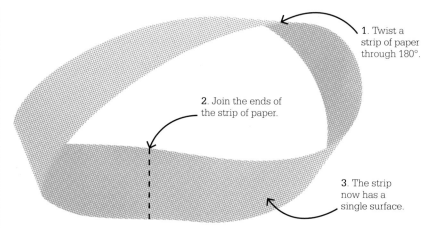

1. Twist a strip of paper through 180°.

2. Join the ends of the strip of paper.

3. The strip now has a single surface.

A Möbius strip can be made from a simple length of paper. It can be colored in with a crayon in one continuous movement without taking the crayon away from the paper. The shape has a single surface; this can be tested by following the surface of the shape with the eye.

Named after 19th-century German mathematician August Möbius, a Möbius strip can be created in seconds by twisting a strip of paper through 180°, then joining its two ends together. The shape that results has some unexpected properties, which have advanced our understanding of complex geometrical figures—a branch of study called topology.

The 19th century was a creative period for mathematics, and the exciting new field of topology spawned many new geometrical shapes. Much of this impetus came from German mathematicians, including Möbius and Johann Listing. In 1858, the two men independently investigated the twisted strip, which Listing is said to have discovered first.

Once formed, the Möbius strip has only one surface—an ant crawling along that surface would be able to cover both sides of the paper in one continuous movement without crossing the edge of

See also: Graph theory 194–95 ▪ Topology 256–59 ▪ Minkowski space 274–75
▪ Fractals 306–11

A Roman mosaic dating from c. 200 CE includes what may be the earliest representation of a Möbius strip, which is thought to represent the eternal nature of time.

the paper. In geometry, it is considered a classic example of a "nonorientable" surface. This means that when you trace your finger around the complete strip, the left and right sides of the paper are reversed. The Möbius strip is the simplest nonorientable, two-dimensional surface that can be created in three-dimensional space.

Experimenting with the Möbius strip produces other unexpected results. For instance, if you draw a line around the center of the strip and then cut along it, the shape does not divide in half. Rather, it produces a longer, continuous twisted loop. Alternatively, draw a line about a third of the way across the width of the strip, then turn the scissors 90° and cut along its length: the result is one twisted loop linked to a second, thinner twisted loop that is twice as long.

Space, industry, and art

The Möbius strip shape sometimes occurs naturally, such as in the movement of magnetically charged particles within the Van Allen radiation belts that surround Earth and in the molecular structure of some proteins. Its properties have been put to use in everyday applications, too. In the early 20th century, the Möbius strip shape was used in continuous-loop recording tapes to provide double the playback time. There are also Möbius strip roller-coasters, such as the Grand National at Blackpool Pleasure Beach in northern England.

The Möbius strip's form has inspired artists and architects. Dutch artist M.C. Escher created a notable woodcut of ants endlessly patrolling the shape. Impressive Möbius strip buildings are being constructed to minimize the impact of the sun's rays. The shape is used in the universal symbol for recycling and also suggested in the mathematical symbol for infinity (∞), echoing the eternity image in the ancient Roman mosaic (above). ■

Our lives are Möbius strips, misery and wonder simultaneously. Our destinies are infinite, and infinitely recurring.
Joyce Carol Oates
American novelist

August Möbius

Born near Naumberg in Saxony, Germany, in 1790, August Ferdinand Möbius was the son of a dance teacher. At the age of 18, he entered the University of Leipzig to study mathematics, physics, and astronomy, and later studied in Göttingen under the great German mathematician Carl Friedrich Gauss. In 1816, Möbius was appointed professor of astronomy at Leipzig and stayed there for the rest of his life, writing treatises on Halley's Comet and other aspects of astronomy.

Möbius is associated with a number of mathematical concepts, including Möbius transformations, the Möbius function, the Möbius plane, and the Möbius inversion formula. He also conjectured a geometrical projection known as a Möbius net. Möbius died in Leipzig in 1868.

Key works

1827 *The Calculus of Centers of Gravity*
1837 *Textbook of Statics*
1843 *The Elements of Celestial Mechanics*

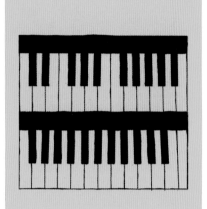

THE MUSIC OF THE PRIMES

THE RIEMANN HYPOTHESIS

IN CONTEXT

KEY FIGURE
Bernhard Riemann
(1826–66)

FIELD
Number theory

BEFORE
1748 Leonhard Euler defines the Euler product, linking a version of what will become the zeta function to the sequence of prime numbers.

1848 Russian mathematician Pafnuty Chebyshev presents the first significant study of the prime counting function $\pi(n)$.

AFTER
1901 Swedish mathematician Helge von Koch proves that the best possible version of the prime counting function relies on the Riemann hypothesis.

2004 Distributed computing is used to prove that the first 10 trillion "nontrivial zeros" lie on the critical line.

It is very **difficult to estimate** how many prime numbers there are between a **pair of numbers**.

The **Riemann hypothesis** states that the **zeta function** (a function in number theory) gives the most **accurate estimate** for the number of primes between two values.

The hypothesis has **not yet been proven**.

I n 1900, David Hilbert listed 23 outstanding mathematical problems. One of them was the Riemann hypothesis, which is still agreed to be one of the most important unsolved problems in mathematics. It concerns the prime numbers—numbers that are only divisible by themselves or 1. Proving the Riemann hypothesis would solve many other theorems.

The most noticeable thing about prime numbers is that the larger they are, the more widely spread out they get. Of the numbers between 1 and 100, 25 are prime (1 in 4); between 1 and 100,000, 9,592 are prime (about 1 in 10). These values are expressed through the prime counting function, $\pi(n)$, but π here is not related to the mathematical constant pi. Inputting n into π gives the number of primes between 1 and n. For example, the number of primes up to 100 gives $\pi(100) = 25$.

Finding the pattern

For centuries, mathematicians' fascination with primes has led them to seek a formula that would

See also: ▪ Mersenne primes 124 ▪ Imaginary and complex numbers 128–31
▪ The complex plane 214–15 ▪ The prime number theorem 260–61

> The failure of the Riemann hypothesis would create havoc in the distribution of prime numbers.
> **Enrico Bombieri**
> **Italian mathematician**

predict the values of this function. Aged just 14, Carl Gauss found a rough answer, and he was soon able to find an improved version of the prime counting function that could predict the number of primes between 1 and 1,000,000 as 78,628, which is accurate to 0.2 percent.

A new formula

In 1859, Bernhard Riemann constructed a new formula for $\pi(n)$, which would give the most accurate estimates possible. One of the inputs needed for this formula is a series of complex numbers defined by what is now called the Riemann zeta function, $\zeta(s)$.

The numbers that are needed to confirm Riemann's formula for $\pi(n)$ are those complex numbers (s) for which $\zeta(s) = 0$. Some of these—the "trivial zeros"—are easy to find; they are all the negative even integers (-2, -4, -6, and so on). Finding the others (the "nontrivial zeros"— all other values for which $\zeta(s) = 0$) is more difficult. Riemann only calculated three. He believed that nontrivial zeros have one thing in common: when they are plotted on

the complex plane, they all lie on "the critical line," where the real part of the number is 0.5. This belief is called the Riemann hypothesis.

A solution

In 2018, British mathematician Michael Atiyah, then aged 89, said he had found a simple proof for the Riemann hypothesis. He died a few months later, the proof unverified.

Although proving the Riemann hypothesis would validate the zeta function's status as the best predictor of the distribution of primes, it still would not allow prime numbers to be fully predicted. Their distribution is to some extent chaotic. But the hypothesis does pin down the blend of predictability and randomness the primes obey. This blend is exactly that exhibited by the energy levels of the nuclei of heavy atoms, according to quantum theory. This profound connection means the hypothesis may one day be proved not by a mathematician, but by a physicist. ▪

The uranium atom is one example of a heavy atom whose nucleus follows the same statistical behavior as prime numbers, making it extremely difficult to predict.

Bernhard Riemann

The son of a pastor, Bernhard Riemann was born in Germany in 1826. Initially fascinated by theology, he was persuaded to change his degree to mathematics by Carl Gauss, under whom he then studied at the University of Göttingen. The result was a series of breakthroughs that remain influential today.

In addition to his work on primes, Riemann helped to formulate the rules for applying calculus to complex functions (functions using complex numbers). His revolutionary understanding of space was used by Einstein in developing relativity theory. Despite his success, Riemann struggled financially. He could finally afford to marry when he was awarded a full professorship by Göttingen in 1862. Just a month later, he fell ill and his health deteriorated until he died of tuberculosis in 1866.

Key work

1868 *Über die Hypothesen, welche der Geometrie zu Grunde liegen* (*On the Hypotheses Which Lie at the Foundation of Geometry*)

SOME INFINITIES ARE BIGGER THAN OTHERS

TRANSFINITE NUMBERS

IN CONTEXT

KEY FIGURE
Georg Cantor (1845–1918)

FIELD
Number theory

BEFORE
450 BCE Zeno of Elea uses a series of paradoxes to explore the nature of infinity.

1844 French mathematician Joseph Liouville proves that a number can be transcendental—have an infinite number of digits arranged with no repeating pattern and without an algebraic root.

AFTER
1901 Bertrand Russell's barber paradox exposes the weakness of set theory's ability to define numbers.

1913 The infinite monkey theorem explains that given infinite time, random input will eventually produce all possible outcomes.

Infinity was a concept that mathematicians had long instinctively mistrusted. It was only in the late 1800s that Georg Cantor was able to explain it with mathematical rigor. He found there was more than one kind of infinity—an infinite variety, in fact—and that some were larger than others. In order to describe these differing infinities, he introduced "transfinite" numbers.

While he was studying set theory, Cantor aimed to create definitions for every number to infinity. This need arose from the discovery of transcendental numbers, such as π and e, which

---

The **infinite** set of **natural numbers** (positive integers) is well ordered and theoretically **can be listed**.

↓

So it is a **countable infinity**.

↓

The **infinite** set of **transcendental numbers**, such as π, **cannot be listed** in any order.

↓

So it is an **uncountable infinity**.

↓

An **uncountable infinity** is **larger** than a **countable infinity**.

↓

Some infinities are bigger than others.

See also: Irrational numbers 44–45 ▪ Zeno's paradoxes of motion 46–47 ▪ Negative numbers 76–79 ▪ Imaginary and complex numbers 128–31 ▪ Calculus 168–75 ▪ The logic of mathematics 272–73 ▪ The infinite monkey theorem 278–79

are irrational, infinitely long, and are not themselves an algebraic root. Between every algebraic number—including the integers, fractions, and certain irrational numbers (such as $\sqrt{2}$)—there is an infinite number of transcendentals.

Counting infinities

To help identify where a number is located, Cantor drew a distinction between two kinds of numbers: cardinals, which are the counting numbers 1, 2, 3… that denote the size of a set; and ordinals, such as 1st, 2nd, or 3rd, which list order.

Cantor created a new transfinite cardinal number—aleph ($\aleph$), the first letter of the Hebrew alphabet—to denote a set containing an infinite number of elements. The set of integers that includes the natural numbers, negative integers, and zero, was given the cardinality of $\aleph_0$, the smallest transfinite cardinal, as these are theoretically countable numbers but are actually impossible to count completely. A set with a cardinality of $\aleph_0$ starts with a first item, and ends with a ω (omega)

Every number within this diagram is real, as opposed to imaginary: it gives a positive result when it is squared.

A transcendental number can never be fully calculated and so cannot be added to a set of numbers in the correct order, thus forming an uncountable set.

Real
Real algebraic
Rational
Integer

Natural
1 2 3

-1
-2
-3
2.25

$^1/_3$

$\sqrt{2}$
$-\sqrt{3}$
$-^2/_3$
$1+\sqrt{5}/_2$

π
e
-2π

Transcendental

The numbers in these two bands are irrational as they cannot be described as fractions containing two integers.

Irrational

These concentric rings show the different types of numbers, which correspond to different types of infinities. Each ring describes a set of numbers. For example, the set of natural numbers is a small subset of rational numbers, which in turn combine with the set of irrational numbers to make the full set of real numbers.

item, a transfinite ordinal number. The number of items in a set with a cardinality of $\aleph_0$ is ω.

Adding to that set makes a new set of $\omega + 1$. A set of all countable ordinals, such as $\omega + 1$, $\omega + 1 + 2$, $\omega + 1 + 2 + 3…$, will contain ω_1 items. This set cannot be counted,

making this infinity larger than countable ones, so it is said to have a cardinality of $\aleph_1$.

The set of all $\aleph_1$ sets contains ω_2 items, with a cardinality of $\aleph_2$. In this way, Cantor's set theory creates infinities nested inside each, expanding forever. ▪

Georg Cantor

Born in St. Petersburg, Russia, in 1845, Georg Cantor moved with his family to Germany in 1856. An outstanding scholar (and violinist), he studied in Berlin and Göttingen. He was later made a professor of mathematics at the University of Halle.

Although much admired by today's mathematicians, Cantor was something of a pariah among his contemporaries. His theory of transfinite numbers clashed with traditional mathematical beliefs and the criticisms of leading mathematicians damaged his career. His work was also

criticized by the clergy, but Cantor, who was deeply religious, saw his research as a glorification of God.

Overwhelmed by depression, Cantor was institutionalized for much of his later life. He began to receive plaudits in the early 1900s, but lived out his old age in poverty. He died of a heart attack in 1918.

Key work

1915 *Contributions to the founding of the theory of transfinite numbers*

A DIAGRAMMATIC REPRESENTATION OF REASONINGS

VENN DIAGRAMS

IN CONTEXT

KEY FIGURE
John Venn (1834–1923)

FIELD
Statistics

BEFORE
c. 1290 Catalan mystic Ramon Llull devises classification systems using devices such as trees, ladders, and wheels.

c. 1690 Gottfried Leibniz creates classification circles.

1762 Leonhard Euler describes the use of logic circles, now known as "Euler circles."

AFTER
1963 American mathematician David W. Henderson outlines the connection between symmetrical Venn diagrams and prime numbers.

2003 In the US, Jerrold Griggs, Charles Killian, and Carla Savage show that symmetrical Venn diagrams exist for all primes.

In 1880, British mathematician John Venn introduced the idea of the Venn diagram in his paper "On the Diagrammatic and Mechanical Representation of Propositions and Reasonings." The Venn diagram is a way of grouping things in overlapping circles (or other curved shapes) to show the relationship between them.

Overlapping circles

The Venn diagram considers two or three different sets or groups of things with something in common, such as all living things, or all planets of the solar system. Each set is given its own circle and the circles are overlapped. Objects in each set are then arranged in the circles so that objects that belong in more than one set are placed where the circles overlap.

Two-circle Venn diagrams can represent categorical propositions, such as "All A are B," "No A are B," "Some A are B," and "Some A are not B." Three-circle diagrams can also represent syllogisms, in which there are two categorical premises

and a categorical conclusion. For example: "All French people are European. Some French people eat cheese. Therefore, some Europeans eat cheese."

As well as being a widely used tool for sorting data in everyday life, in contexts ranging from school classrooms to boardrooms, Venn diagrams are an integral part of set theory, due to their distinctive ability to express relationships. ∎

Great ideas are the ones that lie in the intersection of the Venn diagram of 'is a good idea' and 'looks like a bad idea.'
Sam Altman
American entrepreneur

See also: Syllogistic logic 50–51 ▪ Probability 162–65 ▪ Calculus 168–75 ▪ Euler's number 186–91 ▪ The logic of mathematics 272–73

THE TOWER WILL FALL AND THE WORLD WILL END
THE TOWER OF HANOI

IN CONTEXT

KEY FIGURE
Édouard Lucas (1842–91)

FIELD
Number theory

BEFORE
1876 Édouard Lucas proves that the Mersenne number $2^{127} - 1$ is prime. This is still the largest prime ever found without using a computer.

AFTER
1894 Lucas's work on recreational mathematics is posthumously published in four volumes.

1959 American writer Erik Frank Russell publishes "Now Inhale," a short story about an alien allowed to play a version of the Tower of Hanoi before his execution.

1966 In an episode of the BBC's *Doctor Who*, the villain, The Celestial Toymaker, forces the Doctor to play a ten-disk version of the game.

French mathematician Édouard Lucas is believed to have invented his Tower of Hanoi game in 1883. The aim of the puzzle is simple. The challenger is presented with three poles, one of which holds three disks in order of size, with the largest disk on the bottom. The three disks must be moved one disk at a time so as to recreate the starting arrangement on a different pole using the smallest possible number of moves, with the restriction that players can only place a disk on top of a larger disk or on to an empty pole.

Solving the puzzle

With just three disks, the Tower of Hanoi can be solved in just seven moves. With any number of disks, the formula $2^n - 1$ will give the minimum number of moves (where n is equal to the number of disks). One solution to the challenge employs binary numbers (0 and 1). Each disk is represented by a binary digit, or bit. A value of 0 indicates that a disk is on the

A form of the Tower of Hanoi is a popular toy for small children. Versions with eight disks are often used to test developmental skills of older children.

starting pole; 1 shows that it is on the final pole. The sequence of bits changes at each move.

According to legend, if monks at a certain temple in either India or Vietnam (depending on the version of the tale) succeed in moving 64 disks from one pole to another in line with the rules, the world will end. However, even using the best strategy and moving one disk per second, they would take 585 billion years to complete the game. ∎

See also: Wheat on a chessboard 112–13 ▪ Mersenne primes 124 ▪ Binary numbers 176–77

SIZE AND SHAPE DO NOT MATTER, ONLY CONNECTIONS

TOPOLOGY

IN CONTEXT

KEY FIGURE
Henri Poincaré (1854–1912)

FIELD
Geometry

BEFORE
1736 Leonhard Euler solves the historical topological problem of "The Seven Bridges of Königsberg."

1847 Johann Listing coins the term "topology" as a mathematical subject.

AFTER
1925 Russian mathematician Pavel Aleksandrov establishes the basis for studying the essential properties of topological spaces.

2006 Grigori Perelman's proof of the Poincaré conjecture is confirmed.

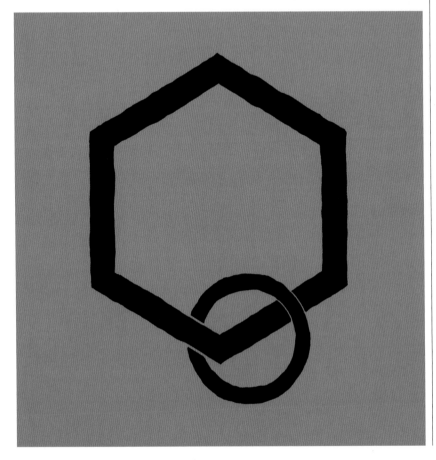

Topology is, in simple terms, the study of shapes without measurements. In classical geometry, if a pair of shapes has equal corresponding lengths and angles, and you can slide, reflect, or rotate one of the shapes into the other, they are congruent— a mathematical way of saying they are identical. To a topologist, however, two shapes are identical— or invariant, in topological terminology—if they can be molded one into the other by continuous stretching, twisting, or bending, but with no cutting, piercing, or sticking together. This has led to topology being called "rubber-sheet geometry."

See also: Euclid's *Elements* 52–57 ▪ Coordinates 144–51 ▪ The Möbius strip 248–49 ▪ Minkowski space 274–75 ▪ Proving the Poincaré conjecture 324–25

Topology is the **study of abstract shapes** without measurements.

↓

Topologically identical **shapes can be molded** into each other through stretching, twisting, or bending.

↓

Shape and size do not matter, only connections (the number of holes).

Henri Poincaré

Born in 1854, in Nancy, France, Henri Poincaré showed such early promise that he was described by a teacher as a "monster of mathematics." He graduated in the subject from the Paris École Polytechnique and earned his doctorate from the University of Paris. In 1886, he was appointed as chair of mathematical physics and probability at the Sorbonne in Paris, where he spent the rest of his career.

In 1887, Poincaré won a prize from King Oscar II of Sweden for his partial solution of the many variables involved in determining the stable orbit of three planets around one another. A self-confessed mistake threw his calculations for the stable orbit into doubt, but in turn paved the way for the study of "chaos theory." He died in 1912.

Key works

1892–99 *Les Méthodes nouvelles de la mécanique céleste (New Methods of Celestial Mechanics)*
1895 *Analysis Situs (Topology)*
1903 *La Science et l'hypothèse (Science and Hypothesis)*

For more than 2,000 years, from the time of Euclid, c. 300 BCE, geometry was concerned with classifying shapes by their lengths and angles. In the 18th and early 19th centuries, some mathematicians began to look at geometric objects differently, considering the global properties of shapes beyond the confines of lines and angles. Out of this grew the mathematical field of topology, which by the early 1900s had moved far from the notion of "shape" to embrace abstract algebraic structures. The most ambitious and influential exponent of this was French mathematician Henri Poincaré, who used complex topology to throw new light on the "shape" of the Universe itself.

Birth of a new geometry

In 1750, Leonhard Euler revealed that he had been working on a formula for polyhedra—three-dimensional figures with four or more planes, such as a cube or pyramid—that involved their vertices, edges, and faces rather than lines and angles. What he postulated became known as Euler's polyhedral formula: »

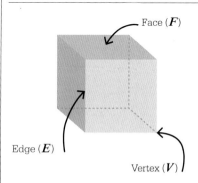

Euler's formula, $V + F - E = 2$, works for most polyhedra, including a cube. Its values of $V = 8$, $F = 6$, and $E = 12$, when fed into the formula, produce the calculation $8 + 6 - 12$ which equals 2.

Algebraic topology allows us to read qualitative forms and their transformations.
Stephanie Strickland
American poet

$V + F - E = 2$, where V is the number of vertices, F the number of faces, and E the number of edges. The formula suggested that all polyhedra shared basic characteristics.

However, in 1813, another Swiss mathematician, Simone L'Huilier, noted that Euler's formula was not true for all polyhedra; it was false for polyhedra with holes and for nonconvex polyhedra—shapes with some diagonals (linked by vertices) not contained within or on the surface. L'Huilier devised a system whereby every shape had its own "Euler characteristic"— $(V - E + F)$—and shapes with the same Euler characteristic were related regardless of how much they might be manipulated.

The term "topology"—derived from the Greek *topos*, meaning "a place"—was introduced to the mathematical world by German mathematician Johann Listing in 1847 in his treatise *Vorstudien zur Topologie* (*Introductory Studies in Topology*), although he had used the word in correspondence at least 10 years earlier. In particular, Listing was interested in shapes that did not satisfy Euler's formula or defied the conventions of having distinct "outside" and "inside" surfaces. He even devised a version of the Möbius strip—a surface that has only one side and one edge—a few months before August Möbius.

Around the same period, another German mathematician, Bernhard Riemann, devised new geometrical coordinate systems that extended beyond the limits of the 2-D and 3-D systems devised by René Descartes. Riemann's new framework enabled mathematicians to explore shapes in four dimensions or higher, including seemingly "impossible" shapes.

One such shape was the "Klein bottle," devised in 1882 by German mathematician Felix Klein. He imagined joining two Möbius strips together to create a shape that has only one surface, is nonorientable (has no "left" or "right"), and, unlike a Möbius strip, has no edge or boundary curve. As it has no intersections, the shape can only truly exist in four-dimensional space. If the shape is represented in 3-D, it has to intersect itself, which is where it starts to look like a bottle. Topologists applied the term "2-manifold" to shapes such as the Möbius strip and Klein bottle to describe their surfaces, which are two-dimensional surfaces embedded within a space of higher dimension (the Möbius strip can exist inside three dimensional space, but the Klein bottle can only exist properly in four).

A universal conjecture

The shape of the Universe has long been a source of speculation. We appear to inhabit a 3-D world, but to make any sense of its shape we need to take ourselves outside this, into four dimensions. In the same way, to gain a sense of the shape of a 2-D surface, we need to look down on it in three dimensions. A starting point would be to imagine that we inhabit a Universe that is a 3-D surface embedded within four dimensions. Taking this one step further, you could consider that this 3-D surface is actually a sphere

To a topologist, a coffee mug is identical in shape to a doughnut, because by pulling, stretching, and bending one, you could mold one into the shape of the other.

Coffee mug

Doughnut

The BlackDog™ robot is designed to carry loads over rough terrain. The robot's moves are computed using algebraic topology that can predict and model the surrounding "space."

embedded in a 4-D space, also known as a "3-sphere." A "2-sphere" is equivalent to a "normal" sphere (such as a ball) in a 3-D space.

In 1904, Henri Poincaré went even further, producing a theory that would help to lay a topological basis for understanding the shape of the Universe. He proposed what became known as the Poincaré conjecture: "every simply connected, closed 3-manifold is homeomorphic to the 3-sphere." A "3-manifold" is a shape that appears 3-D when its surface is enlarged but exists within higher dimensions, and "simply connected" means that it has no holes—like an orange but not a doughnut. A "closed" shape is finite, with no boundaries—like a sphere. Finally, "homeomorphic" describes shapes that can be molded into each other, such as a mug and a doughnut (see opposite). A doughnut and an orange, however, are not homeomorphic because of the hole in the doughnut.

According to Poincaré, if it could be could shown that the Universe did not contain holes, then you could model it as a "3-sphere." To establish whether it contained holes, you could, in theory, conduct an experiment with string. Imagine you are an explorer traveling around the Universe from a set point, and unraveling a ball of string as you go. When you get back to your starting point, you see the end of the string that you started with. You take both ends, and start to gather in the string, pulling both ends. If the Universe is "simply connected," then you would be able to gather in the whole string, like a loop following the smooth contours of a sphere; if you had passed through holes or gaps, then the string could get "snagged." For example, if the Universe were shaped like a doughnut, and, in your travels, you wrapped your string around its girth, the string would get caught. You would not be able to gather in the string without pulling it beyond the Universe.

Shaping the future

Topology developments still continued during the 1900s. In 1905, French mathematician Maurice Fréchet devised the idea of a metric space—a set of points along with a "metric" that defines the distance between them.

Also at the turn of the 20th century, German mathematician David Hilbert invented the idea of a space that took the Euclidean spaces of two and three dimensions and generalized them to infinite dimensions. Mathematics could then be done in any dimension in much the same way as in a 3-D coordinate system. This area of topological mathematics has become known as "infinite-dimensional topology."

The field of topology is now vast, embracing abstract algebraic structures far removed from a simple notion of "shape." It has wide-ranging applications in areas such as genetics and molecular biology, such as helping to unravel the "knots" created around DNA by certain enzymes. ■

Probably no branch of mathematics has experienced a more surprising growth.
Raymond Louis Wilder
American mathematician

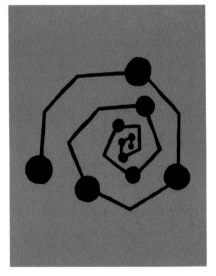

LOST IN THAT SILENT, MEASURED SPACE
THE PRIME NUMBER THEOREM

IN CONTEXT

KEY FIGURE
Jacques Hadamard
(1865–1963)

FIELD
Number theory

BEFORE
1798 French mathematician Adrien-Marie Legendre offers an approximate formula to determine how many prime numbers there are below or equal to a given value.

1859 Bernhard Riemann outlines a possible proof for the prime number theorem, but the necessary mathematics to complete it does not yet exist.

AFTER
1903 German mathematician Edmund Landau simplifies Hadamard's proof of the prime number theorem.

1949 Paul Erdős in Hungary and Atle Selberg in Norway both find a proof of the theorem using only number theory.

The prime numbers—those positive whole numbers that have only two factors, themselves and 1—have long fascinated mathematicians. If the first step was to find them, and they are frequent among the small numbers, the next step was to identify a pattern to describe their distribution. More than 2,000 years before, Euclid had proved that there are infinitely many primes, but it was only at the end of the 1700s that Legendre stated his conjecture—a formula to describe the distribution of primes. This became known as the prime number theorem. In 1896, Jacques Hadamard in France and Charles-Jean de la Vallée Poussin in Belgium both proved the theorem, quite independently.

It is evident that primes decrease in frequency as numbers get larger. Of the first 20 positive whole numbers, eight are prime— 2, 3, 5, 7, 11, 13, 17, and 19. Between the numbers 1,000 and 1,020, there are only three prime numbers (1,009, 1,013, 1,019), and between

There are **25 prime** numbers **from 1 to 100**.

There are **21 prime** numbers **from 101 to 200**.

There are **16 prime** numbers **from 201 to 300**.

Prime numbers become **less common** as numbers **get larger**.

A pattern of primes emerges.

See also: Euclid's *Elements* 52–57 ▪ Mersenne primes 124 ▪ Imaginary and complex numbers 128–31 ▪ The Riemann hypothesis 250–51

1	2	3	4	5	6	7	8	9	10
11	12	13	14	15	16	17	18	19	20
21	22	23	24	25	26	27	28	29	30
31	32	33	34	35	36	37	38	39	40
41	42	43	44	45	46	47	48	49	50

▨ **Primes**

Primes tend to decrease in frequency as numbers get larger. Although there are two primes between 30 and 40, and three between 40 and 50, the accuracy of the prime number theorem increases at higher numbers.

1,000,000 and 1,000,020, the only prime is 1,000,003. This seems reasonable; the higher the number, the more numbers that could be divisors exist below it.

Many notable mathematicians have puzzled over how primes are distributed. In 1859, German mathematician Bernhard Riemann worked toward a proof in his paper *On the Number of Primes Less Than a Given Magnitude*. He believed that complex analysis, a branch of mathematics in which ideas of function are applied to complex numbers (combinations of real numbers, such as 1, and imaginary numbers, such as $\sqrt{-1}$), would lead to a resolution. He was right; the study of complex analysis developed, fueling the proofs of Hadamard and Poussin.

What the theorem says

The prime number theorem is designed to calculate how many primes there are less than or equal to a real number x. It states that $\pi(x)$ is approximately equal to $x \div ln(x)$ as x gets larger and tends to infinity. Here $\pi(x)$ denotes the prime counting function (how many primes) and is unrelated to the number pi, and $ln(x)$ is the

natural logarithm of x. To explain the theorem slightly differently, for a large number x, the average gap between primes from 1 to x is approximately $ln(x)$. Or, for any number between 1 and x, the probability of it being a prime is approximately $1 \div ln(x)$.

The prime numbers are the building blocks for numbers in mathematics, just as the elements are for compounds in chemistry. Fundamental to understanding this is the Riemann hypothesis—an unsolved conjecture—which, if true, could reveal a huge amount more about prime numbers. ∎

The prime numbers… grow like weeds among the natural numbers, seeming to obey no other law than that of chance.
Don Zagier
American mathematician

Jacques Hadamard

Born in Versailles, France, in 1865, Jacques-Salomon Hadamard became interested in mathematics thanks to an inspiring teacher. He obtained his doctorate in Paris in 1892 and the same year won the Grand Prix des Sciences Mathématiques for his work on primes. He moved to Bordeaux to lecture at the university, and there proved the prime number theorem.

In 1894, Alfred Dreyfus, a Jewish relative of Hadamard's wife, was falsely accused of selling state secrets and was sentenced to life in prison. Hadamard, who was also Jewish, worked tirelessly on behalf of Dreyfus and he was eventually freed. Hadamard's brilliant career was marred by personal loss; two of his sons died in World War I, and another in World War II. The death of his grandson Étienne in 1962 was a final blow. Hadamard died a year later.

Key works

1892 *Determination of the Number of Primes Less than a Given Number*
1910 *Lesson on the Calculus of Variations*

MODERN
MATHEM
1900—PRESENT

ATICS

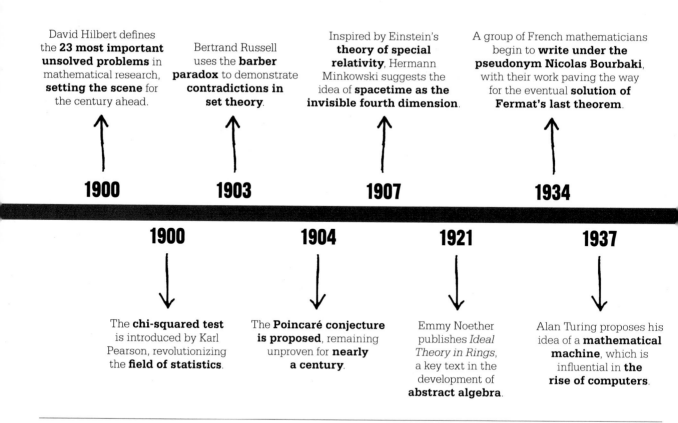

David Hilbert defines the **23 most important unsolved problems** in mathematical research, **setting the scene** for the century ahead.

Bertrand Russell uses the **barber paradox** to demonstrate **contradictions in set theory**.

Inspired by Einstein's **theory of special relativity**, Hermann Minkowski suggests the idea of **spacetime as the invisible fourth dimension**.

A group of French mathematicians begin to **write under the pseudonym Nicolas Bourbaki**, with their work paving the way for the eventual **solution of Fermat's last theorem**.

1900   **1903**   **1907**   **1934**

1900   **1904**   **1921**   **1937**

The **chi-squared test** is introduced by Karl Pearson, revolutionizing the **field of statistics**.

The **Poincaré conjecture is proposed**, remaining unproven for **nearly a century**.

Emmy Noether publishes *Ideal Theory in Rings*, a key text in the development of **abstract algebra**.

Alan Turing proposes his idea of a **mathematical machine**, which is influential in **the rise of computers**.

I n 1900, as the arms race that led to World War I intensified, German mathematician David Hilbert attempted to anticipate the directions that mathematics would take in the 20th century. His list of the 23 unsolved problems he considered crucial was influential in identifying the fields of mathematics that could be fruitfully explored by mathematicians.

New century, new fields

One area of exploration was the foundations of mathematics. In seeking to establish the logical basis of mathematics, Bertrand Russell described a paradox that highlighted a contradiction in Georg Cantor's naive set theory, leading to a reappraisal of the subject. These ideas were taken up by André Weil and others, using

the pseudonym Nicolas Bourbaki. Starting from the basics, they met in the 1930s and 40s, rigorously formalizing all branches of mathematics in terms of set theory.

Others, notably Henri Poincaré, explored the newly established field of topology, the offshoot of geometry dealing with surfaces and space. His famous conjecture concerns the 2-dimensional surface of a 3-dimensional sphere. Unlike many of his peers in the 1900s, Poincaré did not confine himself to any one single field of mathematics. As well as pure mathematics, he made significant discoveries in theoretical physics, including his proposed principle of relativity. Similarly, Hermann Minkowski—whose primary interest was in geometry and the geometrical method applied to problems in number theory—

explored the notion of multiple dimensions, and suggested spacetime as a possible fourth dimension. Emmy Noether, one of the first female mathematicians of the modern era to gain recognition, came to the field of theoretical physics from a perspective of abstract algebra.

The computer age

In the first half of the 1900s, applied mathematics was largely concerned with theoretical physics, especially the implications of Einstein's theories of relativity, but the latter part of the century was increasingly dominated by advances in computer sciences. Interest in computing had begun in the 1930s, in the search for a solution to Hilbert's *Entscheidungsproblem* (decision problem) and the possibility of an

Edward Lorenz publishes his work on **chaos theory** that later becomes **synonymous** with the example of **the "butterfly effect."**

In the US, three mathematicians develop the **RSA algorithm**, using prime numbers **to encrypt information**.

Benoit Mandelbrot creates the **Mandelbrot set**, having coined the term **"fractal."**

Fermat's last theorem is finally declared **to be solved** after British mathematician Andrew Wiles **corrects an error** in his initial proof.

1963   **1977**   **1980**   **1995**

1965   **1977**   **1989**   **2006**

Fuzzy logic is formulated by Lotfi Zadeh and is soon used in a **wide range of technologies**, particularly in Japan.

The **solution to the four-color problem** becomes the first mathematical **theorem to be proved by a computer**.

The **World Wide Web** invented by **Tim Berners-Lee** facilitates the rapid transmission of ideas, including mathematics.

Grigori Perelman's proof of the **Poincaré conjecture** is fully accepted by the mathematical community.

---

algorithm to determine the truth or falsity of a statement. One of the first to tackle the problem was Alan Turing, who went on to develop code-cracking machines during World War II that were the forerunners of modern computers. He later proposed a test of artificial intelligence.

With the advent of electronic computers, mathematics was in demand to provide methods of designing and programming computer systems. But computers also provided a powerful tool for mathematicians. Hitherto unsolved mathematical problems such as the four-color theorem often involved lengthy calculations, which could now be done quickly and accurately by computer. Although Poincaré had laid the foundations of chaos theory, Edward Lorenz was able to establish the principles more firmly with the aid of computer models. His visual images of attractors and oscillators, along with Benoit Mandelbrot's fractals, became icons of these new fields of study.

With the advent of computers, the secure transfer of data became an issue, and mathematicians devised complex cryptosystems using the factorization of large prime numbers. Launched in 1989, the World Wide Web facilitated the rapid transmission of knowledge, and computers became a part of everyday life, especially in the field of information technology.

New logic, new millennium

For a while, it seemed electronic computing could potentially provide answers to almost all problems. But computing science was based on a binary system of logic first proposed by George Boole in the 1800s, and the polar opposites of on-off, true-false, 0-1, and so on could not describe how things are in the real world. To overcome this, Lotfi Zadeh suggested a system of "fuzzy" logic, in which statements can be partly true or false, in a range between 0 (absolutely false) and 1 (absolutely true).

In 2000, 21st-century mathematics was heralded in a similar spirit to that of the 20th century, when the Clay Mathematics Institute announced seven Millennium Prize Problems, offering a $1 million prize for any of their solutions. As yet, only the Poincaré conjecture has been solved; Grigori Perelman's proof was confirmed in 2006. ■

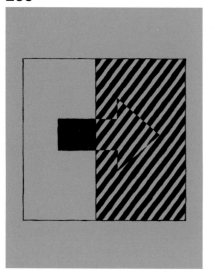

THE VEIL BEHIND WHICH THE FUTURE LIES HIDDEN
23 PROBLEMS FOR THE 20TH CENTURY

In 1900, **David Hilbert** stated **23 problems** that he felt would **occupy mathematicians** for the next **century**.

He felt solving these problems would improve our understanding of a wide range of fields, including number theory, algebra, geometry, and calculus.

10 of the problems have been **resolved**.

Seven have solutions **not universally accepted**.

Four remain **unresolved**.

Two are too vague to ever be **resolved**.

IN CONTEXT

KEY FIGURE
David Hilbert (1862–1943)

FIELDS
Logic, geometry

BEFORE
1859 Bernhard Riemann proposes the Riemann hypothesis, a famous problem that will later be Number 8 on Hilbert's list and remains unresolved today.

1878 Georg Cantor advances the continuum hypothesis, later Number 1 on Hilbert's list.

AFTER
2000 The Clay Institute issues a list of seven Millennium Prize mathematical problems, offering a million dollars for each problem solved.

2008 In a bid to stimulate major new mathematical breakthroughs, the US Defense Advanced Research Projects Agency (DARPA) announces its list of 23 unsolved problems.

It requires a special technical brilliance and self-confidence to predict relevant problems for the next hundred years, but this is what German mathematician David Hilbert did in 1900. Hilbert possessed a substantial grasp of most fields of mathematics. At the International Mathematical Congress in Paris in 1900, he confidently announced his choice of 23 questions that he believed should occupy mathematicians'

thoughts in the decades to come. This proved prescient; the math world rose to the challenge.

The range of problems
Many of Hilbert's questions are highly technical, but some are more accessible. Number 3, for instance, asks if one of any two polyhedra of the same volume can always be cut into finitely many bits that can be reassembled to create the other polyhedron.

The infinite! No other question has ever moved so profoundly the spirit of man.
David Hilbert

This was soon resolved in 1900 by German-born American mathematician Max Dehn, who concluded that it could not.

The continuum hypothesis, the first problem on Hilbert's list, pointed out that the set of natural numbers (the positive integers) was infinite, but so was the set of real numbers between 0 and 1. As a result of the work of German mathematician Georg Cantor, it was agreed that the first infinity was "smaller" than the second.

The continuum hypothesis also stated that there was no infinity lying between these two infinities. Cantor himself was sure this was true, but he could not prove it. In 1940, Austrian–American logician Kurt Gödel showed it could not be proved that such an infinity exists, and, in 1963, American mathematician Paul Cohen showed it could not be proved that such an infinity does not exist. Hilbert's first problem is substantially resolved, although set theory (the study of the properties of sets) is a complex subject, and much more work on it remains to be done. Of Hilbert's 23 problems, 10 are considered resolved, seven have been partially solved, two have been classed as too vague to ever be definitively solved, three remain unsolved, and one (also unsolved) is really a physics problem. Among the unsolved problems is the Riemann hypothesis, which some observers think will remain unsolved for the foreseeable future.

Challenges for the future

Hilbert's remarkable achievement was to accurately predict what would concern mathematicians in the 1900s and beyond. When American mathematician and Fields Medal winner Steve Smale came up with his own list of 18 questions in 1998, it included Hilbert's eighth and 16th problems. Two years later, the Riemann hypothesis was also one of the Clay Institute's Millennium Prize problems. Today's mathematicians face further challenges, but aspects of Hilbert's problems – especially those that are still unsolved – remain relevant. ▪

Problem solving and theory building go hand in hand. That's why Hilbert risked offering a list of unsolved problems instead of presenting new methods or results.
Rüdiger Thiele
German mathematician

David Hilbert

Born in Prussia in 1862 to German parents, David Hilbert entered the University of Königsberg in 1880 and later taught there before becoming professor of mathematics at the University of Göttingen in 1895. In this role, he turned Göttingen into one of the mathematical hubs of the world and taught a number of young mathematicians who later made their own mark.

Hilbert was renowned for his broad understanding of many areas of mathematics, and had a keen interest in mathematical physics, too. Exhausted by anemia, he retired in 1930, and Göttingen's math faculty soon declined after the Nazi purges of Jewish colleagues. Despite his great contribution to mathematics, Hilbert's death in 1943, during World War II, went largely unnoticed.

Key works

1897 *Commentary on Numbers*
1900 "The Problems of Mathematics" (Paris lecture)
1932–35 *Collected Works*
1934–39 *Foundations of Mathematics* (with Paul Bernays)

STATISTICS IS THE GRAMMAR OF SCIENCE

THE BIRTH OF MODERN STATISTICS

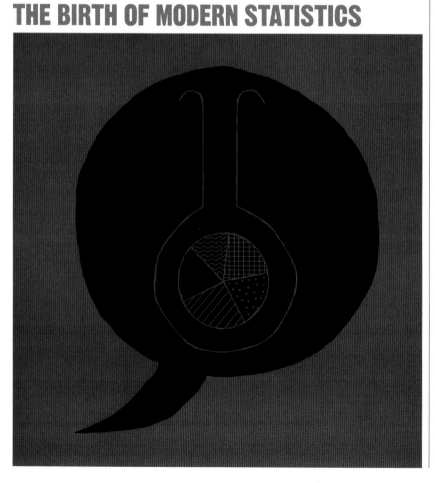

IN CONTEXT

KEY FIGURE
Francis Galton (1822–1911)

FIELD
Number theory

BEFORE
1774 Pierre-Simon Laplace shows the expected pattern of distribution around the norm.

1809 Carl Friedrich Gauss develops the least squares method of finding the best fit line for a scatter of data.

1835 Adolphe Quetelet advocates the use of the bell curve to model social data.

AFTER
1900 Karl Pearson proposes the chi-squared test to determine the significance of differences between expected and observed frequencies.

S tatistics is the branch of mathematics that is concerned with analyzing and interpreting large quantities of data. Its foundations were laid in the late 1800s, principally by British polymaths Francis Galton and Karl Pearson.

Statistics investigates whether the pattern of recorded data is significant or random. Its origins lie in the efforts of 18th-century mathematicians such as Pierre-Simon Laplace to identify observational errors in astronomy. In any set of scientific data, most errors are likely to be very small, and only a few are likely to be very large. So when observations are plotted on a graph, they create a bell-shaped curve with a peak created by the most likely result, or "norm," in the middle. In 1835, Belgian mathematician Adolphe Quetelet posited that characteristics, such as body mass, within a human population follow a bell curve pattern, in which values around the mean are most frequent. Higher and lower values are less frequent. He devised the Quetelet Index (now called the BMI) to indicate body mass.

Typically, plotting two variables, such as height and age, on a graph creates a messy scatter of data points that cannot be linked by a neat line. However, in 1809, German mathematician Carl Friedrich Gauss found an equation to create a "best fit" line, which would show the relationship between the variables. Gauss used a method called "least squares," which involves adding up the squares of the data; this is still used by statisticians. By the 1840s, mathematicians such as Auguste Bravais were looking at the level of error that could be accepted for this line, and tried to pin down the significance of the midpoint or "median" of a set of data.

Correlation and regression

It was first Galton, then Pearson, who began to draw these threads together. Galton was inspired by his cousin Charles Darwin's work on evolution, and his aim was to show how likely it was that factors such as height, physiognomy, and even intelligence and criminal tendencies might be passed from one generation to the next. Galton and Pearson's ideas are tainted by the doctrine of eugenics and racial improvement, but the techniques that they developed have found applications elsewhere. »

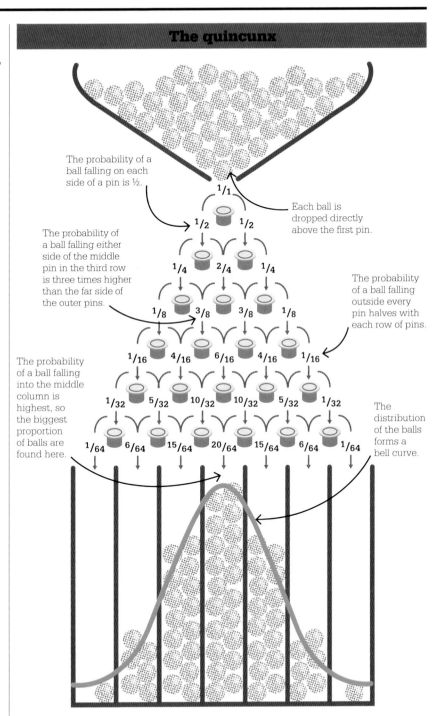

The quincunx

The probability of a ball falling on each side of a pin is ½.

Each ball is dropped directly above the first pin.

The probability of a ball falling either side of the middle pin in the third row is three times higher than the far side of the outer pins.

The probability of a ball falling outside every pin halves with each row of pins.

The probability of a ball falling into the middle column is highest, so the biggest proportion of balls are found here.

The distribution of the balls forms a bell curve.

Francis Galton invented the quincunx (sometimes called the Galton board) to model the bell curve. His original design had beads dropping over pegs.

Galton was a rigorous scientist, determined to analyze data to show mathematically how probable outcomes are. In his innovative 1889 book *Natural Inheritance*, Galton showed how two sets of data can be compared to show if there is a significant relationship

Galton built an "anthropometric laboratory" to collect information on human characteristics, including head size and quality of vision. It generated huge amounts of data that he had to analyze statistically.

between them. His approach involved establishing two related concepts that are now at the heart of statistical analysis: correlation and regression.

Correlation measures the degree to which two random variables, such as height and weight, correspond. It often looks for a linear relationship—that is, a relationship that gives a simple line on a graph, with one variable changing in step with the other. Correlation does not imply a causal relationship between the two variables; it simply means they

vary together. Regression, on the other hand, looks for the best equation for the graph line for two variables, so that changes in one variable can be predicted from changes to the other.

Standard deviation

Although Galton's main interest was human heredity, he created a broad range of data sets. Famously, he measured the size of seeds produced by sweet pea plants grown from seven sets of seeds. Galton found that the smallest pea seeds had larger offspring and the largest seeds produced smaller offspring. He had discovered the phenomenon of "regression to the mean," a tendency for measurements to even out, always drifting toward the mean over time.

Inspired by Galton's work, Pearson set out to develop the mathematical framework for correlation and regression. After exhaustive tests that involved tossing coins and drawing lottery tickets, Pearson came up with the key idea of "standard deviation," which shows how much on average observed values differ from expected. To arrive at this figure, he found the mean, which is the sum of

Regression to the mean

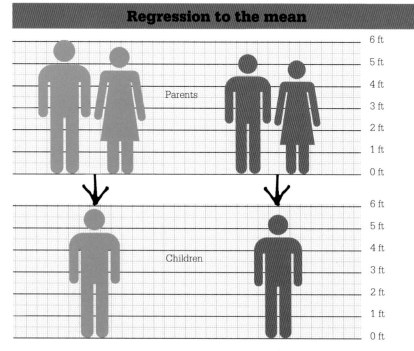

Parents

Children

Galton noticed that very tall parents tend to have children who are shorter than their parents, while very short parents tend to have children who are slightly taller than their parents. The second generation will be closer in height than the first, an example of regression to the mean.

No observational problem will not be solved by more data.
Vera Rubin
American astronomer

all the values divided by how many values there are. Pearson then found the variance—the average of the squared differences from the mean. The differences are squared in order to avoid problems with negative numbers, and the standard deviation is the square root of the variance. Pearson realized that by uniting the mean and the standard deviation, he could calculate Galton's regression precisely.

Chi-squared test

In 1900, after an extensive study of betting data from the gaming tables of Monte Carlo, Pearson described the chi-squared test, now one of the cornerstones of statistics. Pearson's aim was to determine whether the difference between observed values and expected values is significant, or simply the result of chance.

Using his data on gambling, Pearson calculated a table of probability values, called chi-squared (χ^2), in which 0 shows no significant difference from expected (the "null hypothesis"), whereas larger values show a significant difference. Pearson

Francis Galton introduced…		**Karl Pearson** introduced…	
Correlation: the degree to which two variables correspond.	**Regression to the mean**: the tendency of data to even out over time.	**Standard deviation**: the degree to which results differ from the mean.	**The chi-squared test**: for variations between observed and expected data.

Modern statistics was born.

painstakingly worked out his table by hand, but chi-squared tables are now produced using computer software. For each set of data, a chi-squared value can be found from the sum of all the differences between observed and expected values. The chi-squared values are checked against the table to find the significance of the variations in the data within limits set by the researcher and known as "degrees of freedom."

The combination of Galton's correlation and regression, and Pearson's standard deviation and chi-squared test, formed the foundations of modern statistics. These ideas have since been refined and developed, but they remain at the heart of data analysis. This is crucial in many aspects of modern life, from comprehending economic behavior to planning new transportation links and improving public health services. ■

Karl Pearson

Karl Pearson was born in London in 1857. An atheist, freethinker, and socialist, he became one of the greatest statisticians of the 1900s, but he was also a champion of the discredited science of eugenics.

After graduating with a degree in mathematics from Cambridge University, Pearson became a teacher before making his mark in statistics. In 1901, he founded the statistical journal *Biometrika* with Francis Galton and evolutionary biologist Walter F. R. Weldon, followed by the world's first university department of statistics at University College, London, in 1911. His views often led him into disputes. He died in 1936.

Key works

1892 *The Grammar of Science*
1896 *Mathematical Contributions to the Theory of Evolution*
1900 *On the criterion that a given system of deviation from the probable in the case of a correlated system of variables is such that it can be reasonably supposed to have arisen from random sampling.*

A FREER LOGIC EMANCIPATES US

THE LOGIC OF MATHEMATICS

IN CONTEXT

KEY FIGURE
Bertrand Russell
(1872–1970)

FIELD
Logic

BEFORE
c. 300 BCE Euclid's *Elements* contains an axiomatic approach to geometry.

1820s French mathematician Augustin Cauchy clarifies the rules for calculus, inaugurating a new rigor in mathematics.

AFTER
1936 Alan Turing studies the computability of mathematical functions, with a view to analyzing which problems in mathematics can be decided and which cannot.

1975 American logician Harvey Friedman develops the "reverse mathematics" program, which starts with theorems and works backward to axioms.

The **barber paradox** imagines a town in which **all** men must be **clean shaven**.

↓

Any man who does not shave **himself** must be shaved by the **town barber**.

↓

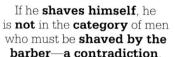

So who shaves the barber?

↓   ↓

If he **shaves himself**, he is **not** in the **category** of men who must be **shaved by the barber—a contradiction**.

If he **does not** shave himself, he **is** in the **category** of men who must be **shaved by the barber—another contradiction**.

The common perception that mathematics is logical, with fixed rules, evolved over millennia, dating back to ancient Greece with the works of Plato, Aristotle, and Euclid. A rigorous definition of the laws of arithmetic and geometry had emerged by the 1800s, with the work of George Boole, Gottlob Frege, Georg Cantor, Giuseppe Peano, and, in 1899, David Hilbert's *Foundations of Geometry*. However, in 1903, Bertrand Russell published *The Principles of Mathematics*, which revealed a flaw in the logic of one area of mathematics. In the book, he explored a paradox, known as Russell's paradox (or the Russell–Zermelo paradox, after German mathematician Ernst Zermelo, who made a similar discovery in 1899).

See also: The Platonic solids 48–49 ▪ Syllogistic logic 50–51 ▪ Euclid's *Elements* 52–57 ▪ The Goldbach conjecture 196 ▪ The Turing machine 284–89

Bertrand Russell

The son of a lord, Bertrand Russell was born in Monmouthshire, Wales, in 1872. He studied mathematics and philosophy at Cambridge University, but was dismissed from an academic post there in 1916 for anti-war activities. A prominent pacifist and social critic, in 1918 he was jailed for six months, during which he wrote his *Introduction to Mathematical Philosophy*.

Russell taught in the US in the 1930s, although his appointment at a college in New York was revoked due to a judicial declaration that his opinions rendered him morally unfit. He was awarded the Nobel Prize in Literature in 1950, and in 1955 he and Albert Einstein released a joint manifesto calling for a ban on nuclear weapons. He later opposed the Vietnam War. Russell died in 1970.

Key works

1903 *The Principles of Mathematics*
1908 *Mathematical Logic as Based on the Theory of Types*
1910–13 *Principia Mathematica* (with Alfred North Whitehead)

The paradox implied that set theory, which deals with the properties of sets of numbers or functions, and was fast becoming the bedrock of mathematics, contained a contradiction. To explain the problem, Russell used an analogy known as the barber paradox in which a barber shaves every man in town aside from those who shave themselves, creating two sets of people: those who shave themselves and those shaved by the barber. However, this begs the question: if the barber shaves himself, to which of the two sets does the barber belong?

Russell's barber paradox contradicted Frege's *Basic Laws of Arithmetic* concerning the logic of mathematics, which Russell had pointed out in a letter to Frege in 1902. Frege declared that he was "thunderstruck," and he never found an adequate solution to the paradox.

A theory of types
Russell went on to produce his own response to his paradox, developing a "theory of types," which placed restrictions on the established model of set theory (known as "naive set theory") by creating a hierarchy so that "the set of all sets" would be treated differently from its constituent smaller sets. In so doing, Russell managed to circumvent the paradox completely. He utilized this new set of logical principles in the momentous *Principia Mathematica*, written with Alfred North Whitehead and published in three volumes from 1910 to 1913.

Logical gaps
In 1931, Kurt Gödel, an Austrian mathematician and philosopher, published his incompleteness theorem (following on from his completeness theorem of a few years earlier). The 1931 theorem concluded that there will always exist some statements regarding numbers that may be true, but can never be proved. Furthermore, expanding mathematics by simply adding more axioms will lead to further "incompleteness." This meant that the efforts of Russell, Hilbert, Frege, and Peano to develop complete logical frameworks for mathematics were destined to have logical gaps, however watertight they tried to make them.

Gödel's theorem also implied that some as-yet unproven theorems in mathematics, such as the Goldbach conjecture, may never be proved. This has not, however, deterred mathematicians in their resolute efforts to prove Gödel wrong. ▪

Every good mathematician is at least half a philosopher, and every good philosopher is at least half a mathematician.
Gottlob Frege

THE UNIVERSE IS FOUR-DIMENSIONAL
MINKOWSKI SPACE

IN CONTEXT

KEY FIGURE
Hermann Minkowski
(1864–1909)

FIELD
Geometry

BEFORE
c. 300 BCE In his book
Elements, Euclid establishes
the geometry of 3-D space.

1904 In his book *The
Fourth Dimension*, British
mathematician Charles Hinton
coins the term "tesseract" for
a four-dimensional cube.

1905 French scientist Henri
Poincaré has the idea of
making time the fourth
dimension in space.

1905 Albert Einstein states
his theory of special relativity.

AFTER
1916 Einstein writes the key
paper outlining his theory of
general relativity, in which he
explains gravity as a curvature
of spacetime.

There are three dimensions
in our familiar view of the
world—length, width, and
height—and they can largely be
described mathematically by the
geometry of Euclid. But in 1907,
German mathematician Hermann
Minkowski delivered a lecture in
which he added time, an invisible
fourth dimension, to create the
concept of spacetime. This has
played a key part in understanding
the nature of the Universe. It has
provided a mathematical framework
for Einstein's theory of relativity,
allowing scientists to develop and
expand this theory.

It was in the 1700s that scientists
first began questioning whether
three-dimensional Euclidean
geometry could describe the entire
Universe. Mathematicians started
to develop non-Euclidean geometric
frameworks, while some considered
time as a potential dimension.
Light provided the mathematical
prompt. In the 1860s, Scottish
scientist James Clerk Maxwell

A black hole occurs when spacetime
warps so much that its curvature
becomes infinite at the hole's center.
Even light is not fast enough to escape
the hole's immense gravitational pull.

See also: Euclid's *Elements* 52–57 ▪ Newton's laws of motion 182–83 ▪ Laplace's demon 218–19 ▪ Topology 256–59 ▪ Proving the Poincaré conjecture 324–25

Stationary object

Moving objects

Object moving at the speed of light

Time / Space

Time / Space — Slower moving object / Faster moving object

Time / Space — As nothing travels faster than light, the worldline of all objects is in this section. — 45°

The worldline for a stationary object is vertical because it is not moving through space.

A slower object has a steeper worldline as it is progressing more slowly along the space axis.

This worldline has a 45° angle, with a 1:1 ratio between the time and space axes.

found that the speed of light is the same whatever the speed of its source. Mathematicians then developed his equations to try to understand how the finite speed of light fit into the coordinate system of space and time.

Mathematics of relativity

In 1904, Dutch mathematician Henrik Lorentz developed a set of equations, called "transformations," to show how mass, length, and time change as a spatial object approaches the speed of light. A year later, Albert Einstein produced his theory of special relativity, which proved that the speed of light is the same throughout the Universe. Time is a relative, not an absolute, quantity—running at different speeds in different places and woven together with space.

Minkowski turned Einstein's theory into mathematics. He showed how space and time are parts of a four-dimensional spacetime, where each point in space and time has a position. He represented movement between positions as a theoretical line, a

"worldline," which could be plotted on a graph, with space and time as the axes. A static object produces a vertical worldline, and the worldline of a moving object is at an angle (see above). The worldline angle of an object moving at the speed of light is 45°. According to Minkowski, no worldline can exceed this angle, but in reality, there are three axes of space, plus the axis of time, so the 45° worldline is really a "hypercone," a 4-dimensional figure. All physical reality is held within it, as nothing can travel faster than light. ∎

Henceforth, space by itself, and time by itself shall fade to mere shadows, and only some union of the two will preserve independent reality.
Hermann Minkowski

Hermann Minkowski

Born in Aleksotas (now in Lithuania) in 1864, Minkowski moved with his family to Königsberg in Prussia in 1872. As a boy, he showed an aptitude for math and began his studies at the University of Königsberg aged 15. By 19, he had won the Paris Grand Prix for mathematics, and at 23, he became a professor at the University of Bonn. In 1897 he taught the young Albert Einstein in Zurich.

Following a move to Göttingen in 1902, Minkowski became fascinated by the mathematics of physics, especially the interaction of light and matter. When Einstein unveiled his theory of special relativity in 1905, Minkowski was spurred on to develop his own theory, in which space and time form part of a four-dimensional reality. This concept inspired Einstein's theory of general relativity in 1915, but by then, Minkowski was dead—killed at 44 years old by a ruptured appendix.

Key work

1907 *Raum und Zeit* (*Space and Time*)

RATHER A DULL NUMBER

TAXICAB NUMBERS

IN CONTEXT

KEY FIGURE
Srinivasa Ramanujan
(1887–1920)

FIELD
Number theory

BEFORE
1657 In France, mathematician Bernard Frénicle de Bessy cites the properties of 1,729, the original "taxicab" number.

1700s Swiss mathematician Leonhard Euler calculates that 635,318,657 is the smallest number that can be expressed as the sum of two fourth powers (numbers to the power of 4) in two ways.

AFTER
1978 Belgian mathematician Pierre Deligne receives the Fields Medal for his work on number theory, including the proof of a conjecture in the theory of modular forms that was first made by Ramanujan.

A "taxicab" number, Ta(n), is the smallest number that can be expressed as the sum of two positive cubed integers (whole numbers) in n (number of) different ways. They owe their name to an anecdote from 1919, when British mathematician G. H. Hardy went to Putney, London, to visit his protégé Srinivasa Ramanujan, who was unwell. Arriving in a cab with the number 1,729, Hardy remarked, "Rather a dull number, don't you think?" Ramanujan disagreed, then explained that 1,729 is the smallest number that is the sum of two positive cubes in two different ways. Hardy's frequent retelling of this story ensured that 1,729 would become one of the best-known numbers in mathematics. Ramanujan was not the first to make note of this number's unique properties; French mathematician Bernard Frénicle de Bessy had also written about them in the 1600s.

Extending the concept
The taxicab story inspired later mathematicians to examine the property that Ramanujan had recognized and to expand its application. The hunt was on for the smallest number that could be expressed as the sum of two

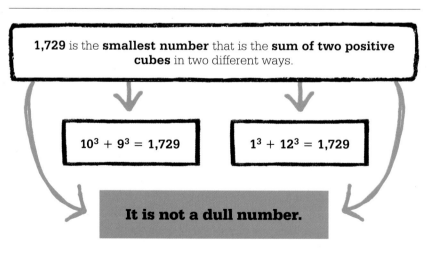

1,729 is the **smallest number** that is the **sum of two positive cubes** in two different ways.

$10^3 + 9^3 = 1,729$

$1^3 + 12^3 = 1,729$

It is not a dull number.

See also: Cubic equations 102–05 ▪ Elliptic functions 226–27 ▪ Catalan's conjecture 236–37 ▪ The prime number theorem 260–61

Does Ta(n) always exist?

The existence of Ta(n) was proved theoretically in 1938 for all values of **n**, but the search is still on for larger taxicab numbers. Even with the benefits of computer calculations, mathematicians have not yet moved beyond Uwe Hollerbach's discovery of Ta(6).

Date	Number	Value	Discoverer
N/A	Ta(1)	2	N/A
1657	Ta(2)	1,729	de Bessy
1957	Ta(3)	87,539,319	Leech
1989	Ta(4)	6,963,472,309,248	Rosenstiel, Dardis, Rosenstiel
1994	Ta(5)	48,988,659,276,962,496	Dardis
2008	Ta(6)	24,153,319,581,254,312,065,344	Hollerbach

Srinivasa Ramanujan

Born in Madras, India in 1887, Ramanujan displayed an extraordinary aptitude for mathematics at an early age. Finding it hard to get full recognition locally, he took the bold step of sending some of his results to G. H. Hardy, then a professor at Trinity College, Cambridge. Hardy declared that they had to be the work of a mathematician "of the highest class," and had to be true, because no one could invent them. In 1913, Hardy invited Ramanujan to work with him in Cambridge. The collaboration was hugely productive: in addition to the taxicab numbers, Ramanujan also developed a formula for obtaining the value of pi to a high level of accuracy.

However, Ramanujan suffered from poor health. He returned to India in 1919 and died a year later—probably as a result of amoebic dysentery contracted years earlier. He left behind several notebooks, which mathematicians are still studying today.

Key work

1927 *Collected papers of Srinivasa Ramanujan*

positive cubes in three, four, or more different ways. A further question was whether Ta(**n**) exists for all values of **n**; in 1938, Hardy and British mathematician Edward Wright proved that it does (an existence proof), but developing a method of finding Ta(**n**) in each case has proved elusive.

Extending the concept further, the expression Ta(**j**, **k**, **n**) seeks the smallest positive number that is the sum of any number of different positive integers (**j**), each to any power (**k**) in **n** distinct ways. For example, Ta(4, 2, 2) requires the smallest number that is the sum of four squares (or two fourth powers) in two different ways: 635,318,657.

Continuing relevance

Taxicab numbers were only one area of Hardy and Ramanujan's work. Their main focus was prime numbers. Hardy was excited by Ramanujan's claim that he had found a function of **x** that exactly represented the number of prime numbers less than **x**; Ramanujan was unable, however, to offer â rigorous proof.

Taxicab numbers have little practical use, but they still inspire scholars as curiosities. Mathematicians now also seek "cabtaxi" numbers: based on the taxicab formula, these allow calculations using both positive and negative cubes. ▪

An equation means nothing to me unless it expresses a thought of God.
Srinivasa Ramanujan

A MILLION MONKEYS BANGING ON A MILLION TYPEWRITERS
THE INFINITE MONKEY THEOREM

IN CONTEXT

KEY FIGURE
Émile Borel (1871–1956)

FIELD
Probability

BEFORE
45 BCE The Roman philosopher Cicero argues that a random combination of atoms forming Earth is highly improbable.

1843 Antoine Augustin Cournot makes a distinction between physical and practical certainty.

AFTER
1928 British physicist Arthur Eddington develops the idea that improbable is impossible.

2003 Scientists at Plymouth University in the UK test Borel's theory with real monkeys and a computer keyboard.

2011 American programmer Jesse Anderson's million virtual monkey software generates the complete works of Shakespeare.

In the early 1900s, French mathematician Émile Borel explored improbability—when events had a very small chance of ever occurring. Borel concluded that events with a sufficiently small probability will never occur. He was not the first to study the probability of unlikely events. In the 4th century BCE, the ancient Greek philosopher Aristotle suggested in *Metaphysics* that Earth was created by atoms coming together entirely by chance. Three centuries later, the Roman philosopher Cicero argued that this was so unlikely that it was essentially impossible.

Defining impossibility
Over the past two millennia, various thinkers have probed the balance between the improbable and the impossible. In the 1760s, French mathematician Jean

In an **infinite amount of time**, an **infinite** number of **events** will happen.

A monkey typing for **infinity** would produce every letter in every **possible combination** an **infinite** number of times.

The **monkey** would therefore produce **every finite text** an **infinite** number of times.

According to mathematical probability, a monkey typing for infinity will eventually type the complete works of Shakespeare.

See also: Probability 162–65 ▪ The law of large numbers 184–85 ▪ Normal distribution 192–93 ▪ Laplace's demon 218–19
▪ Transfinite numbers 252–53

> The physically impossible event is therefore the one that has infinitely small probability, and only this remark gives substance… to the theory of mathematical probability.
> **Antoine Augustin Cournot**

d'Alembert questioned whether it was possible to have a very long string of occurrences in a sequence in which occurrence and non-occurrence are equally likely—for example, whether a person flipping a coin might get "heads" two million times in a row. In 1843, French mathematician Antoine Augustin Cournot questioned the possibility of balancing a cone on its tip. He argued that it is possible but highly unlikely, and made the distinction between a physical certainty—an event that can happen physically, like the balancing cone—and a practical certainty, which is so unlikely that in practical terms it is considered impossible. In what is sometimes known as Cournot's principle, Cournot suggested that an event with a very small probability will not happen.

Infinite monkeys
Borel's law, which he called the law of single chance, gave a scale to practical certainty. For events on a human scale, Borel considered events with a probability of less than 10^{-6} (or 0.000001) to be impossible. He also came up with a famous example to illustrate impossibility: monkeys hitting typewriter keys at random will eventually type the complete works of Shakespeare. This outcome is highly improbable, but mathematically, over an infinite time (or with an infinite number of monkeys), it must happen. Borel

Borel's theory is often applied to stock markets, where the level of chaos means that in some cases random selection performs better than selection based on traditional economic theories.

noted that, while it cannot be mathematically proven that it is impossible for monkeys to type Shakespeare, it is so unlikely that mathematicians should consider it impossible. This idea of monkeys typing the works of Shakespeare captured people's imagination and Borel's law came to be known as the infinite monkey theorem. ▪

Émile Borel

Born in 1871 in Saint-Affrique, France, Émile Borel was a mathematics prodigy and graduated top of his class from the École Normale Supérieure in 1893. After lecturing in Lille for four years, he returned to the École, where he dazzled fellow mathematicians with a series of brilliant papers.

Borel is best known for his infinite monkey theorem, but his lasting achievement was in laying the foundations for the modern understanding of complex functions—what a variable must be altered by to achieve a particular output. During World War I, Borel worked for the War Office and later became minister of the navy. Imprisoned when the Germans invaded France in World War II, he was released and fought for the Resistance, earning himself the Croix de Guerre. He died in 1956 in Paris.

Key works

1913 *Le Hasard* (*Chance*)
1914 *Principes et formules classiques du calcul des probabilités* (*Principles and classic formulas of probability*)

SHE CHANGED THE FACE OF ALGEBRA

EMMY NOETHER AND ABSTRACT ALGEBRA

IN CONTEXT

KEY FIGURE
Emmy Noether (1882–1935)

FIELD
Algebra

BEFORE
1843 German mathematician Ernst Kummer develops the concept of ideal numbers—ideals in the ring of integers.

1871 Richard Dedekind builds on Kummer's idea to formulate definitions of rings and ideals more generally.

1890 David Hilbert refines the concept of the ring.

AFTER
1930 Dutch mathematician Bartel Leendert Van der Waerden writes the first comprehensive treatment of abstract algebra.

1958 British mathematician Alfred Goldie proves that Noetherian rings can be understood and analyzed in terms of simpler ring types.

In the 1800s, analysis and geometry were the leading fields of mathematics, while algebra was considerably less popular. Throughout the Industrial Revolution, applied mathematics was prioritized over areas of study that were more theoretical. This all changed in the early 1900s with the rise of "abstract" algebra, which became one of the key fields of mathematics, largely thanks to the innovations of German mathematician Emmy Noether.

Noether was not the first to focus on abstract algebra. Work on algebra theory had been developed by

My methods are really methods of working and thinking; this is why they have crept in everywhere anonymously.
Emmy Noether

mathematicians such as Joseph-Louis Lagrange, Carl Friedrich Gauss, and British mathematician Arthur Cayley, but gained traction when German mathematician Richard Dedekind began to study algebraic structures. He conceptualized the ring—a set of elements with two operations, such as addition and multiplication. A ring can be broken into parts called "ideals"—a subset of elements. For example, the set of odd integers are an ideal in the ring of integers.

Significant works

Noether began her work on abstract algebra shortly before World War I with her exploration of invariant theory, which explained how some algebraic expressions stay the same while other quantities change. In 1915, this work led her to make a major contribution to physics; she proved that the laws of conservation of energy and mass each correspond to a different type of symmetry. The conservation of electric charge, for example, is related to rotational symmetry. Now called Noether's theorem, it was praised by Einstein for the way it addressed his theory of general relativity.

See also: Algebra 92–99 ▪ The binomial theorem 100–01 ▪ The algebraic resolution of equations 200–01 ▪ The fundamental theorem of algebra 204–09 ▪ Group theory 230–33 ▪ Matrices 238–41 ▪ Topology 256–59

Mathematicians created a system called **abstract algebra** to generalize mathematical objects and the **operations that act on them**.

⬇

A **set** is a collection of **objects** or **elements**, such as integers.

⬇

A **group** is a type of set that **includes an operation** (e.g. addition), and follows certain axioms.

⬇

A **ring** is a **type of group** that includes a **second operation**, often multiplication. It also includes the **axiom of associativity**, whereby each of the operations can be applied in any order without affecting the result.

⬇

Noether's contributions to ring theory furthered our understanding of algebraic structures.

Emmy Noether

Born in 1882, Emmy Noether struggled to find education, recognition, and even basic employment in early 20th century academia as a Jewish woman in Germany. Although her mathematical skill won her a position at the University of Erlangen—where her father also taught mathematics— from 1908 to 1923 she received no pay. She later faced similar discrimination in Göttingen, where her colleagues had to fight to have her officially included in the faculty. In 1933, the rise of the Nazis led to her dismissal, and she moved to the US, working at Bryn Mawr College and at the Institute for Advanced Study until her death in 1935.

Key works

1921 *Idealtheorie in Ringbereichen* (*Ideal Theory in Rings*)
1924 *Abstrakter Aufbau der Idealtheorie im algebraischen Zahlkörper* (*Abstract Construction of Ideal Theory in Algebraic Fields*)

In the early 1920s, Noether's work focused on rings and ideals. In a key paper in 1921, *Idealtheorie in Ringbereichen* (*Ideal Theory in Rings*), she studied ideals in a particular set of "commutative rings," in which the numbers can be swapped around when they are multiplied without affecting the result. In a 1924 paper, she proved that in these commutative rings, every ideal is the unique product of prime ideals. One of the most brilliant mathematicians of her time, Noether laid the foundations for the development of the entire field of abstract algebra with her contributions to ring theory. ▪

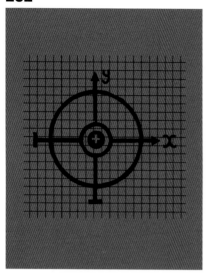

STRUCTURES ARE THE WEAPONS OF THE MATHEMATICIAN
THE BOURBAKI GROUP

IN CONTEXT

KEY FIGURES
André Weil (1906–1998),
Henri Cartan (1904–2008)

FIELDS
Number theory, algebra

BEFORE
1637 René Descartes creates coordinate geometry, allowing points on a flat surface to be described.

1874 Georg Cantor creates set theory, describing how sets and their subsets interrelate.

1895 Henri Poincaré lays the foundations of algebraic topology in *Analysis Situs* (*Analysis of Position*).

AFTER
1960s The New Mathematics movement, which focuses on set theory, becomes popular in American and European schools.

1995 Andrew Wiles publishes his final proof of Fermat's last theorem.

Russian mathematical genius Nicolas Bourbaki was one of the most prolific and influential mathematicians of the 1900s. His monumental work *Éléments de Mathématique* (*Elements of Mathematics*, 1960), occupies a key place in university libraries and countless students of mathematics have learned the tools of their trade from his work.

Bourbaki, however, never existed. He was a fiction created in the 1930s by young French mathematicians who were striving to fill the vacuum left by the

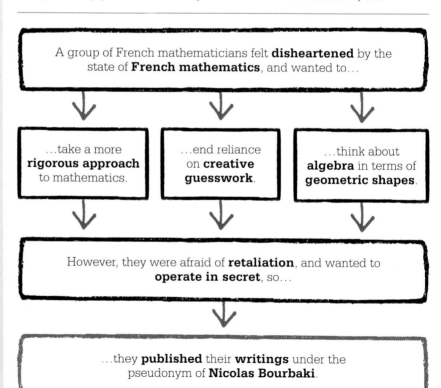

A group of French mathematicians felt **disheartened** by the state of **French mathematics**, and wanted to…

…take a more **rigorous approach** to mathematics.

…end reliance on **creative guesswork**.

…think about **algebra** in terms of **geometric shapes**.

However, they were afraid of **retaliation**, and wanted to **operate in secret**, so…

…they **published** their **writings** under the pseudonym of **Nicolas Bourbaki**.

See also: Coordinates 144–51 ▪ Topology 256–59 ▪ The butterfly effect 294–99 ▪ Proving Fermat's last theorem 320–23 ▪ Proving the Poincaré conjecture 324–25

The Bourbaki group poses for a photo at the first Bourbaki congress in July 1935. Among them are Henri Cartan (standing far left) and André Weil (standing fourth from left).

The group aimed to strip mathematics back to basics and provide a foundation from which it could go forward. While their work sparked a brief fad in the 1960s, it proved too radical for teachers and pupils alike. The group was often at odds with cutting-edge mathematics and physics, and was so focused on pure math that applied math was of little interest to them. Topics containing uncertainty, such as probability, had no place in Bourbaki's work.

Even so, the group made important contributions across a wide range of mathematical topics, particularly in set theory and algebraic geometry. The group, which acts in secrecy and whose members must resign at age 50, still exists, although Bourbaki now publishes infrequently. The most recent two volumes were published in 1998 and 2012. ▪

devastation of World War I. While other countries had kept academics at home, French mathematicians had joined their countrymen in the trenches and a generation of teachers had been killed. French mathematics was stuck with antiquated textbooks and teachers.

Renewing mathematics

Some young teachers believed that French mathematics had fallen victim to a lack of rigor and precision. They were distrustful of the creative guesswork, as they saw it, of older mathematicians such as Henri Poincaré in developing chaos theory and mathematics for physics.

In 1934, two young lecturers at the University of Strasbourg, André Weil and Henri Cartan, took matters into their own hands. They invited six fellow former students from the École Normale Supérieur to lunch in Paris, hoping to persuade them to take part in an ambitious project to write a new treatise that would revolutionize mathematics.

The group—which included Claude Chevalley, Jean Delsarte, Jean Dieudonné, and René de Possel— agreed to create a new body of work that covered all fields of mathematics. Meeting regularly and marshaled by Dieudonné, the group produced book after book, led by *Éléments de Mathématique*. Their work was likely to be controversial, so they adopted the pseudonym Nicolas Bourbaki.

Bourbaki's legacy

Topology and set theory—the meeting between numbers and shapes—were for Bourbaki at the very root of mathematics and lay at the heart of the group's work. René Descartes had first made the link between shapes and numbers in the 1600s with coordinate geometry, turning geometry into algebra. Bourbaki helped make the link the other way, turning algebra into geometry to create algebraic geometry, which is perhaps their lasting legacy. It was at least partly Bourbaki's work on algebraic geometry that led British mathematician Andrew Wiles to finally prove Fermat's last theorem; he published his proof in 1995.

Some mathematicians believe algebraic geometry has great untapped potential for the future. It already has real-world applications such as in programming codes in cell phones and smart cards.

IN CONTEXT

KEY FIGURE
Alan Turing (1912–54)

FIELD
Computer science

BEFORE
1837 In the UK, Charles
Babbage designs the
Analytical Engine, a
mechanical computer using
the decimal system. If it had
been constructed, it would
have been the first "Turing-
complete" device.

AFTER
1937 Claude Shannon designs
electrical switching circuits
that use Boolean algebra to
make digital circuits that
follow rules of logic.

1971 American mathematician
Stephen Cook poses the P
versus NP problem, which
tries to understand why some
mathematical problems can
quickly be verified but would
take billions of years to prove,
despite computers' immense
calculating power.

If a machine is expected
to be infallible, it cannot
also be intelligent.
Alan Turing

Computing the answers to many number problems can be
reduced to an algorithm—a sequence of **mathematical steps**
that are applied in a predefined order.

Some **algorithms**
reach **answers**;
others **loop forever**.

The **Turing machine**
can **process** any
**algorithm, solvable
or not**.

**By inputting algorithms to the
machine, it is possible to prove when an
algorithm has no answer.**

A lan Turing is often cited
as the "father of digital
computing," yet the
Turing machine that earned him
that accolade was not a physical
device but a hypothetical one.
Instead of constructing a prototype
computer, Turing used a thought
experiment in order to solve the
Entscheidungsproblem (decision
problem) that had been posed by
German mathematician David
Hilbert in 1928. Hilbert was
interested in whether logic could
be made more rigorous by being
simplified into a set of rules, or
axioms, in the same way that
arithmetic, geometry, and other
fields of mathematics were thought
possible to simplify at the time.
Hilbert wanted to know if there
was a way to predetermine whether
an algorithm—a method for solving
a specific mathematical problem

using a given set of instructions
in a given order—would arrive
at a solution to the problem.

In 1931, Austrian mathematician
Kurt Gödel demonstrated that
mathematics based on formal
axioms could not prove everything
that was true according to those
axioms. What Gödel called the
"incompleteness theorem" found
that there was a mismatch
between mathematical truth
and mathematical proof.

Ancient roots
Algorithms have ancient origins.
One of the earliest examples is
the method used by the Greek
geometer Euclid to calculate the
greatest common divisor of two
numbers—the largest number
that divides both of them without
leaving a remainder. Another early
example is Eratosthenes' sieve,

See also: Euclid's *Elements* 52–57 ▪ Eratosthenes' sieve 66–67 ▪ 23 Problems for the 20th century 266–67 ▪ Information theory 291 ▪ Cryptography 314–17

A man provided with paper, pencil, and rubber, and subject to strict discipline, is in effect a universal machine.
Alan Turing

attributed to the 3rd-century BCE Greek mathematician. It is an algorithm for sorting primes from composite (not prime) numbers. The algorithms of Eratosthenes and Euclid work perfectly and can be proven always to do so, but they did not conform to a formal definition. It was the need for this that led Turing to create his "virtual machine."

In 1937, Turing published his first paper as a fellow of King's College, Cambridge, "On Computable

Numbers, with an Application to the *Entscheidungsproblem*." It showed that there is no solution to Hilbert's decision problem: some algorithms are not computable, but there is no universal mechanism for identifying them before trying them.

Turing reached this conclusion using his hypothetical machine, which came in two parts. First there was a tape, as long as it needed to be, divided into sections, each section carrying a coded character. This character could be anything, but the simplest version used 1s and 0s. The second part was the machine itself, which read the data from each section of the tape (either by the head or tape moving). The machine would be equipped with a set of instructions (an algorithm) that controlled the behavior of the machine. The machine (or tape) could move left, »

Clerks at work in Hut 8, Bletchley Park, UK, during World War II. At one point, Turing led the work of Hut 8, which deciphered communiqués between Adolf Hitler and his forces.

Alan Turing

Born in London in 1912, Alan Turing was described as a genius by his teachers. After graduating with a first-class degree in mathematics from the University of Cambridge in 1934, he went on to study at Princeton in the US.

Returning to the UK in 1938, Turing joined the Government Code and Cypher School at Bletchley Park. After war broke out in 1939, he and others developed the Bombe, an electromechanical device that deciphered enemy messages. Following the war, Turing worked at Manchester University, where he designed the Automatic Computing Engine (ACE) and developed further digital devices.

In 1952, Turing was convicted of homosexuality, then a crime in the UK. He was also barred from working on codebreaking for the government. To avoid prison, Turing agreed to hormone treatment to reduce his libido. In 1954, he committed suicide.

Key work

1939 "Report on the Applications of Probability to Cryptography"

The Turing machine consists of a head that reads data from an infinitely long tape. The machine's algorithm might either instruct the head or the tape to move—to go left, right, or stay still. The memory keeps track of changes and feeds them back into the algorithm.

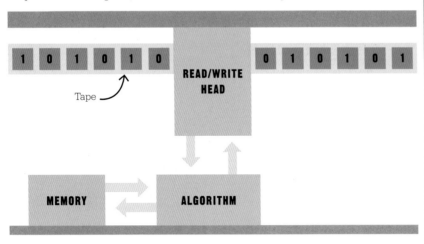

Tape

READ/WRITE HEAD

MEMORY

ALGORITHM

right, or stay where it was, and it could rewrite the data on the tape, switching a 0 to 1 or vice versa. Such a machine could carry out any conceivable algorithm.

Turing was interested in whether any algorithm put into the machine would cause the machine to halt. Halting would signify that the algorithm had arrived at a solution. The question was whether there was a way of knowing which algorithms (or virtual machines), would halt and which would not; if Turing could find out, he would answer the decision problem.

The halting problem

Turing approached this problem as a thought experiment. He began by imagining a machine that was able to say whether any algorithm (A) would halt (provide an answer and stop running) when given an input to which the answer was either Yes or No. Turing was not concerned with the physical mechanics of such a machine. Once he had conceptualized such a machine, however, he could theoretically

take any algorithm and test it using the machine to see if it halted.

In essence, the Turing machine (M) is an algorithm that tests another algorithm (A) to see if it is solvable. It does this by asking: does A halt (have a solution)? M then reaches an answer of Yes or No. Turing then imagined a modified version of this machine (M*), which would be set up so that if the answer was Yes (A does halt), then M* would do the opposite—it would loop forever (and not halt). If the answer was No (A does not halt), then M* would halt.

Turing then took this thought experiment further by imagining that you could use the machine M* to test whether its own algorithm, M*, would halt. If the answer was Yes, the algorithm M* will halt, then the machine M* would not halt. If the answer was No, the algorithm M* never halts, then the machine M* would halt. Turing's thought experiment had, therefore, created a paradox which could be used as a form of mathematical proof. It proved that,

because it was impossible to know if the machine would ever halt or not, then the answer to the decision problem was No: there was no universal test for the validity of algorithms.

Computer architecture

The Turing machine had not finished its job. Turing and others realized that this simple concept could be used as a "computer." At the time, the term "computer" was used to describe a person who carried out complex mathematical calculations. A Turing machine would do so using an algorithm to rewrite an input (the data on the tape) into an output. In terms of computing ability, the algorithms at work in a Turing machine are the strongest type ever devised. Modern computers and the programs that run on them are effectively working as Turing machines, and so are said to be "Turing complete."

As a leading figure in mathematics and logic, Turing made important contributions to the development of real computers, not just virtual ones. However, it

We need to feed [information] through a processor. A human must turn information into intelligence or knowledge. We've tended to forget that no computer will ever ask a new question.
Grace Hopper
American computer scientist

A Turing Bombe, used to decipher coded messages, has been rebuilt at the museum at Bletchley Park, the British code-breaking center during World War II.

was Hungarian mathematician John von Neumann who contrived a real-life version of Turing's hypothetical device using a central processing unit (CPU) that converted an input to an output by calling up information stored in an internal memory and sending back new information to be saved. He proposed his configuration, known as the "von Neumann architecture," in 1945, and today, a similar process is used in almost every computing device.

Binary code

Turing did not initially envisage that his machine would use only binary data. He merely thought it would use code with a finite set of characters. However, binary was the language of the first Turing-complete machine ever built, the Z3. Constructed in 1941 by German engineer Konrad Zuse, the Z3 used electromechanical relays, or switches, to represent 1s and 0s of binary data. Initially referred to as "discrete variables," in 1948 the 1s and 0s in computer code were renamed "bits," short for binary digits. This term was coined by Claude Shannon, a leading figure in information theory—the field of mathematics examining how information could be stored and transmitted as digital codes.

Early computers used multiple bits as "addresses" for sections of memory—showing where the processor should look for data. These chunks of bits became known as "bytes," spelled this way to avoid confusion with "bits." In the early decades of computing, bytes generally contained 4 or 6 bits, but the 1970s saw the rise of Intel's 8-bit microprocessors, and byte became the unit for 8 bits. The 8-bit byte was convenient because 8 bits have 2^8 permutations (256), and can encode numbers from 0 to 255.

Armed with a binary code arranged in sets of eight digits—and later even longer strings—software could be produced for any conceivable application. Computer programs are simply algorithms; the inputs from a keyboard, microphone, or touchscreen are

The popular view that scientists proceed inexorably from well-established fact to well-established fact, never being influenced by any unproved conjecture, is quite mistaken.
Alan Turing

processed by these algorithms into outputs, such as text on a device's screen.

The principles of the Turing machine are still used in modern computers and look set to continue until quantum computing changes how information is processed. A classical computer bit is either 1 or 0, never anything in between. A quantum bit, or "qubit," uses superposition to be both a 1 and 0 at the same time, which boosts computing power enormously. ∎

The Turing test

In 1950, Turing developed a test of a machine's ability to exhibit intelligent behavior equivalent to, or indistinguishable from, that of a human. In his view, if a machine appeared to be thinking for itself, then it was.

The annual Loebner Prize in Artificial Intelligence (AI) was inaugurated in 1990 by American inventor Hugh Loebner and the Cambridge Center for Behavioral Studies, Massachusetts. Every year, computers using AI try to win the prize. The AIs must fool human judges into thinking they are human rather than a computer program. AIs who progress to the final take it in turns to communicate with one of four judges. Each judge is also communicating with a human and must decide whether the AI or the human is most humanlike.

Over the years the test has had many critics, who question its ability to truly judge the intelligence of an AI effectively or see the competition as a stunt that does not advance knowledge in the field of AI.

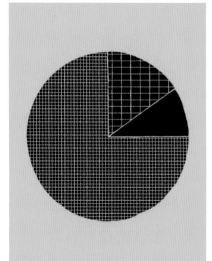

SMALL THINGS ARE MORE NUMEROUS THAN LARGE THINGS
BENFORD'S LAW

IN CONTEXT

KEY FIGURE
Frank Benford (1883–1948)

FIELD
Number theory

BEFORE
1881 Canadian astronomer Simon Newcomb notices that the pages most often referred to in logarithm tables are for numbers starting with 1.

AFTER
1972 Hal Varian, an American economist, suggests using Benford's law to detect fraud.

1995 American mathematician Ted Hill proves that Benford's law can be applied to statistical distributions.

2009 Statistical analysis of the Iranian presidential election results shows that they do not conform to Benford's law, suggesting that the election may have been rigged.

It might be expected that in any large set of numbers, those that start with the digit 3 would occur with roughly the same frequency as those that start with any other digit. However, many sets of numbers—a list of populations for US villages, towns, and cities, for example—show a distinctly different pattern. Often in a set of naturally occurring numbers, around 30 percent of the numbers have a leading digit of 1, around 17 percent have a leading digit of 2, and less than 5 percent have a leading digit of 9. In 1938, American physicist Frank Benford wrote a paper on this phenomenon; mathematicians later referred to it as Benford's law.

Recurring pattern

Benford's law is evident in many situations, from the lengths of rivers to share prices and mortality rates. Some types of data fit the law better than others. Naturally occurring data that extends over several orders of magnitude, from hundreds to millions, for example, fulfils the law better than data that is more closely grouped. The numbers in the Fibonacci sequence follow Benford's law, as do the powers of many integers. Numbers that act as a name or label, such as bus or telephone numbers, do not fit.

When numbers are made up, they tend to have a more equal distribution of leading digits than if they followed Benford's law. This has enabled investigators to use the law to detect financial fraud. ∎

Funnily, of the 20 data sets that Benford collected, six of the sample sizes have leading digit 1. Notice anything strange about that?
Rachel Fewster
Statistical ecologist, New Zealand

See also: The Fibonacci sequence 106–11 ▪ Logarithms 138–41 ▪ Probability 162–65 ▪ Normal distribution 192–93

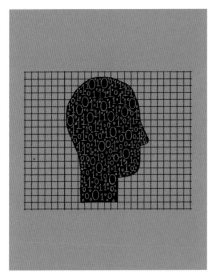

A BLUEPRINT FOR THE DIGITAL AGE
INFORMATION THEORY

IN CONTEXT

KEY FIGURE
Claude Shannon (1916–2001)

FIELD
Computer science

BEFORE
1679 Gottfried Leibniz develops the ancient idea of binary numbering.

1854 George Boole introduces the algebra that will form the basis for computing.

1877 Austrian physicist Ludwig Boltzman develops the link between entropy (measure of randomness) and probability.

1928 In the US, Ralph Hartley, an electronics engineer, sees information as a measurable quantity.

AFTER
1961 German physicist Rolf Landauer shows that the manipulation of information increases entropy.

In 1948, Claude Shannon, an American mathematician and electronics engineer, published a paper called *A Mathematical Theory of Communication*. This launched the information age by unlocking the mathematics of information and showing how it could be transmitted digitally.

At the time, messages could only be transmitted using a continuous, analog signal. The main drawback to this was that waves become weaker the further they travel, and increasing background interference creeps in. Eventually, this "white noise" overwhelms the original message.

Shannon's solution was to divide information into the smallest possible chunks, or "bits" (binary digits). The message is converted into a code made of 0s and 1s— every 0 is a low voltage and every 1 is a high voltage. In creating this code, Shannon drew on binary mathematics, the idea that figures can be represented by just 0s and 1s, which had been developed by Gottfried Leibniz.

Shannon demonstrates Theseus, his electromechanical "mouse," which used a "brain" of telephone relays to find its way around a maze.

Although Shannon was not the first to send information digitally, he fine-tuned the technique. For him, it was not simply about solving the technical problems of transmitting information efficiently. By showing that information could be expressed as binary digits, he launched the theory of information—with implications stretching into every field of science, and into every home or office with a computer. ∎

See also: Calculus 168–75 ▪ Binary numbers 176–77 ▪ Boolean algebra 242–47

WE ARE ALL JUST SIX STEPS AWAY FROM EACH OTHER
SIX DEGREES OF SEPARATION

Most **individuals** have a range of **connections** to people from **different** parts of their **lives**.

These **connections**, in turn, have **links** to other **groups** and **networks** of people.

Further links to individuals who are **three steps removed** (for example, a friend of a friend of a friend) reveal a **wide range** of **people connected** to **each other**.

Studies have suggested that, when connected by our social networks, we are all, on average, six steps away from each other.

Networks are used to model relationships between objects or people in many disciplines, including computer science, particle physics, economics, cryptography, biology, sociology, and climatology. One type of network is a "six degrees of separation" social network diagram, which measures how connected people are to each other.

In 1961, Michael Gurevitch, an American postgraduate student, published a landmark study of the nature of social networks. In 1967, Stanley Milgram studied how many

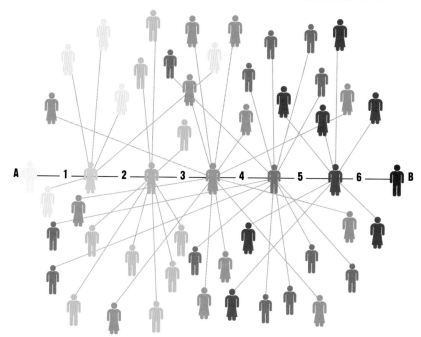

The six degrees of separation theory shows how any two seemingly unconnected people can be connected in no more than six steps by their friends and acquaintances. This number may decrease with the growth of social media.

intermediate acquaintance links were needed to connect strangers in the US. He had people in Nebraska send a letter intended to eventually reach a specific (random) person in Massachusetts. Each recipient then sent the letter on to a person they knew to get it closer to its target destination. Milgram studied how many people each of the letters went through to reach their targets. On average, the letters that reached the target needed six intermediaries.

This "small world theory" predated Milgram. In a 1929 short story *Chains*, Frigyes Karinthy suggested that people's average connection-number across the world might be six when the connecting factor is friendship. Karinthy, who was a writer, not a mathematician, coined the phrase "six degrees of separation." Mathematicians have since tried to model the average degree of separation. Duncan Watts and Steven Strogatz showed that if you have a random network with N nodes, each of which has K links to other nodes, then the average path length between two nodes is $ln N$ divided by $ln K$ (where ln means the natural logarithm). If there are 10 nodes, each with four connections to other nodes, then the average distance between two nodes chosen at random will be $^{ln 10}/_{ln 4} \approx 1.66$.

Other social networks

In the 1980s, friends of Hungarian mathematician Paul Erdős, who was well known for working collaboratively, coined the term "Erdős number" to indicate his degree of separation from other published mathematicians. Erdős's coauthors had an Erdős number of 1, anyone who had worked with one of his coauthors had an Erdős number of 2, and so on. This concept captured the public's imagination following an interview with American actor Kevin Bacon, in which he said he had worked with every actor in Hollywood or with someone who had worked with them. The term "Bacon number" was coined to indicate an actor's degree of separation from Bacon. In rock music, connections to members of the heavy metal group Black Sabbath are indicated by the "Sabbath number." To filter out the truly well-connected, there is the Erdős-Bacon-Sabbath number (the sum of someone's Erdős, Bacon, and Sabbath numbers). Only a few individuals have single-digit EBS numbers.

In 2008, Microsoft conducted research to show that everyone on Earth is separated from every other person by only 6.6 people on average. As social media brings us ever closer, this number may reduce even further. ▪

It is my hope that Six Degrees [a philanthropic project] will… [bring] a social conscience to social networking.
Kevin Bacon

A SMALL POSITIVE VIBRATION CAN CHANGE THE ENTIRE COSMOS

THE BUTTERFLY EFFECT

IN CONTEXT

KEY FIGURE
Edward Lorenz (1917–2008)

FIELD
Probability

BEFORE
1814 Pierre-Simon Laplace ponders the consequences of a deterministic universe where knowing all present conditions can be used to predict the future for all eternity.

1890 Henri Poincaré shows there is no general solution to the three-body problem, which predicts the motion of three celestial bodies kept in contact by gravity. Mostly, the bodies do not move in rhythmic, repeating patterns.

AFTER
1975 Benoit Mandelbrot uses computer graphics to create more complex fractals (shapes that self-repeat). The Lorenz attractor, which revealed the butterfly effect, is a fractal.

I n 1972, Edward Lorenz, an American meteorologist and mathematician, delivered a talk titled "Does the flap of a butterfly's wings in Brazil set off a tornado in Texas?" This was the origin of the term "butterfly effect," which refers to the idea that a tiny change in atmospheric conditions (which could be caused by anything, not just a butterfly) is enough to alter weather patterns somewhere else in the future. If the butterfly had not made its small contribution to the initial conditions,

The idea that a butterfly flapping its wings in one part of the world could alter atmospheric conditions and eventually produce a tornado elsewhere has captured the popular imagination.

then the tornado or other weather event would not have occurred at all, or would have struck some place other than Texas.

The title of the lecture was not chosen by Lorenz himself, but by physicist Philip Merilees, the convener of the American Association for the Advancement

Edward Lorenz

Born in 1917, in West Hartford, Connecticut, Edward Lorenz studied mathematics at Dartford College and Harvard University, gaining a masters degree at Harvard in 1940. After training as a meteorologist, he served with the US Army Air Corps in World War II. After the war, Lorenz studied meteorology at the Massachusetts Institute of Technology and began to develop ways to predict the behavior of the atmosphere. At that time, meteorologists used linear statistical modeling to forecast weather, and they often failed.

In developing a nonlinear model of the atmosphere, Lorenz stumbled across the area of chaos theory that would later be dubbed the butterfly effect. He showed that even the most powerful computers could not produce accurate long-term weather forecasts. Lorenz remained physically and mentally active until just before his death in 2008.

Key work

1963 *Deterministic Nonperiodic Flow*

> A butterfly flaps its wings in the Amazonian jungle, and subsequently a storm ravages half of Europe.
> **Terry Pratchett and Neil Gaiman**
> **British authors**

of Science's annual meeting in Boston. Lorenz had been late to provide information about his proposed talk, so Merilees had improvised, basing his choice of words on what he knew of Lorenz's work and an earlier comment that "one flap of a seagull's wings" could be enough to change the weather forecast.

Chaos theory

The butterfly effect is a popular introduction to chaos theory, which looks at the way complex systems are highly sensitive to initial conditions and are thus extremely unpredictable. Chaos theory has practical relevance to areas such as population dynamics, chemical engineering, and financial markets, and helps in the development of artificial intelligence.

In a Lorenz attractor, small changes in starting conditions result in huge changes to the paths each line takes, yet the lines still fall within the confines of the same shape, providing order within the chaos.

Lorenz began investigating climate modeling in the 1950s. By the early 1960s, he was attracting attention for the unexpected results of a toy climate model ("toy" meaning that it was a simplistic model made to demonstrate processes concisely). The model predicted the way the atmosphere would evolve in terms of three data points, such as air pressure, temperature, and wind speed. Lorenz found that the results were chaotic. He compared two sets of results, each starting with near-identical sets of data, noting that the atmospheric conditions developed along near-identical lines at first, but then changed in completely different ways. He also found that while every starting point in his model rendered unique results, they were all confined within certain limits.

Strange attractor

The computing power available to Lorenz in the early 1960s was unable to plot the modeled atmospheric variables in a three-dimensional space, where the

> The amazing thing is that chaotic systems don't always stay chaotic.
> **Connie Willis**
> **American writer**

values on the x, y, and z axes represented, for example, air temperature, pressure, and humidity (or triplets of other weather data). In 1963, when it became possible to plot this data, the shape created became known as the Lorenz attractor. Each starting point evolves into a looping line that swings from one quadrant of the space to another—indicating, for example, a change from wet »

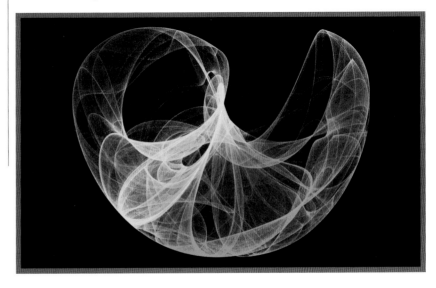

Chaos: when the present
determines the future, but
the approximate present
does not approximately
determine the future.
Edward Lorenz

Many **dynamic systems** of nature appear **deterministic**, following **laws** that **always** apply.

If we know the **initial** conditions **very precisely**, we can **determine** the **future** conditions **exactly**.

However, the system is highly **sensitive**. A small change in the initial conditions will **lead** to a **big difference** in the **future** conditions.

If we only know an **approximation** of initial conditions, our **predictions** will be **imprecise**.

The **system** is **chaotic**.

and windy weather to hot, dry conditions, and all states in between. Each starting point leads to a unique evolution, but all the lines, whatever the start point, fall into the same region of the space. After many iterations, run for long periods, that region becomes a beautiful looping surface. The individual lines within the attractor are highly unstable in their trajectories; those that start in the same area often move far apart at a later point, and lines with very different starting points may end up tracking each other closely for long periods. However, the attractor shows that as a whole, the system is stable. There is no possible starting point within the attractor that can lead to a trajectory that escapes from it. This apparent contradiction is at the heart of chaos theory.

Finding the right path
The roots of chaos theory lie in early attempts to understand and predict motion, especially of heavenly bodies. For example, in the 1600s, Galileo formulated laws about the way pendulums swing and how objects fall; Johannes Kepler showed how planets sweep

through space as they orbit the Sun; and Isaac Newton combined this knowledge with physical laws covering gravity and motion. Along with Gottfried Leibniz, Newton is credited with developing calculus, a system of mathematics designed to analyze and predict the behaviors of more complex systems. Using calculus, the relationships between any complex variables can —in theory—be predicted by solving a particular differential equation.

These physical laws and analytical tools can demonstrate that the Universe is deterministic— if the exact location and condition of an object and all the forces

acting upon it are known, it is possible to determine its future location and condition with perfect accuracy.

The three-body problem
Nevertheless, Newton found a flaw with this deterministic view of the Universe. He reported difficulties in analyzing the movements of three bodies bound together by gravity—even when those bodies were as seemingly stable as the Earth, Moon, and Sun. Later attempts to analyze the movement of the Moon to improve navigation were plagued by inaccuracies. In 1890, French mathematician Henri Poincaré showed that there

was no generalized, predictable way in which three bodies move around each other. In a few cases, where the bodies start in very specific places, the motion is periodic—it repeats the same paths over and over again. Mostly, Poincaré argued, the three bodies do not retrace their paths, and their movement is called aperiodic.

Mathematicians hoping to solve this "three-body problem" have abstracted it to consider imaginary bodies moving around surfaces and spaces with specific curvature. The curvature of an imaginary body can be a mathematical representation of the forces (such as gravity) acting on it. The path the imaginary body takes in each case is called the geodesic path (see below). In a simple case, such as the movement of a pendulum or the orbit of a planet around a star, this imaginary body oscillates (moves back and forth) around a fixed point on the surface, following a repeating path and creating what is called a limit cycle. In the case of a damped

pendulum (one that is losing energy because of friction), the oscillatory motion will diminish until the imaginary body reaches the fixed point—when it stops moving.

When considering the motion of an imaginary body with respect to several others, the geodesic path becomes very complicated. If it were possible to set the start conditions precisely, it would be possible to create every conceivable path. Some would be periodic, repeating a path of whatever complexity over and over again. Others would be unstable initially but would settle into a limit cycle eventually. A third kind would fly off to infinity—perhaps right away, or perhaps after a period of apparent stability.

Approximations

Although it has been studied by physicists and mathematicians alike, the three-body problem is largely theoretical. When it comes to a real physical system, there is no way to be absolutely precise

Determinism was equated with predictability before Lorenz. After Lorenz, we came to see that… in the long run, things could be unpredictable.
Stephen Strogatz
American mathematician

about the starting conditions. This is the essence of chaos theory. Even though the system is deterministic, every measurement of that system is an approximation. Therefore, any mathematical model based on those uncertain measurements will very possibly develop in a different way from the real thing. Even a small uncertainty is enough to create chaos. ■

The geodesic path of a planet

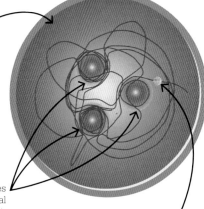

The geodesic path of a planet orbiting a star in a predictable way is shown in the left-hand image. The image on the right shows how the presence of three other celestial bodies—perhaps nearby planets or other stars—complicates the planet's path, making it unpredictable, or chaotic.

The gravity well of a star

A planet with no neighboring bodies

The geodesic path of the planet forms a predictable shape.

Three bodies exert gravitational effects on the planet.

The geodesic path of the planet is scrambled by the proximity of the three bodies.

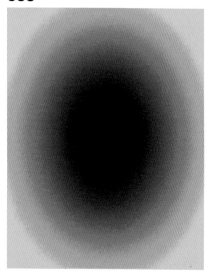

LOGICALLY THINGS CAN ONLY PARTLY BE TRUE
FUZZY LOGIC

IN CONTEXT

KEY FIGURE
Lotfi Zadeh (1921–2017)

FIELD
Logic

BEFORE
350 BCE Aristotle develops a system of logic that dominates Western scientific reasoning until the 1800s.

1847 George Boole invents a form of algebra in which variables can have one of only two values (true or false), paving the way for symbolic, mathematical logic.

1930 Polish logicians Jan Łukasiewiecz and Alfred Tarski define a logic with infinitely many truth values.

AFTER
1980s Japanese electronics companies use fuzzy logic control systems in industrial and domestic appliances.

The binary logic of any computer is clear: given valid inputs, it will provide appropriate outputs. However, binary computer systems are not always well suited for dealing with real-world inputs that are ambiguous or unclear. In the case of handwriting recognition, for example, a binary system would not be sufficiently subtle. A system controlled by fuzzy logic, however, allows for degrees of truth that can better analyze complex phenomena, including human actions and thought processes. Fuzzy logic is an offshoot of the fuzzy set theory

The classes of objects encountered in the real physical world do not have precisely defined criteria of membership.
Lotfi Zadeh

developed in 1965 by Lotfi Zadeh, an Iranian–American computer scientist. Zadeh claimed that as a system becomes more complex, precise statements about it become meaningless; the only meaningful statements about it are imprecise. Such situations demand a many-valued (fuzzy) reasoning system.

Standard set theory allows an element to either belong or not belong to a set, but fuzzy set theory allows degrees of membership or a continuum. Similarly, fuzzy logic allows a range of truth values for a proposition—not just completely true or completely false, the two values of Boolean logic. Fuzzy truth values also require fuzzy logical operators—for example, the fuzzy version of the AND operator of Boolean algebra is the MIN operator, which outputs the minimum of the two inputs.

Creating fuzzy sets
A basic computer program that mimics the simple human task of soft-boiling an egg might apply a single rule: boil the egg for five minutes. A more sophisticated program would, like a human, take the weight of the egg into account. It might divide eggs

See also: Syllogistic logic 50–51 ▪ Binary numbers 176–77 ▪ Boolean algebra 242–47 ▪ Venn diagrams 254 ▪ The logic of mathematics 272–73 ▪ The Turing machine 284–89

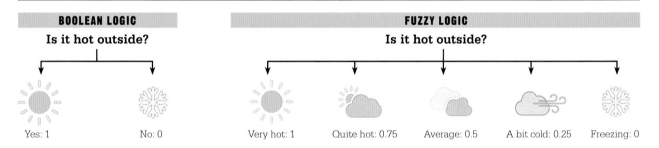

BOOLEAN LOGIC			FUZZY LOGIC				
Is it hot outside?			**Is it hot outside?**				
Yes: 1		No: 0	Very hot: 1	Quite hot: 0.75	Average: 0.5	A bit cold: 0.25	Freezing: 0

Fuzzy logic recognizes a continuum of truth values instead of the Boolean binary values of "yes" (1) or "no" (0). These fuzzy values resemble probabilities, but are fundamentally quite distinct—they indicate the degree to which a proposition is true, not how likely it is.

into two sets—small eggs of 1.76 oz (50 g) or less, and large ones over 1.76 oz—and boil the former for four minutes, and the latter for six. Fuzzy logicians call these crisp sets: each egg either does or does not belong.

To achieve a perfectly cooked egg, however, the boiling time must be adjusted to match the weight of the egg. While an algorithm could use traditional logic to divide a set of eggs into precise weight ranges and assign exact cooking times, fuzzy logic achieves this result with a more

general approach. The first step is to make the data fuzzy—every egg is regarded as both large and small, belonging to both sets to different degrees. For example, a 1.76 oz egg would have a membership degree of 0.5 for both sets, while an 2.82 oz (80 g) egg would be "large" with degree nearly 1, and also "small" with degree nearly 0. A fuzzy rule is then applied, with large eggs boiled for six minutes and small eggs for four. Through a process called fuzzy inference, the algorithm applies the rule to each egg based on its fuzzy set membership. The system will

deduce that an 2.82 oz egg should be boiled for both four and six minutes (with degrees of almost 0 and almost 1 respectively). This output is then defuzzified to give a crisp logical output that can be used by the control system. As a result, the 2.82 oz egg would be assigned a boiling time of nearly 6 minutes.

Fuzzy logic is now a ubiquitous part of computer-controlled systems. It has many applications, from forecasting weather to trading stocks, and plays a vital role in programming artificial intelligence systems. ▪

A humanoid robot using AI works at the front desk of a Henn-na hotel in Tokyo, which claims to be the world's first hotel with robotic staff.

Artificial intelligence

Fuzzy control systems can work effectively with uncertainties in the everyday world, and are therefore used in artificial intelligence (AI) systems. The fuzziness of AI helps to give the illusion of a self-directing intelligence, but in reality fuzzy logic processes data to smooth out uncertainty. AI is therefore entirely the product of a pre-programmed set of rules.

Techniques such as machine learning, in which AIs program themselves by a process of trial

and error, and expert systems, in which the AI draws upon a database of knowledge provided by human programmers, have greatly extended the abilities of AI. Nevertheless most AI is "narrow," in that it is tasked with doing one job very well, generally better than a human can, but it cannot learn to do anything else and is unaware of what it does not know. A general AI that can direct its own learning in the same way as evolved intelligence (such as human intelligence) is the next goal of computer science.

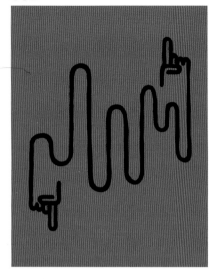

A GRAND UNIFYING THEORY OF MATHEMATICS
THE LANGLANDS PROGRAM

IN CONTEXT

KEY FIGURE
Robert Langlands (1936–)

FIELD
Number theory

BEFORE
1796 Carl Gauss proves the quadratic reciprocity theorem, relating the solvability of quadratic equations to prime numbers.

1880–84 Henri Poincaré develops the concept of automorphic forms—tools that allow us to keep track of complicated groups.

1927 Austrian mathematician Emil Artin extends the reciprocity theorem to groups.

AFTER
1994 Andrew Wiles uses a special case of Langlands' conjectures to translate Fermat's last theorem from a problem in number theory to one in geometry, enabling him to solve it.

I n 1967, the young Canadian–American mathematician Robert Langlands suggested a set of profound links between two major and seemingly unconnected areas of mathematics—number theory and harmonic analysis. Number theory is the mathematics of integers, in particular prime numbers. Harmonic analysis (in which Langlands specialized) is the mathematical study of waveforms, exploring how they can be broken down to sine waves. These fields seem fundamentally different: while sine waves are continuous, integers are discrete.

Langlands' letter
In a 17-page handwritten letter to number theorist André Weil in 1967, Langlands offered several conjectures linking number theory and harmonic analysis. Realizing its significance, Weil had the letter typed up and circulated among number theorists through the late 1960s and '70s. Once they had

Number theory deals with the **properties of** and **relationship between** integers.

Harmonic analysis **analyzes complicated functions**, breaking them into groups of **sine waves**.

The Langlands Program **joins together** these seemingly **disparate branches** of mathematics.

The Program can be described as "a grand unifying theory of mathematics."

been made public, Langlands' conjectures became influential across mathematics, and continue to shape research 50 years later.

Uncovering links

Langlands' ideas involve highly technical mathematics. In basic terms, his areas of interest are Galois groups and functions called automorphic forms. Galois groups turn up in number theory and are a generalization of the groups that Évariste Galois used in order to study roots of polynomials.

Langlands' conjectures are significant in that they allowed problems from number theory to be reframed in the language of harmonic analysis. The Langlands Program has been described as a mathematical Rosetta Stone, helping to translate ideas from one area of mathematics into another. Langlands himself has helped to develop the means for working on the Program, including generalizing functoriality—a way of comparing the structures of different groups.

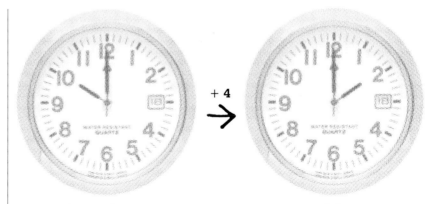

Modular ("clock") arithmetic involves number systems with finite sets of numbers. On a 12-hour clock, for example, if you count on four hours from 10 o'clock, you get 2 o'clock; 10 + 4 = 2, because the remainder of 14 ÷ 12 is 2. In the Langlands program, numbers are usually manipulated by modular arithmetic.

Langlands' marriage of harmonic analysis and number theory could lead to a wealth of new tools, just as the 19th-century unification of electricity and magnetism into electromagnetism provided a new understanding of the physical world. By finding new links between mathematical fields that seem profoundly different, the Program has revealed some of the structures at the heart of mathematics. In the 1980s, Ukrainian mathematician Vladimir Drinfel'd expanded the Program's scope to show that there might be a Langlands-type connection between specific topics within harmonic analysis and others within geometry. In 1994, Andrew Wiles used one of Langlands' conjectures to help solve Fermat's last theorem. ▪

Robert Langlands

Born near Vancouver, Canada, in 1936, Robert Langlands did not plan to go to study at a university until a teacher "took up an hour of class time" to publicly implore him to make use of his talents. He was also a gifted linguist, but at 16, he enrolled at the University of British Columbia, Canada, to study mathematics. He later moved to the US, where he was awarded a doctorate from Yale University in 1960. Langlands taught at Princeton, Berkeley, and Yale before moving to the Institute for Advanced Study (IAS), where he still occupies Einstein's old office.

Langlands began studying the relationship between integers and periodic functions as part of research into patterns in prime numbers. He was awarded the Abel Prize in 2018 for his "visionary" Program.

Key works

1967 *Euler Products*
1967 *Letter to André Weil*
1976 *On the Functional Equations Satisfied by Eisenstein Series*
2004 *Beyond Endoscopy*

ANOTHER ROOF, ANOTHER PROOF
SOCIAL MATHEMATICS

IN CONTEXT

KEY FIGURE
Paul Erdős (1913–96)

FIELD
Number theory

BEFORE
1929 Hungarian author
Frigyes Karinthy postulates
the concept of six degrees
of separation in his short
story, *Láncszemek* (*Chains*).

1967 American social
psychologist Stanley Milgram
conducts experiments on
the interconnectedness
of social networks.

AFTER
1996 The Bacon number is
introduced on an American TV
show. It indicates the number
of degrees of separation an
actor has from American actor
Kevin Bacon.

2008 Microsoft conducts
the first experimental study
into the effects of social
media on connectedness.

Hungarian mathematician Paul Erdős wrote and cowrote around 1,500 academic papers in his lifetime. He worked with more than 500 others in the global mathematical community across different branches of mathematics, including number theory (the study of integers) and combinatorics—a field of mathematics concerned with the number of permutations that are possible in a collection of objects. His motto, "Another roof, another proof," referred to his habit of staying at the homes of fellow mathematicians in order to "collaborate" for a while.

The Erdős number, first used in 1971, indicates how far a mathematician is removed from Erdős in their published work. To qualify for an Erdős number, a person has to have written a mathematical paper—someone who coauthored a paper with Erdős would have an Erdős number of 1. Someone who worked with a coauthor (but not with Erdős directly) would have an Erdős

number of 2, and so on. Albert Einstein has an Erdős number of 2; Paul Erdős's number is 0.

Oakland University runs the Erdős Number Project, which analyzes collaboration among research mathematicians. The average Erdős number is around 5. The rarity of an Erdős number higher than 10 indicates the degree of collaboration within the mathematical community. ∎

Erdős has an amazing ability to match problems with people. Which is why so many mathematicians benefit from his presence.
Béla Bollobás
Hungarian–British mathematician

See also: Diophantine equations 80–81 ▪ Euler's number 186–91 ▪ Six degrees of separation 292–93 ▪ Proving Fermat's last theorem 320–23

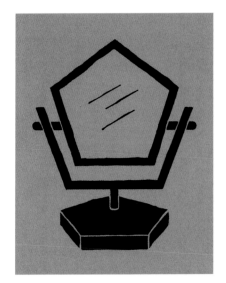

PENTAGONS ARE JUST NICE TO LOOK AT
THE PENROSE TILE

IN CONTEXT

KEY FIGURE
Roger Penrose (1931–)

FIELD
Applied geometry

BEFORE
3000 BCE Sumerian buildings incorporate tessellations into wall decorations.

1619 Johannes Kepler conducts the first documented study of tessellations.

1891 Russian crystallographer Evgraf Fyodorov proves there are only 17 possible groups that form periodic tilings of the plane.

AFTER
1981 Dutch mathematician Nicolaas Govert de Bruijn explains how to construct Penrose tilings from five families of parallel lines.

1982 Israeli engineer Dan Shechtman discovers quasicrystals whose structure is similar to Penrose tilings.

ile patterns have been a feature of art and construction for millennia, especially in the Islamic world. The need to fill two-dimensional space as efficiently as possible led to the study of tessellations—the fitting together of polygons with no gaps or overlap. Some natural structures, such as a honeycomb, tessellate.

There are three regular shapes that tessellate on their own, without the need for another shape: the square, equilateral triangle, and regular hexagon. However, many irregular shapes also tessellate, and semiregular tessellations involve more than one regular shape. The pattern of such tessellations usually repeats. This is known as a "periodic tessellation."

Nonperiodic tessellations, in which the pattern does not repeat, are harder to find, although some regular shapes can be combined to create nonperiodic tessellations. British mathematician Roger Penrose investigated whether any polygons could only lead to nonperiodic tessellations. In 1974,

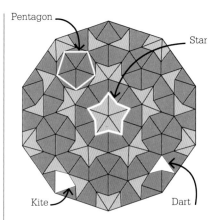

Penrose tiling consists of kites and darts, producing a nonperiodic tessellation. Shapes with five-fold symmetry, such as pentagons and stars, can also be identified.

he created tiles using kite and dart shapes. The kite and dart must be exactly the same shape as the ones shown (above); the area of the kite to that of the dart is expressed by the golden ratio. Although no part of the tiling matches another part exactly, the pattern does repeat on a larger scale in a similar way to a fractal. ■

See also: The golden ratio 118–23 ▪ The problem of maxima 142–43 ▪ Fractals 306–11

ENDLESS
VARIETY AND
UNLIMITED
COMPLICATION
FRACTALS

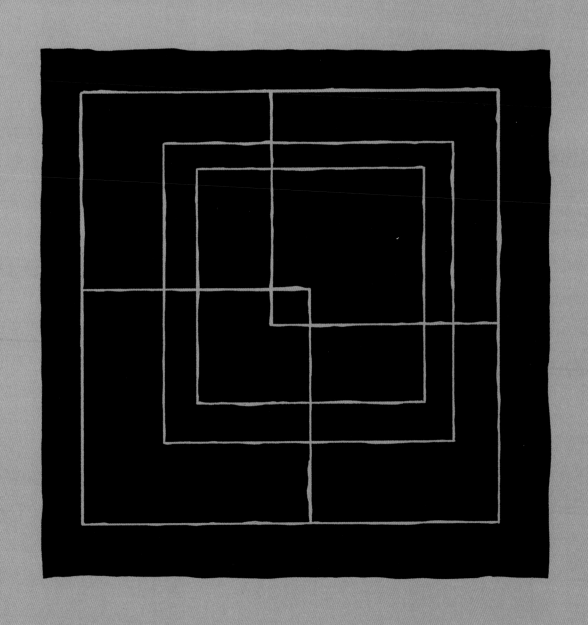

IN CONTEXT

KEY FIGURE
Benoit Mandelbrot
(1924–2010)

FIELDS
Geometry, topology

BEFORE
c. 4th century BCE Euclid
sets out the foundations of
geometry in *Elements*.

AFTER
1999 The study of "allometric
scaling" applies fractal growth
to metabolic processes within
biological systems, leading to
valuable medical applications.

2012 In Australia, the largest
3-D map of the sky suggests
that the Universe is fractal up
to a point, with clusters of
matter within larger clusters,
but ultimately matter is
distributed evenly.

2015 Fractal analysis is
applied to electrical power
networks, leading to the
modeling of the frequency
of power failure.

A geometry able to include
mountains and clouds now
exists… Like everything
in science this new
geometry has very, very
deep and long roots.
Benoit Mandelbrot

A computer graphic shows a
fractal pattern derived from the
Mandelbrot set. Mesmerizingly
beautiful, such images produced
with fractal-generating software
make popular screen savers.

After Euclid, scholars and
mathematicians modeled
the world in terms of
simple geometry: curves and
straight lines; the circle, ellipse,
and polygons; and the five Platonic
solids—the cube, the tetrahedron,
the octahedron, the dodecahedron,
and the icosahedron. For much of
the past 2,000 years, the prevailing
assumption has been that most
natural objects—mountains, trees,
and so on—can be deconstructed
into combinations of these shapes
to ascertain their size. However,
in 1975, Polish-born mathematician
Benoit Mandelbrot drew attention
to fractals—nonuniform shapes that
echo larger and smaller shapes in
a structure such as a jagged
mountaintop. Fractals, a word
derived from the Latin *fractus*,
meaning "broken," would eventually
lead to the topic of fractal geometry.

A new geometry

Although it was Mandelbrot who
brought fractals to the attention of
the world, he was building on the
findings of earlier mathematicians.
In 1872, German mathematician
Karl Weierstrass had formalized
the mathematical concept of
"continuous function," meaning
that changes in the input result
in roughly equal changes in the
output. Composed entirely of
corners, the Weierstrass function
has no smoothness anywhere,
however much it is magnified.
This was seen at the time as a
mathematical abnormality that,
unlike the sensible Euclidean
shapes, had no real-world relevance.

In 1883, another German
mathematician, Georg Cantor, built
on work by British mathematician
Henry Smith to demonstrate how
to create a line that is nowhere
continuous and has zero length.
He did so by drawing a straight

See also: The Platonic solids 48–49 ▪ Euclid's *Elements* 52–57 ▪ The complex plane 214–15 ▪ Non-Euclidean geometries 228–29 ▪ Topology 256–59

line, removing the middle third (leaving two lines and a gap), and then repeating the process ad infinitum. The result is a line composed entirely of disconnected points. Like the Weierstrass function, this "Cantor set" was considered unsettling by the mathematical establishment, who branded these new shapes "pathological"—meaning "lacking usual properties."

In 1904, Swedish mathematician Helge von Koch constructed a shape known as the Koch curve or "Koch snowflake," which repeated a triangular motif at an ever smaller size. This was followed in 1916 by the Sierpinski triangle, or Sierpinski gasket, composed entirely of triangular holes.

All these shapes possess self-similarity, which is a key property of fractal geometry. This means that enlargement of a portion of the shape reveals smaller replicas with equal detail. Mathematicians realized that this was a fundamental property of natural growth—a repetition of a pattern on many scales, from the macro to the micro. In 1918, German mathematician Felix Hausdorff proposed the existence of fractional dimensions. Whereas the simple line, plane, and solid occupy one, two, and three dimensions respectively, these new shapes could be given non-whole-number dimensions. For example, the British coastline could, in theory, be measured with a one-dimensional rope, but inlets would require string, and crevices require thread. This implies that the coastline cannot be measured in one dimension. The British coastline has a Hausdorff dimension of 1.26, like the Koch curve.

Dynamic self-similarities

French mathematician Henri Poincaré found that dynamical systems (systems that change over time) also had fractal properties of self-similarity. By their nature, dynamical states are "nondeterministic": two systems that are nearly identical can lead to very different behaviors even when the initial conditions are also »

Benoit Mandelbrot

Born into a Jewish family in Warsaw in 1924, Benoit Mandelbrot left Poland in 1936 to escape the Nazis. His family went first to Paris and then to the south of France. After World War II, Mandelbrot gained scholarships to study in France and then the US, before returning to Paris, where he was awarded a doctorate in mathematical sciences from the city's university in 1952.

In 1958, Mandelbrot joined IBM in New York, where his role as a researcher gave him the space and facilities to develop new ideas. In 1975, he coined the term "fractal," and in 1980 he unveiled the Mandelbrot set, a structure that became synonymous with the new science of fractal geometry. The topic gained popular appeal in 1982 with the publication of his book *The Fractal Geometry of Nature*. Mandelbrot received many honors and prizes for his work, including France's Légion d'honneur in 1989. He died in 2010.

Key works

1982 *The Fractal Geometry of Nature*

The **Mandelbrot set** has an extraordinarily **elaborate structure**.

Its boundary is **highly complex** and **infinitely squiggly**.

Magnifying **any part** of it, however small, reveals a **replica of the set itself**.

No one can fully comprehend the set's endless variety and unlimited complication.

Timeline of fractals

1883
Cantor set
Constructed by repeatedly removing the middle third from a succession of lines, the Cantor set creates a series of intervals.

1872
Weierstrass function
Composed of corners, the Weierstrass function will never appear smooth however much it is magnified.

1904
Koch snowflake
The shape becomes infinitely more intricate as more triangles are added.

almost identical. This phenomenon is popularly known as the "butterfly effect," after the frequently cited example of the massive effect a single butterfly can theoretically have on a weather system when it causes a small disturbance by flapping its wings. The differential equations devised by Poincaré to prove his theory implied the existence of dynamical states that possess self-similarity much like fractal structures. Large-scale weather systems, such as major cyclonic flows, for instance, repeat themselves on much smaller scales, right down to gusts of wind.

In 1918, French mathematician Gaston Julia, a former student of Poincaré, explored the concept of self-similarity when he began to map the complex plane (the coordinate system based on complex numbers) under a process called iteration—entering a value into a function, obtaining an output, and then plugging that back into the function. Along with George Fatou, who undertook similar research independently, Julia found that by taking a complex number, squaring it, adding a

constant (a fixed number or a letter standing for a fixed number) to it, and then repeating the process, some initial values would diverge to infinity while others would converge to a finite value. Julia and Fatou mapped these different values on a complex plane, noting which ones converged and which ones diverged. The boundaries between these regions were self-replicating, or fractal. With

the limited computational power available at the time, Julia and Fatou were unable to see the true significance of their discovery, but they had found what would become known as the Julia set.

The Mandelbrot set
In the late 1970s, Benoit Mandelbrot used the term "fractal" for the first time. Mandelbrot had become interested in the work of Julia and Fatou while working at the IT company IBM. With the computer facilities available at IBM, he was able to analyze the Julia set in great detail, noting that some values of the constant (c) gave "connected" sets, in which each of the points is joined to another, and others were disconnected. Mandelbrot mapped each value of c on the complex plane, coloring the connected sets and the disconnected sets in different colors. This led, in 1980, to the creation of the

Infinite complexity is suggested by the self-similarities of a Romanesco cauliflower. The natural world is full of fractals, from ferns and sunflowers to ammonites and seashells.

1916
Sierpinski triangle
Repeatedly adding triangles
within triangles creates an
infinitely lacy pattern.

1980
Mandelbrot set
Infinitely squiggly, the Mandelbrot
set looks more elaborate the more it
is magnified.

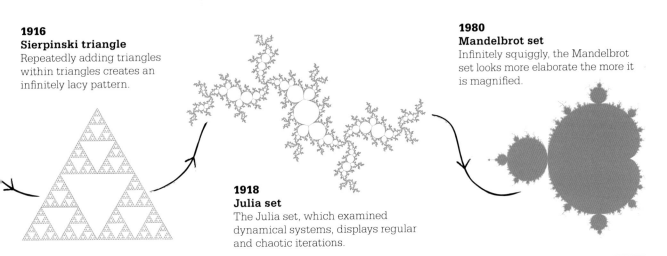

1918
Julia set
The Julia set, which examined
dynamical systems, displays regular
and chaotic iterations.

Mandelbrot set. Beautifully
complex, the Mandelbrot set
displays self-similarity at all scales:
magnification reveals smaller
replicas of the Mandelbrot set itself.
In 1991, Japanese mathematician
Mitsuhiro Shishikura proved that the
boundary of the Mandelbrot set has
a Hausdorff dimension of 2.

Application of fractals
Fractal geometry has allowed
mathematicians to describe the
irregularity of the real world.

Many natural objects exhibit self-
similarity, including mountains,
rivers, coastlines, clouds, weather
systems, blood circulatory
systems, and even cauliflowers.
Being able to model these diverse
phenomena using fractal
geometry enables us to better
understand their behavior and
evolution, even if that behavior
is not entirely deterministic.

Fractals have applications
in medical research, such as
understanding the behavior of

viruses and the development of
tumors. They are also used in
engineering, particularly in the
development of polymer and
ceramic materials. The structure
and evolution of the Universe can
also be modeled on fractals, as
can the fluctuations of economic
markets. As the range of
applications grows, along with ever-
increasing computational capacity,
fractals are becoming integral to
our understanding of the seemingly
chaotic world in which we live. ∎

Fractals and the arts

Under the Wave off Kanagawa by
Japanese artist Katsushika Hokusai
(1760–1849) employs the concept of
self-similarity to dramatic effect.

Self-similarity on infinite scales
is explored in philosophy and the
arts, often to produce a meditative
effect. It is a key tenet of Buddhist
meditation and mandalas (symbols
used in rituals to represent the
Universe), and is also used to
suggest the infinite nature of God
in Islamic decoration, such as
tilework. Self-similarity is even
suggested in the poem "Auguries
of Innocence" by the 19th-century
British poet William Blake, which
begins with the line "To see a
world in a grain of sand."

The work of the Japanese artist
Katsushika Hokusai, with its
swirling repeated motifs, is often
cited as an example of fractal
use in art, as is the architecture
of Catalan artist Antoni Gaudí.

The musical "rave" scene
in the US and UK in the late
1980s and early '90s was
linked to a surge of interest
in fractal art. Nowadays there
are many fractal-generating
computer programs, making it
possible for the general public
to create fractals.

FOUR COLORS BUT NO MORE

THE FOUR-COLOR THEOREM

IN CONTEXT

KEY FIGURES
Kenneth Appel (1932–2013),
Wolfgang Haken (1928–)

FIELD
Topology

BEFORE
1852 South African law
student Francis Guthrie asserts
that four colors are needed to
color a map so that adjacent
areas are not the same color.

1890 British mathematician
Percy Heawood proves that
five colors are sufficient to
color any map.

AFTER
1997 In the US, Neil Robertson,
Daniel P. Sanders, Robin
Thomas, and Paul Seymour
provide a simpler proof
of the four-color theorem.

2005 Microsoft researcher
Georges Gonthier proves
the four-color theorem with
general purpose theorem-
proving software.

How many colors do you need **to color a map** so that no two countries with a common border have the same color?

It **can't be done** with just **two or three** colors.

In 1890, it is proved that any map can be colored with **five colors**.

In 1976, a computer is used to prove that no more than **four colors** are necessary.

Four colors are enough to color a map.

Cartographers have long known that any map, however complicated, can be colored in with just four colors, so that no two nations or regions sharing a border are the same color. Although five colors can seem to be necessary, there is always a way of recoloring the map using only four colors. Mathematicians searched for a proof for this deceptively simple theorem for more than 120 years, making it one of the most enduring unsolved theorems in mathematics.

The first person to formulate the four-color theorem is thought to have been Francis Guthrie, a South

See also: Euler's number 186–91 ▪ Graph theory 194–95 ▪ The complex plane 214–15 ▪ Proving Fermat's last theorem 320–23

Any combination of shapes in a plane, however complex the pattern, can be colored in using just four colors so that no two adjacent shapes have the same color.

African law student. He had colored a map of the English counties using just four colors and believed that the same could be done with any map, however complex. In 1852, he asked his brother Frederick, who was studying under mathematician Augustus De Morgan in London, if his theory could be proved. Admitting that he could not prove the theorem, De Morgan shared it with Irish mathematician William Hamilton. Hamilton went on to attempt to prove the theorem himself, but did not succeed.

False start

In 1879, British mathematician Alfred Kempe claimed a proof for the four-color theorem in the scientific journal *Nature*. Kempe received plaudits for this work, and two years later became a Fellow of the Royal Society partly on the strength of his proof. However, in 1890, fellow British mathematician Percy Heawood found a hole in Kempe's proof, and Kempe himself acknowledged that

he had made a mistake that he could not rectify. Heawood did prove correctly that no more than five colors were needed to color any map.

Mathematicians continued to work on the problem, and gradual progress was made. In 1922, Philip Franklin proved that any map with 25 regions or fewer was four-colorable. The figure of 25 was slowly increased; Norwegian mathematician Øystein Ore and American mathematician Joel Stemple together achieved 39 in 1970, and Frenchman Jean Mayer lifted the figure to 95 in 1976.

New hope

The introduction of supercomputers, computers capable of handling huge amounts of data, in the 1970s revived interest in solving the four-color theorem. Although German mathematician Heinrich Heesch suggested a method for doing this, he did not have sufficient access to a supercomputer to test it. Wolfgang Haken, a former student of Heesch's, became interested in the problem, and began to make progress after meeting computer programmer Kenneth Appel at the University of Illinois. The pair finally cracked the problem in 1977. Relying completely on computing power—the first proof in the history of mathematics to do so—they examined around 2,000 cases, involving billions of calculations and using 1,200 hours of computer time. ▪

Computer proofs

When Appel and Haken proved the four-color theorem in 1977, it was the first time that a computer had been used to prove a mathematical theorem. This was controversial among mathematicians, who were used to solving problems through logic that could be checked by their peers. Appel and Haken had used the computer to carry out a proof by exhaustion—all possibilities were meticulously checked one by one, a feat that would have been impossible to do manually. The question was whether a long calculation that could not be checked by humans, followed by a simple verdict of "yes, the theorem has been proved," could be accepted. Many mathematicians argued that it could not. Proof by computers remains controversial, but advances in technology have increased confidence in their reliability.

The IBM System/370 computer c. 1970 was one of the first computers to use virtual memory, a working memory system that allowed it to process large amounts of data.

SECURING DATA WITH A ONE-WAY CALCULATION

CRYPTOGRAPHY

IN CONTEXT

KEY FIGURES
Ron Rivest (1947–),
Adi Shamir (1952–),
Leonard Adleman (1945–)

FIELD
Computer science

BEFORE
9th century CE Al-Kindi
develops frequency analysis.

1640 Pierre de Fermat
states his "little theorem"
(on primality), which is still
used as a test when searching
for primes to use in public
key encryption.

AFTER
2004 Elliptic curves are first
used in cryptography; they use
smaller keys but offer the same
security as the RSA algorithm.

2009 An anonymous computer
scientist mines the first Bitcoin,
a cryptocurrency without a
central bank. All transactions
are encrypted but public.

Cryptography is the
development of means
of secret communication.
It has become a ubiquitous feature
of modern life, with almost every
connection between one digital
device and another starting with a
"handshake," in which the devices
agree on a way of securing their
connection. That handshake is
often the result of the work of three
mathematicians: Ron Rivest, Adi
Shamir, and Leonard Adleman.
In 1977, they developed the RSA
algorithm (named for their initials),
an encryption procedure that won
them the Turing Award in 2002.
The RSA algorithm is special

See also: Group theory 230–33 ▪ The Riemann hypothesis 250–51 ▪ The Turing machine 284–89 ▪ Information theory 291 ▪ Proving Fermat's last theorem 320–23

The work did not really need mathematics, but mathematicians tended to be good at it.
Joan Clarke
British cryptanalyst

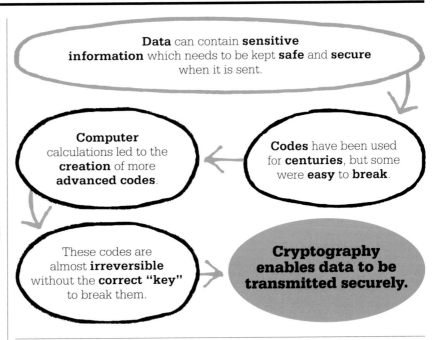

Data can contain **sensitive information** which needs to be kept **safe** and **secure** when it is sent.

Computer calculations led to the **creation** of more **advanced codes**.

Codes have been used for **centuries**, but some were **easy** to **break**.

These codes are almost **irreversible** without the **correct "key"** to break them.

Cryptography enables data to be transmitted securely.

because it ensures that any third party monitoring the connection will be completely unable to figure out any private details.

One main reason people have needed to encrypt communications is to ensure financial transactions can happen without banking information falling into the wrong hands. However, encryption is used against all kinds of third-party "adversary"—a rival company, an enemy power, or a security service. Cryptography is an ancient practice. Mesopotamian clay tablets from c. 1500 BCE were often encrypted to protect recipes for pottery glazes and other such commercially valuable information.

Cipher and key
The term "cryptography" comes from the Greek for "hidden writing study." For much of history it was used to secure written messages. The unencrypted message is known as the plaintext, while the encrypted version is called the ciphertext. For example, "HELLO" might become "IFMMP." Going from plaintext to this ciphertext requires a cipher and a key. A

cipher is an algorithm (a systematic, repeatable method)—in this case, to substitute each letter with one in another position in the alphabet. The key is +1, because each of the letters in plaintext is substituted with the letter +1 along in the alphabet. If the key were −6, then the cipher would turn the same plaintext "HELLO" into "BZFFI." This simple substitution system

is known as the Caesar cipher (or Caesar shift) after the Roman dictator Julius Caesar, who used it in the 1st century BCE. The Caesar cipher is an example of symmetric encryption, as the same cipher and key are used (in reverse) to decipher the message.

Deciphering processes
Given enough paper and time, it is relatively easy to figure out a Caesar cipher by trying out every possible substitution. In modern terms this is known as a "brute force" technique. More complex ciphers and keys make brute force more time-consuming—and, before computers, effectively unworkable for messages long enough to hold »

Cipher wheels, such as this British example from 1802, sped up the decryption of Caesar ciphers. Once the key was uncovered, the two individual wheels could be set accordingly.

large amounts of information. Longer messages were vulnerable to another decryption technique called frequency analysis. Initially developed by the Arab mathematician al-Kindi in the 9th century, this technique made use of the frequency of each letter of the alphabet in a particular language. The most common letter in the English language is "e," so a cryptanalyst would find the most common letter in the ciphertext and designate that as e. The next most common letter is "t," then "a," and so on. Common groupings of letters, such as "th" and "ion" could also provide a way into revealing the cipher. Given a large enough ciphertext, this system worked on any substitution cipher, no matter how elaborate the encryption.

There are two ways of combatting frequency analysis. The first is to obscure the plaintext by using a "code." Cryptography uses a specific definition of this term. A code changes an entire word or phrase in the plaintext before it is encrypted. An encoded plaintext might read "buy lemons

on Thursday," where "buy" is code for "kill" and "lemons" is code for a particular target on a hit list—perhaps with all targets encoded as fruits. Without the list of code words, deciphering the message's full meaning is impossible.

The Enigma code

Another method of increasing the security of encryption is to use a polyalphabetic cipher, where a letter in plaintext can be substituted for several different letters in ciphertext, thus removing the possibility of frequency analysis. Such ciphers were first developed in the 1500s, but the most famous one was the encryption produced by the Enigma machines used by the Axis forces in World War II.

The Enigma machine was a formidable encryption device. In essence, it was a battery connected to 26 lightbulbs, or lamps—one for each letter of the alphabet. When a signaler pressed a letter on the keyboard, a corresponding letter lit up on the lampboard. Pressing the same key a second time always lit a different lamp (never the same letter as the key) because the connections between battery and lampboard were altered by three rotors that clicked around with every key press. Added complexity was introduced by the plugboard, which swapped 10 pairs of letters, thus scrambling the message further. To encrypt and decrypt an Enigma message, both machines needed to be set up in the correct way. This involved the correct three rotors being inserted and set to the right starting positions, and the

The Enigma machine was used in German espionage between 1923 and 1945. The three rotor wheels sit behind the lampboard, and the plugboard is at the front.

Computer technology is on the verge of providing the ability for [people] to communicate and interact with each other in a totally anonymous manner.
Peter Ludlow
American philosopher

10 plugs being connected correctly on the board. The settings became the encryption key. A three-rotor Enigma had over 158,962,555,217 billion possible settings, which were changed daily.

Enigma's flaw was that it could not encrypt a letter as itself. This allowed Allied codebreakers to try frequently used phrases, such as "Heil Hitler" and "Weather Report" to attempt to figure out that day's key. Ciphertext without any of the letters in those words was a potential ciphertext of them. Allied codebreakers used the Turing Bombe, an electromechanical device that mimicked Enigma machines to break the encryption by brute force, using shortcuts developed by British mathematician Alan Turing and others. The British encryption device, the Typex, was a modified version of Enigma that could encode a letter as itself. The Nazis gave up trying to crack it.

Asymmetric encryption

With symmetric encryption, messages are only as secure as the key. This must be exchanged by physical means—written in a

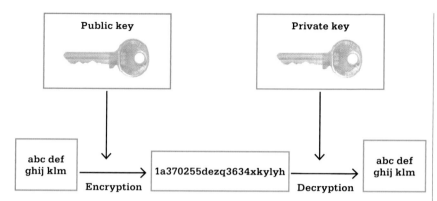

Public key	Private key

| abc def ghij klm | → | 1a370255dezq3634xkylyh | → | abc def ghij klm |

Encryption **Decryption**

Public key encryption scrambles data with an encryption key available to anyone. The data can only be unscrambled with a private key, known only to its owner. This method is effective for small amounts of data, but is too time-consuming for large amounts.

military code book or whispered in the ear of a spy at a secluded rendezvous. If a key falls into the wrong hands, the encryption fails.

The rise of computer networks has allowed people to communicate easily over great distances without ever meeting. However, the most commonly used network, the internet, is public, so any symmetric key shared over a connection would be available to unintended parties, making it useless. The RSA algorithm was an early development in building asymmetric encryption,

where a sender and receiver use two keys: one private and the other public. If two people, Alice and Bob, wish to communicate in secret, Alice can send Bob her public key. It is made up of two numbers, n and a. She keeps a private key, z, to herself. Bob uses n and a to encrypt a plaintext message (M), which is a string of numbers (or letters ciphered into numbers). Each plaintext number is raised to the power of a, and then divided by n. The division is a modulo operation (abbreviated to

mod_n), meaning the answer is just the remainder. So, for example, if n were 10 and M^a were 12, that would give the answer 2. If M^a were 2, it would also give an answer of 2, because 10 goes into 2 zero times with a remainder of 2. The answer to $M^a \, mod_n$ is the ciphertext (C), and in this example it is 2. Someone spying could know the public key, n and a, but would have no idea whether M is 2, 12, or 1,002 (all divisible by 10 with a remainder of 2). Only Alice can find out using her private key, z, because $C^z \, mod_n = M$.

The crucial number in this algorithm is n, which is formed by multiplying two prime numbers: p and q. Then a and z are calculated from p and q using a formula which ensures that the modulo calculations work. The only way to crack the code is to figure out what p and q are and then calculate z. To do that, a codebreaker must figure out the prime factors of n, but today's RSA algorithms use values for n with 600 digits or more. It would take a supercomputer thousands of years to work out p and q by trial and error, making RSA and similar protocols practically unbreakable. ■

Lava lamps can be hooked up to computers in order to generate a selection of random numbers based on their movements.

Finding primes in random ways

The RSA algorithm and other public key encryption systems require a large collection of primes to act as p and q. If the system relies heavily on too few primes, then it is possible for attackers to figure out some of the values for p and q being used in everyday encryption. The solution is to have a supply of fresh primes. These are found by generating random numbers and testing their primality with Pierre de Fermat's "little theorem": if a number (p) is prime, when another number (n)

is raised to the power of p, and n is subtracted from the result, the answer is a multiple of p.

Computers are not easily programmed to create truly random sequences of numbers, so companies use physical phenomena to generate them. Computers are programmed to follow the movements of lava lamps, measure radioactive decay, or listen to white noise made by radio transmissions, turning that input into random numbers to use for encryption.

JEWELS STRUNG ON AN AS-YET INVISIBLE THREAD

FINITE SIMPLE GROUPS

IN CONTEXT

KEY FIGURE
Daniel Gorenstein
(1923–92)

FIELD
Number theory

BEFORE
1832 Évariste Galois defines
the concept of a simple group.

1869–89 Camille Jordan, a
French mathematician, and
Otto Hölder, a German, prove
that all finite groups can be
built from finite simple groups.

1976 Croatian mathematician
Swonimir Janko introduces the
sporadic simple group Janko
Group 4, the last finite simple
group to be discovered.

AFTER
2004 American
mathematicians Michael
Aschbacher and Stephen
D. Smith complete the
classification of finite
simple groups begun by
Daniel Gorenstein.

Simple groups have been described as algebra's atoms. The Jordan-Hölder theorem, proven around 1889, asserts that, just as all positive integers can be constructed from prime numbers, so all finite groups can be built from finite simple groups. In mathematics, a group is not simply a collection of things, but a specification of how the group members can be used to generate more members, for example, by multiplication, subtraction, or addition. In the early 1960s, American mathematician Daniel Gorenstein began to pioneer the classification of groups and issued his complete classification of finite simple groups in 1979.

There are similarities between simple groups and symmetry in geometry. Just as a cube rotated through 90 degrees looks the same as it did before it was rotated, the transformations (rotational and

A **group** is a **set of elements** (numbers, letters, or shapes) that are combined with other elements of the same group through an **operation** (for example, addition, subtraction, or multiplication).

A group is **finite** if it has a **finite number of elements**.

A group is **simple** if it **cannot be broken down** into smaller groups.

Finite simple groups are the fundamental building blocks for all finite groups.

See also: The Platonic solids 48–49 ▪ Algebra 92–99 ▪ Projective geometry 154–55 ▪ Group theory 230–33 ▪ Cryptography 314–17 ▪ Proving Fermat's last theorem 320–23

reflexive) associated with a regular 2-D or 3-D shape can be arranged into a type of simple group known as a symmetry group.

Infinite and finite groups

Some groups are infinite, as in the group of all integers under addition, which is infinite because numbers can be added infinitely. However, the numbers –1, 0, and 1 with the multiplication operation form a finite group; multiplying any members of the group produces only –1, 0, or 1. All the members of a group and the rules for generating it can be visualized using a Cayley graph (see right).

A group is simple if it cannot be broken down into smaller groups. While the number of simple groups is infinite, the number of types of simple group is not—at least, not when simple groups of finite size are considered. In 1963, American mathematician John G. Thompson proved that, with the exception of trivial groups (for example, 0 + 0 = 0, or 1 × 1 = 1), all simple groups have an even number of elements. This led Daniel Gorenstein to propose a more difficult task: the classification of every finite simple group.

The Monster

There are precise descriptions of 18 families of finite simple groups, with each family related to symmetries of certain types of geometrical structure. There are also 26 individual groups called sporadic groups, the largest of which is called the Monster, which has 196,883 dimensions and approximately 8×10^{53} elements. Every finite simple group either belongs to one of the 18 families or is one of the 26 sporadic groups. ▪

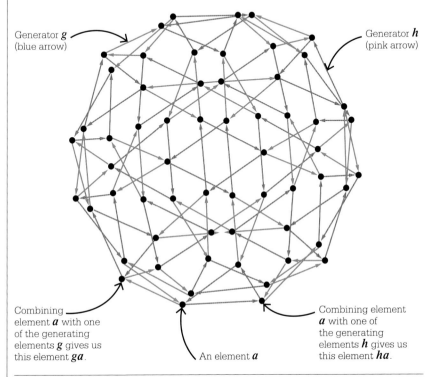

This Cayley graph shows all 60 elements (different orientations) of the group A5 (the group of rotational symmetries of a regular icosahedron, a three-dimensional shape with 20 faces), and how they relate to each other. Since A5 has a finite number of elements, it is a finite group. A5 is also a simple group. It has two generators (elements that can be combined to give any other element of the group).

Generator **g** (blue arrow)

Generator **h** (pink arrow)

Combining element **a** with one of the generating elements **g** gives us this element **ga**.

An element **a**

Combining element **a** with one of the generating elements **h** gives us this element **ha**.

Daniel Gorenstein

Born in Boston, Massachusetts, in 1923, Daniel Gorenstein had taught himself calculus by the age of 12 and later attended Harvard University. There, he became acquainted with finite groups, which would become his life's work. After graduating in 1943, he stayed at Harvard for several years, first to teach mathematics to military personnel during World War II, then to earn his PhD under mathematician Oscar Zariski.

In 1960–61, Gorenstein attended a nine-month program in group theory at the University of Chicago, which inspired him to propose a classification of finite simple groups. He continued to work on this project until his death in 1992.

Key works

1968 *Finite groups*
1979 "The classification of finite simple groups"
1982 *Finite simple groups*
1986 "Classifying the finite simple groups"

A TRULY MARVELOUS PROOF

PROVING FERMAT'S LAST THEOREM

IN CONTEXT

KEY FIGURE
Andrew Wiles (1953–)

FIELD
Number theory

BEFORE
1637 Pierre de Fermat states that there are no sets of positive whole numbers x, y, and z that satisfy the equation $x^n + y^n = z^n$, where n is greater than 2. However, he does not provide the proof.

1770 Swiss mathematician Leonhard Euler shows that Fermat's last theorem is true when $n = 3$.

1955 In Japan, Yutaka Taniyama and Goro Shimura propose that every elliptic curve has a modular form.

AFTER
2001 The Taniyama–Shimura conjecture is established. It becomes known as the modularity theorem.

When he died in 1665, French mathematician Pierre de Fermat left behind a well-thumbed copy of *Arithmetica* by the 3rd-century CE Greek mathematician Diophantus, its margins marked with Fermat's ideas. All the questions posed in Fermat's marginal scribbles were later solved, except for one. He left a tantalizing note in the margin: "I have discovered a truly marvelous proof, which this margin is too small to contain here."

Fermat's note related to Diophantus's discussion of Pythagoras's theorem—that in a right-angled triangle the square

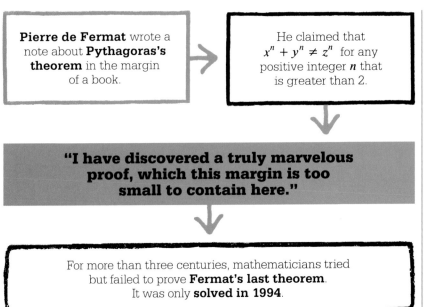

Pierre de Fermat wrote a note about **Pythagoras's theorem** in the margin of a book.

He claimed that $x^n + y^n \neq z^n$ for any positive integer n that is greater than 2.

"I have discovered a truly marvelous proof, which this margin is too small to contain here."

For more than three centuries, mathematicians tried but failed to prove **Fermat's last theorem**. It was only **solved in 1994**.

of the hypotenuse (the side opposite the right angle) is equal to the sum of the squares on the other two sides, or $x^2 + y^2 = z^2$. Fermat knew that this equation had an infinity of integer solutions for x, y, and z, such as 3, 4, and 5 (9 + 16 = 25) and 5, 12, and 13 (25 + 144 = 169), known as "Pythagorean triples." He then wondered if other triples could be found to the power of 3, 4, or any integer beyond 2. The conclusion Fermat reached was that no integer greater than 2 could stand for n. Fermat wrote: "It is impossible for a cube to be the sum of two cubes, a fourth power [number to the power of 4] to be the sum of two fourth powers, or in general for any number that is a power greater than the second to be the sum of two like powers." Fermat never revealed the proof he claimed to have for his theory and so it remained unsolved, becoming known as Fermat's last theorem.

Many mathematicians attempted to reconstruct Fermat's claimed proof after his death, or to find their own. But despite the seeming simplicity of the problem, no one was successful, although a century later Leonhard Euler did prove the theory where $n = 3$.

Finding a solution

Fermat's last theorem remained one of the great unsolved problems in mathematics for more than 300 years, until it was proved by British mathematician Andrew Wiles in 1994. Wiles had first read about Fermat's challenge when he was ten. He had been amazed that he, just a boy, could make sense of it, and yet the best mathematical minds in the world had failed to prove it. It made him want to study mathematics at the University of Oxford, and then to get his PhD at Cambridge. There, he chose elliptic curves as the area of study for his

doctoral thesis—a subject that seemed to have little to do with his interest in Fermat. Yet it was this branch of mathematics that would enable Wiles to prove Fermat's last theorem.

In the mid-1950s, Japanese mathematicians Yutaka Taniyama and Goro Shimura had made the bold step of linking two apparently unrelated branches of mathematics. They claimed that every elliptic curve (an algebraic structure) could be associated with a unique modular form, one of a class of highly symmetrical structures belonging to number theory.

The potential importance of their conjecture was gradually understood over the next three decades and it became part of an ongoing program to link different mathematical disciplines. However, no one had any idea how to prove it.

In 1985, German mathematician Gerhard Frey made a link between the conjecture and Fermat's »

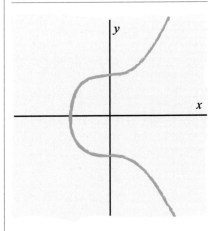

Wiles's investigation of Fermat's last theorem began with his study of elliptic curves, which are described by the equation $y^2 = x^3 + Ax + B$, where A and B are constants (fixed).

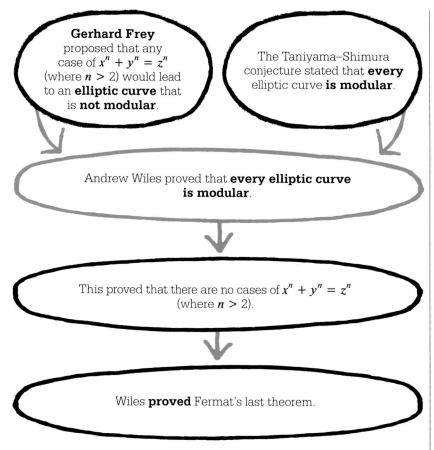

Gerhard Frey proposed that any case of $x^n + y^n = z^n$ (where $n > 2$) would lead to an **elliptic curve** that is **not modular**.

The Taniyama–Shimura conjecture stated that **every** elliptic curve **is modular**.

Andrew Wiles proved that **every elliptic curve is modular**.

This proved that there are no cases of $x^n + y^n = z^n$ (where $n > 2$).

Wiles **proved** Fermat's last theorem.

last theorem. Working from a hypothetical solution to the Fermat equation, he constructed a curious elliptic curve that appeared not to be modular. He argued that such a curve could only exist if the Taniyama–Shimura conjecture were false, in which case Fermat's last theorem would also be false. On the other hand, if the Taniyama–Shimura conjecture were true, Fermat's last theorem would follow. In 1986, Ken Ribet, a professor at Princeton University, in New Jersey, managed to prove Frey's conjectured link.

Proving the unprovable
Ribet's proof electrified Wiles. Here was the chance he had been waiting for—if he could prove the seemingly impossible Taniyama–Shimura conjecture, then he would also prove Fermat's last theorem. Unlike most mathematicians, who like to work collaboratively, Wiles decided to pursue this goal on his own, telling no one except his wife. He felt that to talk openly about working on Fermat would stir up excitement in the mathematics community, and perhaps lead to unwanted competition. However, as the proof reached its final stages, in the seventh year of working on it, Wiles realized he needed help.

At the time, Wiles was employed at the Institute for Advanced Study (IAS) in Princeton, home to some of the world's finest mathematicians. These colleagues were completely astounded when

Wiles revealed that he had been working on Fermat while still carrying out his daily tasks of lecturing, writing, and teaching.

Wiles recruited the help of these colleagues for the final step in compiling his proof. He turned to American mathematician Nick Katz to check his reasonings. Katz could find no errors, so Wiles decided to go public. In June 1993, at a conference at the University of Cambridge, Wiles delivered his results. Tension rose as he piled his results one on top of the other, with only one end in view. He delivered his final line, "Which proves Fermat's last theorem," smiled, and added, "I think I'll leave it there."

Fixing an error
The next day, the world's press was full of the story, transforming Wiles into the world's most famous mathematician. Everyone wanted to know how this problem had finally been solved. Wiles was delighted, but then came a twist; there was a problem with his proof.

The results had to be verified before they could be published— and Wiles's proof covered scores of pages. Among the reviewers was

Some mathematics problems look simple. There's no reason why these problems shouldn't be easy, and yet they turn out to be extremely intricate.
Andrew Wiles

> I had this rare privilege of being able to pursue in my adult life what had been my childhood dream.
> **Andrew Wiles**

Wiles's friend Nick Katz. For a whole summer Katz went through the proof line by line, querying and questioning until the meaning was clear. One day, he thought he had spotted a hole in the argument. He emailed Wiles, who replied, but not to Katz's satisfaction. More emails followed, before the truth emerged—Katz had found a flaw at the heart of Wiles's work. A vital point in the proof contained an error that undermined Wiles's method.

Suddenly Wiles's approach was brought into question. Had he worked with others rather than alone, the error might have been identified earlier. The world believed that Wiles had resolved Fermat's last theorem, and it was waiting for the finished, published proof. Wiles was under immense pressure. His mathematical achievements so far had been impressive, but his reputation was at stake. Day after day, Wiles tried different approaches to the problem, which proved futile—as his fellow IAS mathematician Peter Sarnak said, "It was like pinning down a carpet in one corner of a room, only for the carpet to pop up in another." Eventually, Wiles turned to a friend, British algebra specialist Richard Taylor, and they worked together on the proof for the next nine months.

Wiles was close to having to admit that he had claimed a proof prematurely. Then, in September 1994, he had a revelation. If he took his present problem-solving method and added its strengths to an earlier approach of his, then one might fix the other, allowing him to solve the problem. It seemed a small insight, but it made all the difference. Within weeks, Wiles and Taylor had plugged the gap in the proof. Nick Katz and the wider mathematical community were now convinced there were no mistakes, and Wiles emerged for a second time as the conqueror of Fermat's last theorem—this time on solid ground.

After the theorem

Fermat was amazingly far-sighted in his original conjecture, but it is unlikely that the "marvelous proof" he claimed to have discovered existed. The idea that every mathematician since the 1600s could have missed a proof that a mathematician from Fermat's time could have discovered is inconceivable. In addition, Wiles solved the theorem using advanced mathematical tools and ideas invented long after Fermat.

In many ways, it is not the proving of Fermat's last theorem that has significance, but rather the proofs used by Wiles. A seemingly impossible problem about integers had been solved by marrying number theory to algebraic geometry, using new and existing techniques. This in turn opened up new ways of looking at how to prove many other mathematical conjectures. ∎

Andrew Wiles

The son of an Anglican priest who later became a professor of divinity, Wiles was born in Cambridge in 1953, and was a passionate problem-solver in mathematics from an early age. Awarded his first degree in mathematics at Merton College, Oxford, and his doctorate at Clare College, Cambridge, he took up a post at the Institute for Advanced Study in Princeton in 1981, and was appointed professor there the following year.

While in the US, Wiles made contributions to some of the most elusive problems in his field, including the Taniyama–Shimura conjecture. He also began his long solo attempt to prove Fermat's last theorem. His eventual success led to him receiving the Abel Prize—the highest honor in mathematics—in 2016.

Wiles has also taught in Bonn and Paris, and at the University of Oxford, where he was appointed Regius Professor of Mathematics in 2018. A new mathematics building at Oxford—as well as an asteroid—9999 Wiles—have been named after him.

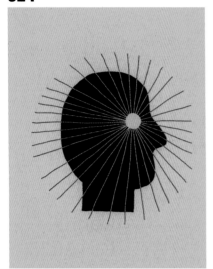

NO OTHER RECOGNITION IS NEEDED

PROVING THE POINCARÉ CONJECTURE

The **"3-sphere"** is a 3-D spherical surface **existing within 4 dimensions**.

⬇

Poincaré argued that **any 3-D shape** without holes can be **distorted to form** the 3-sphere.

⬇

His conjecture can be **extended** to **any number** of **dimensions**.

⬇

Perelman's proof of the Poincaré conjecture was confirmed in 2006.

I n 2000, the Clay Mathematics Institute in the US celebrated the millennium with seven prize problems. Among them was the Poincaré conjecture, which had challenged mathematicians for nearly a century. Within a few years, it was solved—by a little-known Russian mathematician, Grigori Perelman.

Poincaré's conjecture, conceived by the French mathematician in 1904, is stated as: "Every simply connected, closed 3-manifold is homeomorphic to the 3-sphere." In topology, a field that studies the geometrical properties, structure, and spatial relations of shapes, a sphere (a 3-D object in geometry) is said to be a 2-manifold with a 2-D surface existing within a 3-D space—a solid ball, for example. A 3-manifold, such as the 3-sphere, is a purely theoretical concept: it has a 3-D surface and exists in a 4-D space. The description "simply connected" means that the figure has no holes, unlike a bagel or hoop shape (torus), and "closed" means the shape is limited by boundaries, unlike the open endlessness of an infinite plane. In topology, two figures are homeomorphic if they can be distorted or stretched into

A 3-sphere is the 3-D equivalent of a spherical surface, that is a two-dimensional surface, or 2-sphere, such as the ball shown here. To appreciate the shape of the ball, it has to be viewed in 3-D space. To see a 3-sphere requires 4-D space.

Grigori Perelman

Born in 1966 in St. Petersburg, Grigori Perelman developed a passion for mathematics from his mother, who taught the subject. Aged 16, he won a gold medal at the International Mathematical Olympiad in Budapest, achieving a perfect score. A successful academic career followed, including a spell at several research institutes in the US, where he solved a major geometry problem called the Soul conjecture. While there, he met Richard Hamilton, whose work influenced his proof of the Poincaré conjecture.

The reclusive Perelman did not enjoy the fame his proof brought him. He turned down the two greatest accolades for a mathematician: the Fields Medal in 2006 and the Clay Mathematics Institute prize (and its $1 million award) in 2010, saying it belonged as much to Hamilton.

Key works

2002 "The entropy formula for the Ricci flow and its geometric applications"
2003 "Finite extinction time for the solutions to the Ricci flow on certain 3-manifolds"

the same shape. While the question of whether every closed 3-manifold could be deformed to take the shape of a 3-sphere is hypothetical, Perelman has claimed that it holds the key to understanding the shape of the Universe.

Finding a solid proof

Initially, it proved easier to substantiate the conjecture for manifolds of the fourth, fifth, and higher dimensions than it was for 3-manifolds. In 1982, American mathematician Richard Hamilton attempted to prove the conjecture using Ricci flow, a mathematical process that potentially allows any 4-D shape to be distorted to an increasingly smooth version, and ultimately to a 3-sphere. However, the flow failed to handle spikelike "singularities"—deformities including "cigars" and infinitely dense "necks."

Perelman, who learned much from Hamilton during a two-year fellowship at Berkeley in the early 1990s, continued to study Ricci flow and its application to the Poincaré conjecture when he returned to Russia. He masterfully overcame the limitations that Hamilton encountered by using a

technique called surgery, in effect cutting out the singularities, and was able to prove the conjecture.

Surprising the math world

Perelman had achieved success quietly. Unconventionally, he posted his first 39-page paper on the subject online in 2002, emailing a summary to 12 mathematicians in the US. He published two more installments a year later. Others reconstructed his results and explained them in the *Asian Journal of Mathematics*. Finally, his proof was fully accepted by the mathematical community in 2006.

Since then, Perelman's work has been closely studied, fuelling new developments in topology, including a more powerful version of his and Hamilton's technique for using Ricci flow to smooth singularities. ▪

Perelman's proof… solved a problem that for more than a century was an indigestible seed at the core of topology.
Dana Mackenzie
American science writer

DIRECTO

RY

DIRECTORY

In addition to the mathematicians covered in the preceding chapters of this book, many other men and women have made an impact on the development of mathematics. From the ancient Egyptians, Babylonians, and Greeks to the medieval scholars of Persia, India, and China and the city-state rulers of Renaissance Europe, those looking to build, trade, fight wars, and manipulate money realized that measuring and calculating were crucial. By the 19th and 20th centuries, mathematics had become a global discipline, with its practitioners involved in all the sciences. Math remains crucial in the 21st century as space exploration, medical innovations, artificial intelligence, and the digital revolution press ahead, and more secrets about the Universe are revealed.

THALES OF MILETUS
c. 624–c. 545 BCE

Thales lived in Miletus, an ancient Greek city in what is now Turkey. A student of mathematics and astronomy, he broke with the tradition of using mythology as a way of explaining the world. Thales used geometry to calculate the height of pyramids and the distance of ships from the shore. The theorem named after him states that where the longest side of a triangle contained within a circle is the diameter of the circle, that triangle has to be a right-angled triangle. The astronomical discoveries attributed to Thales include his forecast of the 585 BCE solar eclipse.
See also: Pythagoras 36–43 ▪ Euclid's *Elements* 52–57 ▪ Trigonometry 70–75

HIPPOCRATES OF CHIOS
c. 470–c. 410 BCE

Originally a merchant on the Greek island of Chios, Hippocrates later moved to Athens, where he first studied, then practiced mathematics. References by later scholars suggest that he was responsible for the first systematic compilation of geometrical knowledge. He was able to calculate the area of crescent-shaped figures contained within intersecting circles (lunes).

The Lune of Hippocrates, as it was later called, is bounded by the arcs of two circles, the smaller of which has as its diameter a chord spanning a right angle on the larger circle.
See also: Pythagoras 36–43 ▪ Euclid's *Elements* 52–57 ▪ Trigonometry 70–75

EUDOXUS OF CNIDUS
c. 390–c. 337 BCE

Eudoxus lived in the Greek city of Cnidus (now in Turkey). He developed the "method of exhaustion" to prove statements about areas and volumes by successive approximations. For example, he was able to show that the areas of circles relate to each other according to the squares of their radii; that the volumes of spheres relate to each other according to the cubes of their radii; and that the volume of a cone is one-third that of a cylinder of the same height.
See also: The Rhind papyrus 32–33 ▪ Euclid's *Elements* 52–57 ▪ Calculating pi 60–65

HERO OF ALEXANDRIA
c. 10–c. 75 CE

A native of Alexandria in the Roman province of Egypt, Hero (or Heron) was an engineer, inventor, and mathematician. He published descriptions of a steam-powered device called an aeolipile, a wind wheel that could operate an organ, and a vending machine that dispensed "holy" water. His mathematical accomplishments included describing a method for computing the square roots and cubic roots of numbers. He also devised a formula for finding the area of a triangle from the lengths of its sides.
See also: Euclid's *Elements* 52–57 ▪ Trigonometry 70–75 ▪ Cubic equations 102–05

ARYABHATA
476–550 CE

A Hindu mathematician and astronomer, Aryabhata worked in Kusumapara, an Indian center of learning. His verse treatise *Aryabhatiya* contains sections on algebra and trigonometry, including an approximation for pi (π) of 3.1416, accurate to four decimal places. Aryabhata also correctly believed pi to be irrational. He calculated Earth's circumference as a distance close to the current accepted figure. He also defined some trigonometric functions, produced complete and accurate sine and cosine tables, and calculated solutions to simultaneous quadratic equations.
See also: Quadratic equations 28–31 ▪ Calculating pi 60–65 ▪ Trigonometry 70–75 ▪ Algebra 92–99

BHASKARA I
c. 600–c. 680

Little is known about Bhaskara I, although he may have been born in the Saurastra region on India's west coast. He became one of the most important scholars of the astronomy school founded by Aryabhata (see page 328), and wrote a commentary, *Aryabhatiyabhasya*, on Aryabhata's earlier *Aryabhatiya* treatise. Bhaskara I was the first person to write numbers in the Hindu-Arabic decimal system with a circle for zero. In 629, he also found a remarkably accurate approximation of the sine function.
See also: Trigonometry 70–75
▪ Zero 88–91

IBN AL-HAYTHAM
c. 965–c. 1040

Also known as Alhazen, Ibn al-Haytham was an Arab mathematician and astronomer, born in Basra, now in Iraq, who worked at the court of the Fatimid Caliphate in Cairo. He was a pioneer of the scientific method that maintained that hypotheses should be tested by experiment and not just assumed to be true. Among his achievements, he established the beginnings of a link between algebra and geometry, building on the work of Euclid and trying to complete the lost eighth volume of Apollonius of Perga's *Conics*.
See also: Euclid's *Elements* 52–57
▪ Conic sections 68–69

BHASKARA II
1114–85

One the greatest of the medieval Indian mathematicians, Bhaskara II was born in Vijayapura, Karnataka, and is believed to have become the head of the astronomical observatory at Ujjain in Madhya Pradesh. He introduced some preliminary concepts of calculus; established that dividing by zero yields infinity; found solutions to quadratic, cubic, and quartic equations (including negative and irrational solutions); and suggested ways to unlock Diophantine equations of the second order (to the power of two), which would not be solved in Europe until the 1700s.
See also: Quadratic equations 28–31
▪ Diophantine equations 80–81 ▪ Cubic equations 102–05

NASIR AL-DIN AL-TUSI
1201–74

Born in Tus, the Persian mathematician al-Tusi devoted his life to study after he lost his father at a young age. He became one of the great scholars of his day, making important discoveries in math and astronomy. He established trigonometry as a discipline, and in his *Commentary on the Almagest*—an introduction to trigonometry—described methods for calculating sine tables. Although taken prisoner by invading Mongols in 1255, al-Tusi was appointed a scientific advisor by his captors and later established an astronomical observatory in the Mongol capital Maragheh, now in Iran.
See also: Trigonometry 70–75

KAMAL AL-DIN AL-FARISI
c. 1260–c. 1320

Al-Farisi was born in Tabriz, Persia (now Iran). He was a student of polymath Qutb al-Din al-Shirazi, himself a pupil of Nasir al-Din al-Tusi (see above), and, like them, was a member of the Maragheh school of mathematician–astronomers. His explorations of number theory included amicable numbers and factorization. He also applied the theory of conic sections (circles, ellipses, parabolas, and hyperbolas) to solve optical problems, and explained that the different colors of a rainbow were produced by the refraction of light.
See also: Conic sections 68–69
▪ The binomial theorem 100–01

NICOLE ORESME
c. 1320–82

Born in Normandy, France, probably to a peasant family, Oresme studied at the College of Navarre, where pupils from poor backgrounds were subsidized by the royal estate. He later became dean of Rouen Cathedral. Oresme devised a coordinate system with two axes to represent the change of one quality with respect to another—for example, how temperature changes with distance. He worked on fractional exponents and infinite series and was the first to prove the divergence of harmonic series, but his proof was lost and the theory was not proven again until the 1600s. He also argued that Earth could be rotating in space, rather than the Church-approved view that the celestial bodies circled around Earth.
See also: Algebra 92–99 ▪ Coordinates 144–151 ▪ Calculus 168–75

NICCOLÒ FONTANA TARTAGLIA
1499–1557

As a child, Tartaglia was attacked by French soldiers invading Venice. He survived, but with serious facial injuries and a speech impediment, which earned him the nickname "Tartaglia," or stutterer. Essentially self-taught, he became a civil engineer, designing fortifications. Tartaglia realized that an understanding of the trajectory of cannonballs was critical for his designs, which led him to pioneer the study of ballistics. His published mathematical works included a formula for solving cubic equations, an encyclopedic math treatment—*Treatise on Numbers and Measures*—and translations of Euclid and Archimedes.
See also: The Platonic solids 48–49
▪ Trigonometry 70–75 ▪ Cubic equations 102–05 ▪ The complex plane 214–15

GEROLAMO CARDANO
1501–76

A contemporary of Niccolò Tartaglia, Cardano was born in Lombardy and became an outstanding physician, astronomer, and biologist, as well as a renowned mathematician. He studied at the universities of Pavia and Padua in what is now Italy, was awarded a doctorate in medicine, and worked as

a physician before becoming a teacher of mathematics. Cardano published a solution to cubic and quartic equations, acknowledged the existence of imaginary numbers (based on the square root of −1), and is alleged to have forecast the exact date of his own death.
See also: Algebra 92–99 ▪ Cubic equations 102–05 ▪ Imaginary and complex numbers 128–31

JOHN WALLIS
1616–1703

Although Wallis studied medicine at Cambridge University and was later ordained a priest, he retained the interest in arithmetic he first developed as a schoolboy in Kent, England. A supporter of the Parliamentarian cause, Wallis deciphered Royalist dispatches during the English Civil War. In 1644, he was appointed professor of geometry at the University of Oxford and became a champion of arithmetic algebra. His contributions toward the development of calculus include originating the idea of the number line, introducing the symbol for infinity, and developing standard notation for powers. He was one of the small group of scholars whose meetings led to the establishment of the Royal Society of London in 1662.
See also: Conic sections 68–69 ▪ Algebra 92–99 ▪ The binomial theorem 100–01 ▪ Calculus 168–75

GUILLAUME DE L'HÔPITAL
1661–1704

Born in Paris, l'Hôpital was interested in math from a young age and was elected to the French Academy of Sciences in 1693. Three years later, he published the first textbook on infinitesimal calculus: *Analyse des infiniment petits pour l'intelligence des lignes courbes* (*Analysis of the Infinitesimally Small for the Understanding of Curved Lines*). Although l'Hôpital was an accomplished mathematician, many of his ideas were not original. In 1694, he had offered the Swiss mathematician Johann Bernoulli 300 *livres* a year for information on his latest discoveries and an agreement that

he would not share them with other mathematicians. Many of these ideas were published by l'Hôpital in *Infinitesimal Calculus*.
See also: Calculus 168–75

JEAN LE ROND D'ALEMBERT
1717–83

The illegitimate son of a celebrated Paris hostess, d'Alembert was brought up by a glazier's wife. Funded by his estranged father, he studied law and medicine, then turned to mathematics. In 1743, he stated that Newton's third law of motion is as true for freely moving bodies as it is for fixed bodies (d'Alembert's principle). He also developed partial differential equations, explained the variations in the orbits of Earth and other planets, and researched integral calculus. Like other French *philosophes*, such as Voltaire and Jean-Jacques Rousseau, d'Alembert believed in the supremacy of human reason over religion.
See also: Calculus 168–75 ▪ Newton's laws of motion 182–83 ▪ The algebraic resolution of equations 200–01

MARIA GAETANA AGNESI
1718–99

Born in Milan, then under Austrian Hapsburg rule, Agnesi was a child prodigy who, as a teenager, lectured friends of her father on a wide range of scientific subjects. In 1748, Agnesi became the first woman to write a math textbook, the two-volume *Instituzioni analitiche* (*Analytical Institutions*), which covered arithmetic, algebra, trigonometry, and calculus.
Two years later, recognizing her achievement, Pope Benedict XIV awarded her the chair of mathematics and natural philosophy at the University of Bologna, making her the first woman professor of math at any university. The equation describing a particular bell-shaped curve called the "witch of Agnesi" is named in her honor, although "witch" was a mistranslation from the Italian word for "curve."
See also: Trigonometry 70–75 ▪ Algebra 92–99 ▪ Calculus 168–75

JOHANN LAMBERT
1728–77

Lambert was a Swiss-German polymath, born in Mulhouse (now in France), who taught himself math, philosophy, and Asian languages. He worked as a tutor before becoming a member of the Munich Academy in 1759 and the Berlin Academy five years later. Among his mathematical achievements, he provided rigorous proof that pi is an irrational number, and introduced hyperbolic functions into trigonometry. He produced theorems on conic sections, simplified the calculation of the orbits of comets, and created several new map projections. Lambert also invented the first practical hygrometer, used to measure the humidity of air.
See also: Calculating pi 60–65 ▪ Conic sections 68–69 ▪ Trigonometry 70–75

GASPARD MONGE
1746–1818

The son of a merchant, by the age of 17, Monge was teaching physics in Lyon, France. He later worked as a draftsman at the École Royale, Mézières, and in 1780 became a member of the Academy of Sciences. Monge was active in public life, embracing the ideals of the French Revolution. He was appointed Minister of the Marine in 1792, and also worked to reform France's education system, helping to found the École Polytechnique in Paris in 1794 and contributing to the founding of the metric system of measurement in 1795. Described as the "father of engineering drawing," Monge invented descriptive geometry, the mathematical basis of technical drawing, and orthographic projection.
See also: Decimals 132–37 ▪ Projective geometry 154–55 ▪ Pascal's triangle 156–61

ADRIEN-MARIE LEGENDRE
1752–1833

Legendre taught physics and math at the École Militaire in Paris from 1775 to 1780. During this period, he also worked on the Anglo-French Survey, using trigonometry to calculate the distance between the Paris Observatory and London's Royal Greenwich Observatory. During the French Revolution, he lost his private fortune, but in 1794 he published *Eléments de géométrie* (*Elements of Geometry*), which remained a key geometry textbook for the next century, and he was then appointed a math examiner at the École Polytechnique. In number theory, he conjectured the quadratic reciprocity law and the prime number theorem. He also produced the least-squares method for estimating a quantity based on consideration of measurement errors, and gave his name to three forms of elliptic integrals—the Legendre transform, transformation, and polynomials.

See also: Calculus 168–75 ▪ The fundamental theorem of algebra 204–09 ▪ Elliptic functions 226–27

SOPHIE GERMAIN
1776–1831

During the chaos of the French Revolution, 13-year-old Sophie Germain was confined to her wealthy father's house in Paris and began to study the mathematics books in his library. As a woman she was ineligible to study at the École Polytechnique, but she obtained lecture notes and corresponded with the mathematician Joseph-Louis Lagrange. In her work on number theory, Germain also corresponded with Adrien-Marie Legendre (see above) and Carl Gauss, and her ideas on Fermat's last theorem helped Legendre to prove the theorem where $n = 2$. In 1816, she was the first woman to win a prize from the Academy of Sciences in Paris, for a paper on the elasticity of metal plates.

See also: The fundamental theorem of algebra 204–09 ▪ Proving Fermat's last theorem 320–23

NIELS ABEL
1802–29

Abel was a Norwegian mathematician who died tragically young. After graduating from the University of

Christiana (now Oslo) in 1822, he traveled widely in Europe, visiting leading mathematicians. He returned to Norway in 1828, but died from tuberculosis the following year at the age of 26, days before a letter arrived offering him a prestigious math professorship at the University of Berlin. Abel's most important mathematics contribution was to prove that there is no general algebraic formula for solving all quintic (fifth-degree) equations. To make his proof, he invented a type of group theory where the order of the elements within a group is immaterial. This is now known as an abelian group. The annual Abel Prize for mathematics is awarded in his honor.

See also: The fundamental theorem of algebra 204–09 ▪ Elliptic functions 226–27 ▪ Group theory 230–33

JOSEPH LIOUVILLE
1809–82

Born in northern France, Liouville graduated from the École Polytechnique, Paris, in 1827 and took up a teaching post there in 1838. His academic work spanned number theory, differential geometry, mathematical physics, and astronomy, and in 1844 he was the first to prove the existence of transcendental numbers. Liouville wrote more than 400 papers and in 1836 founded the *Journal de Mathématiques Pures et Appliquées* (*Journal of Pure and Applied Mathematics*), the world's second-oldest mathematical journal, which is still published monthly.

See also: Calculus 168–75 ▪ The fundamental theorem of algebra 204–09 ▪ Non-Euclidean geometries 228–29

KARL WEIERSTRASS
1815–97

Born in Westphalia, Germany, Weierstrass developed an interest in mathematics at an early age. His parents wanted their son to have a career in administration, so he was sent to study law and economics at his university, but left without gaining a degree. He then trained as a teacher, ultimately

becoming a professor of mathematics at the Humboldt University of Berlin. Weierstrass was a pioneer in the development of mathematical analysis and in the modern theory of functions, and rigorously reformulated calculus. An influential teacher, he included among his pupils the young Russian émigré and pioneering mathematician Sofya Kovalevskaya (see page 332).

See also: Calculus 168–75 ▪ The fundamental theorem of algebra 204–09

FLORENCE NIGHTINGALE
1820–1910

Named after her Italian birthplace, Florence Nightingale was a British social reformer and pioneer of modern nursing, who based much of her work on the use of statistics. In 1854, after the outbreak of the Crimean War, Nightingale went to work among wounded soldiers at The Barrack Hospital in Scutari, Turkey. There, she campaigned tirelessly for better hygiene, earning the nickname "The lady with the lamp." Back in Britain, Nightingale became an innovator in the use of graphs to display statistical data. She developed the Coxcomb chart, a variation on the pie chart, using circle segments of different sizes to display variations in data, such as the causes of mortality among soldiers. Her actions helped to establish a Royal Commission on health in the army in 1856. In 1907, she was the first woman to receive the Order of Merit, Britain's highest civilian honor.

See also: The birth of modern statistics 268–71

ARTHUR CAYLEY
1821–95

Born in Richmond, Surrey, Cayley was probably the leading British pure mathematician of the 1800s. Graduating from Trinity College, Cambridge, he embarked on a career as a conveyancing lawyer. In 1860, however, he gave up his lucrative law practice to take up a pure math professorship at Cambridge, on a far more modest salary. Cayley was a pioneer of group theory and matrix

algebra, devised the theories of higher singularities and invariants, worked in higher-dimensional geometry, and extended the quaternions of William Hamilton to create octonions.
See also: Non-Euclidean geometries 228–29 ▪ Group theory 230–33 ▪ Quaternions 234–35 ▪ Matrices 238–41

RICHARD DEDEKIND
1831–1916

Dedekind was one of Carl Gauss's students at the University of Göttingen, Germany. After graduating, he worked as an unsalaried lecturer before teaching at the Zurich Polytechnic, Switzerland. Returning to Germany, in 1862 he started work at the Technical High School in Braunschweig, where he remained for the rest of his working life. He proposed the Dedekind cut, now a standard definition of real numbers, and defined concepts of set theory, such as similar sets and infinite sets.
See also: The fundamental theorem of algebra 204–09 ▪ Group theory 230–33 ▪ Boolean algebra 242–47

MARY EVEREST BOOLE
1832–1916

Mary Everest's love of math began young when she studied the books in the study of her clergyman father, whose friends included polymath Charles Babbage, the inventor of the Difference Engine. At 18, Mary met renowned mathematician George Boole (who, like her, was self-taught) in Ireland. They married five years later, but George died soon after the birth of their fifth child. In 1864, with five daughters to raise and no financial support, Mary returned to London, where she worked as a librarian at Queen's College, a girls' school, and later gained a reputation as an eminent children's teacher. She also wrote books that made math more accessible to young students, including *Philosophy and Fun of Algebra* (1909).
See also: Algebra 92–99 ▪ The fundamental theorem of algebra 204–09

GOTTLOB FREGE
1848–1925

The son of the principal of a girls' school in Wismar, northern Germany, Frege studied mathematics, physics, chemistry, and philosophy at the universities of Jena and Göttingen. He then spent his whole working life teaching mathematics in Jena. He lectured in all areas of mathematics, specializing in calculus, but wrote mostly on the philosophy of the subject, bringing the two disciplines together to almost single-handedly invent modern mathematical logic. He once commented that "Every good mathematician is at least half a philosopher, and every good philosopher at least half a mathematician." Frege mixed little with students or colleagues and remained largely unrecognized in his lifetime, although he was a major influence on the work of Bertrand Russell, Ludwig Wittgenstein, and other mathematical logicians.
See also: The logic of mathematics 272–73 ▪ Fuzzy logic 300–01

SOFYA KOVALEVSKAYA
1850–91

Moscow-born Kovalevskaya was the first woman in Europe to gain a doctorate in mathematics, the first woman to join the editorial board of a scientific journal, and the first woman to be appointed a professor of math. She achieved all this despite being barred from a university education in her native Russia because of her gender. Aged 17, Sofya eloped with paleontologist Vladimir Kovalevsky to Germany, where she studied at the University of Heidelberg and then Berlin, where she received tuition from German mathematician Karl Weierstrass (see page 331). Her doctorate was awarded for three papers, the most significant being on partial differential equations. Sofya ended her career as a professor of math at the University of Stockholm, where she died of influenza aged just 41.
See also: Calculus 168–75 ▪ Newton's laws of motion 182–83

GIUSEPPE PEANO
1858–1932

Brought up on a farm in the northern Italian region of Piedmont, Peano studied at the University of Turin, where he gained his doctorate in math in 1880. Almost immediately, he began to teach infinitesimal calculus at the same institution, where he was appointed a full professor in 1889. Peano's first textbook, on calculus, was published in 1884, and in 1891 he began work on the five-volume *Formulario Mathematico* (*Formulation of Mathematics*), which contained the fundamental theorems of math in a symbolic language largely developed by him. Many of the symbols and abbreviations are still in use today. He devised axioms for natural numbers (Peano axioms), developed natural logic and set theory notation, and contributed to the modern method of mathematical induction, used as a proof technique.
See also: Calculus 168–75 ▪ Non-Euclidean geometries 228–29 ▪ The logic of mathematics 272–73

HELGE VON KOCH
1870–1924

Born in Stockholm, Sweden, Koch studied at the universities of Stockholm and Uppsala, later becoming professor of mathematics at Stockholm. He is best known for the fractal—Von Koch's "snowflake" curve—he described in a 1906 paper. This fractal is constructed from an equilateral triangle in which the central third of each side is replaced by the base of another equilateral triangle, with this process continuing indefinitely. If all the triangles face outward, the resulting curve takes on the appearance of a snowflake.
See also: Fractals 306–11

ALBERT EINSTEIN
1879–1955

Einstein was an outstandingly gifted physicist and mathematician. Born in Germany, he moved with his family to Italy when young and studied in

Switzerland. In 1905, he was awarded his doctorate by the University of Zurich and published groundbreaking papers on Brownian motion, the photoelectric effect, special and general relativity, and the equivalence of matter and energy. In 1921, he was awarded the Nobel Prize for his contribution to physics, and he continued to develop the understanding of quantum mechanics in the years that followed. Because of his Jewish background, he did not return to Germany after Hitler came to power in 1933, but settled in the United States, becoming a citizen there in 1940.
See also: Newton's laws of motion 182–83 ▪ Non-Euclidean geometries 228–29 ▪ Topology 256–59 ▪ Minkowski space 274–75

L. E. J. BROUWER
1881–1966

Luitzen Egbertus Jan Brouwer (known to his friends as "Bertus") was born in Overschie, Netherlands. He graduated in mathematics in 1904 from the University of Amsterdam, where he also taught from 1909 to 1951. Brouwer criticized the logical foundations of mathematics as espoused by David Hilbert and Bertrand Russell and helped to found mathematical intuitionism, based on the view of math governed by self-evident laws. He also transformed the study of topology by associating it with algebraic structures, in his fixed-point theorem.
See also: Topology 256–59 ▪ 23 problems for the 20th century 266–67 ▪ The logic of mathematics 272–73

EUPHEMIA LOFTON HAYNES
1890–1980

Born in Washington, DC, Lofton Haynes was the first African-American woman to gain a doctorate in mathematics. After graduating from Smith College, Massachusetts, with a math degree in 1914, she then embarked on a teaching career, and in 1930 established the math department at Miner Teachers College, which later merged with the University of the District of Columbia.

Her doctorate was awarded by The Catholic University of America in 1943 for a dissertation on set theory. In 1959, Lofton Haynes received a Papal medal for her contributions to education and community activism, and in 1966 she was the first woman to chair the District of Columbia State Board of Education.
See also: The logic of mathematics 272–73

MARY CARTWRIGHT
1900–98

The daughter of an English country vicar, Cartwright was one of the first mathematicians to investigate what would later be known as chaos theory. She graduated from the University of Oxford in 1923 with a degree in mathematics. Seven years later, her doctoral thesis was examined by mathematician John E. Littlewood, with whom she would have a long academic collaboration, especially on the study of functions and differential equations. In 1947, Cartwright became the first female mathematician to be elected a Fellow of the Royal Society in London. She had a long association with Girton College, Cambridge, from 1930 to 1968, during which time she taught, researched, and served as Mistress of the college.
See also: The butterfly effect 294–99

JOHN VON NEUMANN
1903–57

The son of affluent Jewish parents in Budapest, Hungary, von Neumann was a child prodigy, able to divide eight-digit numbers in his head at the age of six. He began to publish major mathematical papers in his late teens and started teaching math at the University of Berlin aged 24. In 1933, he moved to the United States to take up a post at the Institute of Advanced Study, Princeton, New Jersey, and became a US citizen in 1937. During a lifetime of mathematical study, von Neumann contributed to virtually every area of the discipline. He was a pioneer of game theory, based on the "two-person zero-sum game," whereby one side wins what the other loses. The

theory provided insights into complex systems in daily life such as economics, computing, and the military. He also created a design model for modern computer architecture, and worked in quantum and nuclear physics, contributing to the atomic bomb during World War II.
See also: The logic of mathematics 272–73 ▪ The Turing machine 284–89

GRACE HOPPER
1906–92

Born Grace Murray in New York City, Hopper was a pioneering computer programmer. After gaining a doctorate from Yale University in 1934, she taught for several years before the outbreak of World War II. When her application to enlist in the US Navy was rejected, she joined the Naval Reserve and began her transition to computer science. After the war, while employed as a senior mathematician at a computer company, she developed the Common Business-Oriented Language (COBOL), which became the most widely used programming language. Hopper retired from the Navy Reserve in 1966, but was called back on active duty the following year, not retiring until 1986, by which time she held the rank of rear admiral. She coined the word "bug" for a computer glitch after a moth flew into circuits on which she was working.
See also: The mechanical computer 222–25 ▪ The Turing machine 284–89

MARJORIE LEE BROWNE
1914–79

Only the third African-American woman to earn a PhD in math, Browne was born in Tennessee at a time when it was hard for women of color to pursue an academic career. With the support of her father, a railroad clerk, she graduated from Howard University, Washington DC, in 1935, and, after teaching briefly in New Orleans, continued her studies at the University of Michigan, gaining her doctorate in 1949. Two years later, she was appointed chair of the mathematics department at North Carolina Central

University. Marjorie gained a reputation for being an excellent teacher, and for her research, especially in topology.
See also: Topology 256–59

JOAN CLARKE
1917–96

London-born Clarke achieved a double first in math at the University of Cambridge on the eve of World War II but was denied a full degree because of her gender. Her mathematical prowess had been recognized, however, and when the Bletchley Park project was established to decipher the German Enigma Code, Clarke was recruited. At Bletchley, she became one of the leading cryptanalysts, working closely with Alan Turing, to whom she was engaged for a short time. Although they did the same work as the male code-breakers, Clarke and the other Bletchley women were paid less. The Bletchley Park operation was hugely successful, cutting short the length of the war and saving countless lives. After the war, Clarke worked at the British government's surveillance center, GCHQ. Because so much of Clarke's work was secret, the full extent of her accomplishments is still unknown.
See also: The Turing machine 284–89 ▪ Cryptography 314–17

KATHERINE JOHNSON
1918–

A child math prodigy, Katherine Johnson (born Coleman) was a pioneer of computing and the American space program. Her calculations on flight trajectories were critical in enabling Alan Shepard to become the first American in space (1961), John Glenn to be the first American to orbit Earth (1962), Apollo 11 to land on the Moon (1969), and the Space Shuttle program to launch (1981). Johnson graduated in 1937 from West Virginia State College and became one of the first African-Americans to enroll in a graduate program at West Virigina University. She worked for the National Advisory Committee for Aeronautics (NACA) from 1953 as part of a group of African-

American women mathematicians known as the West Area Computers, who later inspired the film *Hidden Figures* (2016). Johnson then worked for the National Aeronautics and Space Administration (NASA) from 1958 as part of its Space Task Group. In 2015, President Obama awarded Johnson the Presidential Medal of Freedom.
See also: Calculus 168–75 ▪ Newton's laws of motion 182–83 ▪ Non-Euclidean geometries 228–29

JULIA BOWMAN ROBINSON
1919–85

Born Julia Bowman in St. Louis, Missouri, Robinson gained her mathematics doctorate at the University of California, Berkeley, in 1948. She developed a fundamental theorem of elementary game theory (see John von Neumann, page 333) in 1951, but is best known for her work on solving the tenth of David Hilbert's list of 23 mathematical problems, drawn up in 1900—whether there is an algorithm that could find a solution to any Diophantine equation (one that uses whole numbers and finite unknowns). Robinson proved, along with other mathematicians, such as Yuri Matiyasevich (see opposite), that such an algorithm could not exist. Robinson was appointed a professor at Berkeley in 1975, and in 1976 she was the first woman to be elected to the American National Academy of Sciences.
See also: Diophantine equations 80–81 ▪ 23 problems for the 20th century 266–67

MARY JACKSON
1921–2005

An aerospace engineer, Mary Jackson (born Winston) worked on the US space program and campaigned for better opportunities in engineering for women and people of color. After graduating in math and physical sciences from Hampton University, Virginia, Jackson taught for a while, then in 1951 started work in the West Area Computing Unit of the National Advisory Committee for Aeronautics (NACA). The unit,

known as the West Area Computers, comprised female African-American mathematicians, including Katherine Johnson (see left). From 1958—when Jackson became NASA's first female black engineer—to 1963, she worked on Project Mercury, the program that put the first Americans into space.
See also: Calculus 168–75 ▪ Newton's laws of motion 182–83 ▪ Non-Euclidean geometries 228–29

ALEXANDER GROTHENDIECK
1928–2014

Considered by many to be the greatest pure mathematician of the second half of the 20th century, Grothendieck was unorthodox in every respect. Born in Germany to anarchist parents, at the age of 10 he fled the Nazi regime as a refugee to France, where he spent most of his life. His huge output—much of it never published—included revolutionary advances in algebraic geometry, the devising of the theory of schemes, and contributions to algebraic topology, number theory, and category theory. Grothendieck's radical political activities included delivering math lectures just outside Hanoi while the city was being bombed during the Vietnam War.
See also: Non-Euclidean geometries 228–29 ▪ Topology 256–59

JOHN NASH
1928–2015

American mathematician John Nash is best known for establishing the mathematical principles of game theory (see John von Neumann, page 333). After graduating from Carnegie Mellon University in 1948 and being awarded a doctorate from Princeton University in 1950, he joined the Massachusetts Institute of Technology (MIT), where he researched partial differential equations and began the work on game theory that won him the Nobel Prize for Economics in 1994. For much of his life, Nash fought paranoid schizophrenia, as dramatized in the film *A Beautiful Mind* (2001).
See also: Calculus 168–75 ▪ The logic of mathematics 272–73

PAUL COHEN
1934–2007

New Jersey-born Cohen was awarded the Fields Medal (the mathematical equivalent of a Nobel Prize) in 1966 for solving the first of David Hilbert's list of 23 unresolved mathematical problems—that there is no set whose number of elements is between that of the integers and that of the real numbers. Cohen graduated and later received his doctorate, in 1958, from the University of Chicago before moving to the Massachusetts Institute of Technology (MIT), Princeton University, and finally Stanford University, where he became professor emeritus in 2004.
See also: 23 problems for the 20th century 266–67

CHRISTINE DARDEN
1942–

Along with Katherine Johnson and Mary Jackson (see page 334), Darden is one of the African-American women whose work as mathematicians made key contributions to the work of NASA's space programs. After graduating from Hampton University, Darden taught at Virginia State University before moving in 1967 to NASA's Langley Research Center. There, she built her reputation as an aeronautical engineer, specializing in supersonic flight. In 1989, she was appointed leader of the Sonic Boom Team, working on designs to reduce noise pollution and other negative effects of supersonic flight.
See also: Calculus 168–75 ▪ Newton's laws of motion 182–83 ▪ Non-Euclidean geometries 228–29

KAREN KESKULLA UHLENBECK
1942–

In 2019, Uhlenbeck became the first woman to be awarded the Abel Prize for Mathematics. Born in Cleveland, Ohio, in 1942, she gained a PhD in mathematics from Brandeis University, Waltham, Massachusetts in 1968, and went on to achieve notable breakthroughs in mathematical physics, geometrical analysis, and topology. A champion of gender equality in science and mathematics, in 1990 she became the first woman since Emmy Noether to give a plenary speech at the International Congress of Mathematics. In 1994, she founded the Women and Mathematics Program at the Institute of Advanced Study in Princeton, New Jersey.
See also: Topology 256–59

EVELYN NELSON
1943–87

The Krieger–Nelson Prize, awarded by the Canadian Mathematical Society for outstanding research by a female mathematician, is named in honor of Evelyn Nelson and fellow Canadian Cecilia Krieger. Nelson began a career of teaching and research at McMaster University after obtaining her doctorate there in 1970. She published more than 40 research papers in a 20-year career that was cut short by cancer. Her main contributions were to universal algebra (the study of algebraic theories and their models) and algebraic logic, applying these to the field of computer science.
See also: The fundamental theorem of algebra 204–09 ▪ The logic of mathematics 272–73

YURI MATIYASEVICH
1947–

While studying for his doctorate at the Steklov Institute of Mathematics in Leningrad (now St. Petersburg), Matiyasevich became fascinated by the challenge of solving David Hilbert's tenth problem. Just as he was about to give up, he read the paper "Unsolvable Diophantine problems" (1969) by American mathematician Julia Robinson (see page 334), and a solution fell into place. In 1970, Matiyasevich provided the final proof that the tenth problem is unsolvable because there is no general method of determining whether Diophantine equations have a solution. In 1995, he was appointed professor at St. Petersburg University, first as chair of software engineering and later as chair of algebra and number theory.
See also: Diophantine equations 80–81 ▪ 23 problems for the 20th century 266–67

RADIA PERLMAN
1951–

Virginia-born Perlman has been described as the "mother of the internet." While a student at the Massachusetts Institute of Technology (MIT), she worked on a program that introduced children as young as three to computer programming. After graduating with a masters degree in mathematics in 1976, Perlman worked for a government contractor that developed software. Then, in 1984, while working for the Digital Equipment Corporation (DEC), she invented the Spanning Tree Protocol (STP), which ensures there is only one active path between two network devices; this would later prove crucial for the development of the internet. Perlman has taught at MIT and the universities of Washington and Harvard, and continues to work on computer network and security protocols.
See also: The mechanical computer 222–25 ▪ The Turing machine 284–89

MARYAM MIRZAKANI
1977–2017

At the age of 17, Mirzakani became the first Iranian woman to win a gold medal in the International Mathematical Olympiad. She graduated from Tehran's Sharif University of Technology, before earning a doctorate from Harvard University in 2004 and taking up a professorship at Princeton University. Ten years later, Mirzakani was both the first woman and the first Iranian to receive the Fields Medal—for her contribution to the study of Riemann surfaces. She was working at Stanford University when she died of breast cancer, aged 40.
See also: Non-Euclidean geometries 228–29 ▪ The Riemann hypothesis 250–51 ▪ Topology 256–59

GLOSSARY

In this glossary, terms defined within another entry are identified with *italic* type.

Abstract algebra The branch of *algebra*, developed mainly in the 1900s, that investigates abstract mathematical structures such as *groups* and *rings*.

Acute angle An angle that is less than 90 *degrees*.

Algebra A branch of mathematics that involves the use of letters to stand for unknown or *variable* numbers in calculations.

Algebraic geometry The use of *graphs* to plot lines and curves that represent algebraic *functions*, such as $y = x^2$.

Algebraic numbers All the rational numbers and those *irrational numbers* that can be obtained by calculating the *roots* of a *rational number*. An irrational number that is not algebraic (such as *pi* or *e*) is called a *transcendental number*.

Algorithm A defined sequence of mathematical or logical instructions, or rules, devised to solve a class of problems. Algorithms are widely used in mathematics and computer science for calculation, organizing data, and a multitude of other tasks.

Amicable numbers Any pair of *whole numbers*, where the *factors* of each one add up to form the other. The smallest pair are 220 and 284.

Analysis The branch of mathematics that studies *limits* and handles infinitely large and small quantities, especially to solve problems in *calculus*.

Analytic geometry See *algebraic geometry*.

Apex The *vertex* that is furthest from the base in a 3-D shape.

Applied mathematics The use of mathematics to solve problems in science and technology. It includes techniques for solving particular kinds of *equations*.

Arc A curved line that forms part of the *circumference* of a circle.

Area The amount of space inside any 2-D shape. Area is measured in square units, such as square inches (in^2).

Associative law This states that if you add, for example, $1 + 2 + 3$, the numbers can be added in any order. The law works for ordinary addition and multiplication, but not for subtraction or division.

Average The typical or middle value of a set of data. For the different kinds of averages, see *mean*, *median*, and *mode*.

Axiom A rule, especially one that is fundamental to an area of mathematics.

Axis (plural **axes**) A fixed reference line, such as the vertical y-axis and horizontal x-axis on a *graph*.

Base (1) In a *number system*, the base is the number around which the system is organized. The main number system we use today is the base-10 or decimal system, where the numerals 0 to 9 are used and the next number is written 10, indicating one ten and no units. See also *place value system*. (2) In *logarithms*, a fixed base (usually 10 or Euler's number *e*) is used; the logarithm of any given number x is the *power* to which that base must be raised to produce x.

Binary notation Writing numbers using the binary system, in which the only numerals used are 0 and 1. For example, the number 6 is written as 110 in the binary system. Here, the leftmost 1 has the value of 4 (2 × 2), the middle 1 means one 2, and the zero means no single units: $4 + 2 + 0$ makes 6.

Binomial An *expression* consisting of two *terms* added together, e.g. $x + y$. When a binomial expression is raised to a *power,* for example $(x + y)^3$, the result when multiplied out gives (in this case) $x^3 + 3x^2y + 3xy^2 + y^3$. This process is called binomial expansion, and the numbers multiplying the terms (3s in this case) are called binomial *coefficients*. The binomial theorem is a rule for working out binomial coefficients in complex cases. See also *polynomial*.

Calculus A branch of mathematics that deals with continuously changing quantities. It includes differential calculus, which is concerned with rates of change, and integral calculus, which calculates *areas* and *volumes* under curves or curved surfaces.

Cardinal numbers Numbers that denote a quantity, such as 1, 2, 3 (in contrast to *ordinal numbers*).

Chord A straight line that cuts across a circle, but does not go through its center.

Cipher Any systematic method of coding messages so that they cannot be understood without being deciphered first.

Circumference The distance all the way around the outside edge of a circle.

Coefficient A number or *expression*, usually a *constant*, that is placed before another number (especially a *variable*) and multiplies it. For example, in the expressions ax^2 and $3x$, a and 3 are coefficients.

Coincident In *geometry*, two or more lines or figures that, when superimposed on each other, share all points and occupy exactly the same space.

Combinatorics A branch of mathematics that studies the ways in which sets of numbers, shapes, or other mathematical objects can be combined.

Commutative law The law that states that $1 + 2 = 2 + 1$, for example, and that the order in which the numbers are set down doesn't matter. It works for ordinary addition and multiplication, but not for subtraction and division.

Complex number A number that is a combination of a *real number* and an *imaginary number*.

Complex plane The infinite 2-D *plane* on which *complex numbers* can be plotted.

Composite number A *whole number* that is not *prime,* but can be created by multiplying together smaller numbers.

Cone A 3-D shape with a circular base and a side that narrows upward toward a point (*apex*).

Congruent Having the same size and shape. (Used when comparing geometrical shapes.)

Conjecture A mathematical statement or claim that has not yet been proved or disproved. A pair of related conjectures can be strong or weak: if the strong conjecture is proved, then the weak conjecture is also proved, but not vice versa.

Constant A quantity in a mathematical *expression* that does not vary—often symbolized by a letter such as a, b, or c.

Convergence A property of some infinite mathematical *series* where not only is each *term* smaller than the last, but the terms, when added up, approach a finite answer. The value of numbers such as *pi* can be estimated using convergent series.

Coordinates Combinations of numbers that describe the position of a point, line, or shape on a *graph* or a geographical position on a map. In mathematical contexts, they are written (for a 2-D case) in the form (x,y), where x is the horizontal position and y the vertical position.

Cosine (abbreviation **cos**) A *function* in *trigonometry* similar to a *sine*, except that it is defined as the ratio of the length of the side of a *right-angled* triangle adjacent to a given angle to the length of the triangle's *hypotenuse*.

Cube A 3-D geometrical figure whose *faces* are six identical squares. A cube number is one that is obtainable by multiplying a smaller number together twice— for example 8, which is $2 \times 2 \times 2$ (2^3). This multiplication resembles the way the volume of a cube is calculated, by multiplying its length $\times$ height $\times$ depth.

Cubic equation An *equation* containing at least one *variable* multiplied by itself twice (for example, $y \times y \times y$, also written as y^3), but no variable multiplied more times than this.

Cubit A measure of length used in the ancient world, based on the length of the human forearm.

Cylinder A 3-D shape, such as a tin can, with two identical circular ends joined by one curved surface.

Deduction A process by which a problem is solved by drawing on known or assumed mathematical principles. See also *induction*.

Degree (1) A measure of angle in *geometry*: rotating a full circle involves turning 360 degrees. (2) The degree or *order* of a *polynomial* is the highest-*power* term within it: for example, a polynomial is "of degree 3" or "of order 3" if it contains a cubed term, such as x^3, as its highest power. Similarly, in *differential equations*, the term that has been differentiated most times in a given equation determines its degree or order.

Denominator The lower number in a fraction, such as the 4 in ¾.

Derivative See *differentiation*.

Diameter A straight line touching two points on the edge of a circle and passing through the center.

Differential equation An *equation* that represents a *function* including the *derivative*(s) of a given *variable*.

Differentiation In *calculus*, the process of working out the rate of change of a given mathematical *function*. The result of the calculation is another function called the differential or *derivative* of the first function.

Divergence A term applied mainly to infinite *series* that do not approach closer and closer to an end-number. See also *convergence*.

Divisor The number or quantity by which another number or quantity is being divided.

Dodecahedron A 3-D *polyhedron* made up of 12 pentagonal (5-sided) *faces*. A regular dodecahedron is one of the five *Platonic solids*.

Ellipse A shape like a circle, but stretched out symmetrically in one direction.

Encryption The process of converting data or a message to a secure, coded form.

Equation A statement that two mathematical *expressions* or

quantities are equal to each other. An equation is the usual way of expressing a mathematical *function*. When an equation is true of all the values of a *variable* (for example, the equation $y \times y \times y = y^3$), it is called an identity.

Equilateral triangle A triangle in which all three sides are the same length and all three angles the same size.

Existence proof A mathematical *proof* that something exists, obtained either by constructing an example or by general *deduction*.

Expansion In *algebra*, the expansion of an *expression* is the opposite of *factorization*. For example, $(x + 2)(x + 3)$ can be expanded to $x^2 + 5x + 6$, by multiplying each term in the first pair of parentheses by each term in the second pair of parentheses.

Exponent The superscript number that indicates the *power* to which a number or quantity has been raised, such as the 2 in x^2 ($x \times x$). Also called an *index*.

Exponential function A mathematical *function* where, as a quantity gets larger, its rate of increase also gets faster. The result is often called exponential growth.

Expression Any meaningful combination of mathematical symbols, such as $2x + 5$.

Face A flat surface of any 3-D shape.

Factor A number or *expression* that divides exactly into another number or expression. For example, 1, 2, 3, 4, 6, and 12 are all factors of 12.

Factorial The *product* of any positive *integer* and all the positive integers that are smaller than it. For example, factorial 5, also written 5! (with an exclamation mark) is $5 \times 4 \times 3 \times 2 \times 1 = 120$.

Factorization Expressing a number or mathematical *expression* in terms of *factors* that when multiplied together give the original number or expression.

Formula A mathematical rule that describes a relationship between quantities.

Fractals Self-similar curves or shapes of different sizes that form complex patterns that have the same general appearance at any magnification. Many natural phenomena, such as clouds and rock formations, approximate to fractals.

Function A mathematical relationship where the value of one *variable* is worked out uniquely from the value of other numbers, using a particular rule. For example, in the function $y = x^2 + 3$, the value of y is calculated by squaring x and then adding 3. The same function can also be written $f(x) = x^2 + 3$, where $f(x)$ stands for "function of x."

Geometry The branch of mathematics that studies shapes, lines, points, and their relationships. See also *non-Euclidean geometries*.

Gradient The *slope* of a line.

Graph (1) A chart on which data is plotted using, for instance, lines, points, curves, or bars. (2) In *graph theory*, a graph is a collection of points, called *vertices*, and lines, called edges, that can be used to model theoretical and real networks, relations, and processes in a range of scientific and social fields.

Graph theory A branch of mathematics that studies how *graphs* made up of points and lines are connected.

Group A mathematical *set*, together with an *operation* which, when performed on members of the set, yields an answer that is still a member of the set. For example, the set of *integers* forms a group when addition is the operation. Groups can be finite or *infinite*, and their study is called group theory.

Harmonic series The mathematical *series* $1 + \frac{1}{2} + \frac{1}{3} + \frac{1}{4} + \frac{1}{5} + \ldots$. The individual terms in the series define the different ways that a stretched string, for example, or air in a tube, can vibrate to produce sound. The resulting series of musical pitches forms the basis of the musical scale.

Hyperbola A mathematical curve that looks something like a *parabola*, but in which the two extensions of the curve approach two imaginary straight lines at angles to each other without ever touching or crossing the lines.

Hypotenuse The longest side of a right-angled triangle, located on the opposite side from the *right angle*.

Icosahedron A 3-D *polyhedron* made up of 20 triangular *faces*. A regular icosahedron is one of the five *Platonic solids*.

Ideal In *abstract algebra,* a mathematical *ring* that is a component of a larger ring.

Identity element In a *set* of numbers or other mathematical objects, an *operation* carried out on the set, such as multiplication or addition, always has an identity element—a number or *expression* that leaves other *terms* unchanged after the operation has been carried out. The identity element in ordinary multiplication, for example, is 1, as $1 \times x = x$, and in the addition of *real numbers*, it is 0, as $0 + x = x$.

Imaginary number Any number that is a multiple of $\sqrt{-1}$, which does not exist as a *real number*. It is expressed as the symbol *i*.

Incommensurable Something that cannot be measured exactly in terms of something else.

Index (plural **indices**) Another word for an *exponent*.

Induction A way of obtaining a general conclusion in mathematics by establishing that if a statement is true for one step in a process, it is also true for the next step in a process and all those that follow. See also *deduction*.

Infinite Indefinitely large and without limit. In mathematics, there are different kinds of infinity: the *set* of *natural numbers*, for example, is countably infinite (countable one by one, even though the end is never reached), while the *real numbers* are uncountably infinite.

Infinite series A mathematical series with an infinite number of *terms*: see *series*.

Infinitesimal calculus Another term for *calculus*, generally used in the past when calculus was viewed

as involving the adding up of infinitesimals (infinitely small but nonzero quantities).

Input Any *variable*, which when combined with a *function*, produces an *output*.

Integer Any of the negative or positive *whole numbers*. (Fractions are not integers.)

Integral (1) Relating to *integers*. (2) A mathematical *expression* used in integral *calculus*, or the result of performing an *integration*.

Integration The process of performing a calculation in integral *calculus*.

Inverse A mathematical *expression* or *operation* that is the opposite of another one and undoes it. For example, division is the inverse of multiplication.

Irrational number Any number that cannot be expressed as one *whole number* divided by another and is not an *imaginary number*.

Isosceles triangle A triangle with two sides the same length and two angles the same size.

Iteration Performing the same *operation* again and again to achieve a desired result.

Limit The end number that is approached as certain calculations are iterated to infinity.

Linear equation An equation that contains no *variable* multiplied by itself (for example, no x^2 or x^3). Linear equations result in straight lines when they are plotted as *graphs*.

Linear transformation Also called linear mapping, a *mapping* between *vector* spaces.

Logarithm The logarithm of a number is the *power* to which another number (called the *base*—usually either 10 or Euler's number *e*)—must be raised to give the original number. For example, $10^{0.301} = 2$, and so 0.301 is the logarithm (to the base 10) of 2. A logarithm to the base *e* (2.71828…) is called a *natural logarithm* and is indicated by the prefix *ln* or $\log_e$. The advantage of logarithms is that when numbers need to be multiplied, the calculation can be simplified by adding their logarithms instead.

Logic The study of reasoning, that is, how conclusions can be deduced correctly from given starting information (premises) by following valid rules.

Manifold A kind of abstract mathematical space that in any particular small region resembles ordinary 3-D space. It is a concept within *topology*.

Mapping Establishing a relationship between members of one mathematical *set* and another. It is often but not always used to mean a one-to-one mapping, where each member of one set is associated with one member of the other set, and vice versa.

Matrix (plural **matrices**) A square or rectangular array of numbers or other mathematical quantities that can be treated as a single object in calculations. Matrices have special rules for addition and multiplication. Their many uses include solving several *equations* simultaneously, describing *vectors*, calculating *transformations* in the shape and position of geometrical figures, and representing real-world data.

Mean An *average* found by adding up the values of a set of data and dividing by the number of values. For example, the mean of the four numbers 1, 4, 6, and 13 is 1 + 4 + 6 + 13 = 24 divided by 4 = 6.

Median The middle value of a set of data, when the values are put in order from lowest to highest.

Meridian An imaginary line on Earth's surface joining the North Pole and South Pole through any given locality. Lines of longitude are meridians.

Mode The value that occurs most often in a set of data.

Modular arithmetic Also called clock arithmetic, a form of arithmetic where, after counting up to a certain point, 0 is reached, and the process is repeated.

Natural logarithm See *logarithm*.

Natural number Any of the positive *whole numbers*. See also *integer*.

Non-Euclidean geometries A key *postulate* of traditional geometry, as described by Euclid in ancient times, is that *parallel* lines never meet (often expressed as meeting at infinity). Geometries in which this and other Euclidean postulates are not valid are called non-Euclidean.

Number line A horizontal line with numbers written on it that is used for counting and calculating. The lowest numbers are on the left, the highest on the right. All *real numbers* can be placed on a number line.

Number system Any system of writing down and expressing numbers. The Hindu–Arabic system used today is based on the ten numerals 0 to 9: when 10 is reached, 1 is written again, but with a 0 after it. This system is both a *place value system* and a *base*-10 or decimal system.

Number theory A branch of mathematics that studies the properties of numbers (especially *whole numbers*), their patterns, and their relationships. It includes the study of *prime numbers*.

Numerator The upper number in a fraction, such as the 3 in ¾.

Obtuse angle An angle between 90 and 180 *degrees*.

Octahedron A 3-D *polyhedron* made up of eight triangular *faces*. A regular octahedron is one of the five *Platonic solids*.

Operation Any standard mathematical procedure such as addition or multiplication. The symbols used for such operations are called operators.

Order See *degree*.

Ordinal numbers Numbers that denote a position, such as 1st, 2nd, or 3rd. See also *cardinal numbers*.

Origin The point at which the *x* and *y* axes of a *graph* intersect.

Oscillation A regular to-and-fro movement between one position or value to another and back again.

Output The result when an *input* is combined with a *function*.

Parabola A curve that is similar to one end of an *ellipse*, except that the arms of a parabola diverge.

Parabolic Relating to a parabola, or to a *function* based on it, such as a quadratic function, which produces a parabola-shaped *graph*.

Parallel Of a line, going in exactly the same direction as another line.

Parallelogram A *quadrilateral* where each side has the same length as the side opposite to it and the two sides are also parallel. A square, rectangle, and *rhombus* are types of parallelogram.

Partial differential equation A *differential equation* containing several *variables*, in which the *differentiation* is applied to only one of the variables at a time.

Periodic function A *function* whose value repeats periodically, as seen, for example, in the *graph* of a *sine* function, which is in the form of a repeating series of waves.

Perpendicular At *right angles* to something else.

Pi (π) The ratio of a circle's *circumference* to its *diameter*, approximately $^{22}/_7$, or 3.14159. It is a fundamental *transcendental number* that appears in many branches of mathematics.

Place value system The standard system for writing numbers, where the value of a digit depends on its place in a larger number. The 2 in 120, for example, has a place value of 20, but in 210 it stands for 200.

Placeholder A numeral, usually zero, used in a *place value system* to differentiate 1 from 100, for example, but that does not necessarily imply an exact measurement as in phrases such as "about 100 miles away."

Plane A flat surface.

Plane geometry The *geometry* of 2-D figures on a flat surface.

Platonic solid One of the five *polyhedra* that form completely regular and symmetrical shapes: each face is an identical *polygon* and all the angles between the *faces* are the same. The five Platonic solids are the *tetrahedron*, *cube*, *octahedron*, *dodecahedron*, and *icosahedron*.

Polygon Any flat shape with three or more straight sides, such as a triangle or pentagon.

Polyhedron Any 3-D shape whose faces are *polygons*.

Polynomial A mathematical *expression* made up of two or more *terms* added together. A polynomial expression usually includes different *powers* of a *variable*, together with *constants*, for example, $x^3 + 2x + 4$.

Positional number An individual numeral whose value depends on its position within a larger number. See *place value system*.

Postulate In mathematics, a statement whose truth is taken for granted or thought to be obvious, but is not supported by a *proof*.

Power The number of times a quantity or number has been multiplied by itself. For example,

four ys multiplied together ($y \times y \times y \times y$) is called "$y$ raised to the power of 4" and written y^4.

Power series A mathematical *series* where each *term* has a greater *power* than the previous one, such as $x + x^2 + x^3 + x^4 + \ldots$.

Prime number A *natural number* that can be divided exactly only by itself and 1.

Probability The branch of mathematics that studies the likelihood of different outcomes occurring in the future.

Product The result of one number or quantity being multiplied by another.

Proof Any method of showing beyond doubt that a mathematical statement or result is true. There are different kinds, including proof by *induction* and *existence proofs*.

Proportion The relative size of something compared with something else. For example, if two quantities are in inverse proportion, the larger one of them becomes, the smaller the other one will become; for example, if one quantity is multiplied by 3, the other is divided by 3.

Pure mathematics Topics in mathematics that are studied for their own sake rather than for any practical application. See also *applied mathematics*.

Quadratic equation An equation containing at least one *variable* multiplied by itself once (for example $y \times y$, also written y^2), but containing no variables raised to higher *powers*.

Quadrilateral Any flat 2-D shape with four straight sides.

Quartic Referring to *equations* or *expressions* of the fourth *degree*, where the highest *power* contained in them is 4—for example, x^4.

Quaternion A mathematical object that is a development of the idea of a *complex number,* but uses four components added together, rather than just two.

Quintic Referring to *equations* or *expressions* of the fifth *degree*, where the highest *power* contained in them is 5, for example, x^5.

Quotient The result that is obtained when one number is divided by another.

Radian A measure of angles that is an alternative to *degrees* and is based on the length of the *radius* and *circumference* of a circle. Turning around by 2 × *pi* (2π) radians is the same as turning 360 degrees (that is, in a complete circle).

Radius Any straight line from the center of a circle or sphere to its *circumference*.

Rational number A number that can be expressed as a fraction of one *whole number* over another. See also *irrational number*.

Real number Any number that is either a *rational number* or an *irrational number*. Real numbers include fractions and negative numbers, but not *imaginary* or *complex numbers*.

Reciprocal A number or *expression* that is the *inverse* of another one, meaning that the

result of multiplying them together is 1. For example, the reciprocal of 3 is $\frac{1}{3}$.

Recurring Any number that is repeated without limit. For example, $\frac{1}{3}$ expressed in decimals is 0.333333… , which can also be described in words as "zero point three recurring."

Rhombus A *quadrilateral* with all four sides the same length; informally, a diamond shape. A square is a special kind of rhombus, with all angles 90 *degrees*.

Right angle An angle that is 90 *degrees* (a quarter turn), such as the angle between vertical and horizontal lines.

Ring A mathematical structure that is like a *group* except that it includes two *operations* rather than one. For example, the *set* of all *integers* forms a ring when taken together with the operations addition and multiplication, because performing these operations on members of the set produces an answer that is still a member of the set.

Root (1) The root of a number, which is another number that when multiplied gives the original number. For example, 4 and 8 are roots of 64, with 8 being the square root (8 × 8 = 64) and 4 the cube root (4 × 4 × 4 = 64). (2) The root of an *equation* is its solution.

Scalar A quantity that has magnitude (size), but not direction, in contrast to a *vector*.

Scalene triangle A triangle where none of the sides and none of the angles are the same size.

Segment (1) Part of a line, with definite end points. (2) In a circle, the area between a *chord* and the outside edge (*circumference*).

Sequence An arrangement of numbers or mathematical *terms* placed one after the other and usually following a set pattern.

Series A list of mathematical *terms* added together. Series usually follow a mathematical rule, and even if the series is *infinite*, it may add up to a finite number. See also *sequence*.

Set Any collection of numbers, or mathematical structures based on numbers. Sets can be finite or *infinite* (for example, the set of all *integers*).

Set theory The theory of *sets* and a branch of mathematics which now forms the underlying basis of many other branches of mathematics.

Sexagesimal A *number system* used by the ancient Babylonians based on the number 60, and still used in a modified form for time, angles, and geographic *coordinates*.

Simultaneous equations A set of several *equations* that include the same unknown quantities, such as x, y, and z. Usually, the equations must be calculated together to solve the value of the unknowns.

Sine (abbreviation **sin**) An important *function* in *trigonometry*, and defined as the ratio of the length of the side opposite a given angle in a *right-angled* triangle to the length of the triangle's *hypotenuse*. This ratio starts at 0 and varies with the size of the angle, repeating its pattern after

360 *degrees*. The *graph* of a sine function is also the shape of many waves, including light waves.

Slope The angle of a line to the horizontal, or an angle of a *tangent* to a curve to the horizontal.

Square number A *whole number* that can be formed by multiplying a smaller whole number by itself once. For example, 25 is a square number as it is 5 × 5 (5^2).

Statistics (1) Measurable data collected in an orderly way for any purpose. (2) The branch of mathematics that develops and applies methods for analyzing and studying such data.

Surd An *expression* that includes a *root* that is an *irrational number* such as $\sqrt{2}$. It is left in root form as it cannot be simplified or written exactly as a decimal.

Surface area The *area* of a flat or curved surface, or of the outside of a 3-D object.

Tangent (1) A line which grazes the outside of a curve, just touching it at one point. (2) In *trigonometry*, the tangent *function*, abbreviated as tan, is defined as the ratio of the side length opposite a given angle to the side length adjacent to that angle, in a *right-angled* triangle.

Term In an algebraic *expression*, one or more numbers or *variables*, usually separated by a plus (+) or minus (−) sign, or in a *sequence*, by a comma. In $x + 4y - 2$, for example, x, $4y$, and 2 are all terms.

Tessellation A pattern that is formed on a flat 2-D surface by repeated copies of one or more

regular geometrical shapes that cover the surface without any gaps in between. This is also called a tiling.

Tesseract A 4-D shape with four edges at every *vertex*, whereas a *cube* has three edges at every vertex, and a square has two.

Tetrahedron A 3-D *polyhedron* that is made up of four triangular *faces*. A regular tetrahedron is one of the five *Platonic solids*.

Theorem A significant proven result on a mathematical topic, especially one that is not self-evident. An unproved statement is called a *conjecture*.

Topology The branch of mathematics that studies surfaces and objects by examining how their parts are connected rather than according to their exact geometrical shapes. For example, a doughnut and a teacup are topologically similar because they are both shapes that have one hole going through them (going through the handle, in the case of the teacup).

Transcendental number Any *irrational number* that is not an *algebraic number*. The number *pi* (π) and Euler's number e are both transcendental numbers.

Transfinite number Another term for an *infinite* number. It is used particularly when infinities of different sizes or infinite collections of objects are compared.

Transformation The conversion of a given shape or mathematical *expression* into another related one, using a particular rule.

Translation A *function* that moves an object a certain distance in a direction without affecting its shape, size, or orientation.

Trigonometry Originally, the study of the way the ratios between different sides of a *right-angled* triangle change when other angles in the triangle change, and later extended to all triangles. The way the ratios change is described by trigonometric *functions*, which are now fundamental to many branches of mathematics.

Variable A mathematical quantity that can take on different values, often symbolized by a letter such as x or y.

Vector A mathematical or physical quantity that has both magnitude and direction. In diagrams, vectors are often represented by bold arrows.

Vector space A complex abstract mathematical structure that involves the multiplication of *vectors* by each other and by *scalars*.

Venn diagram A diagram that shows sets of data as overlapping circles. The overlaps show what the sets have in common.

Vertex (plural **vertices**) A corner or angle, where two or more lines, curves, or edges meet.

Volume The amount of space inside a 3-D object.

Whole number Any of the negative and positive counting numbers. For example, −1, 0, 19, 55, and so on. It is another term for *integer*.

INDEX

Page numbers in **bold** refer to main entries;
those in *italics* refer to illustrations and captions.

QUOTATIONS

The following primary quotations are attributed to people who are not the key figure for the relevant topic.

ACKNOWLEDGMENTS

Dorling Kindersley would like to thank Gadi Farfour, Meenal Goel, Debjyoti Mukherjee, Sonali Rawat, and Garima Agarwal for design assistance; Rose Blackett-Ord, Daniel Byrne, Kathryn Hennessy, Mark Silas, and Shreya Iyengar for editorial assistance; and Gillian Reid, Amy Knight, Jacqueline Street-Elkayam, and Anita Yadav for production assistance.

PICTURE CREDITS